Applied Agricultural Practices for Mitigating Climate Change

Applied Agricultural Practices for Mitigating Climate Change

Edited by
Rohitashw Kumar, Vijay P Singh,
Deepak Jhajharia, and Rasoul Mirabbasi

CRC Press is an imprint of the
Taylor & Francis Group, an **informa** business

CRC Press
Taylor & Francis Group
6000 Broken Sound Parkway NW, Suite 300
Boca Raton, FL 33487-2742

© 2020 by Taylor & Francis Group, LLC
CRC Press is an imprint of Taylor & Francis Group, an Informa business

No claim to original U.S. Government works

Printed on acid-free paper

International Standard Book Number-13: 978-0-367-34529-7 (Hardback)

Visit the Taylor & Francis Web site at
http://www.taylorandfrancis.com

and the CRC Press Web site at
http://www.crcpress.com

Contents

Preface

Agriculture is the backbone of the economy in most countries and has a special role in the food chain. In many parts of the world, agriculture is climate restricted and many factors affect crop production, such as water availability and quality, climate, soil productivity and management, degree of farm mechanization, crop variety, and fertilizers. In recent years, climate change has been causing spatial and temporal changes in temperature and precipitation and, hence affecting agricultural production. Another important issue is the effect of cultivation of soil, which may reduce the productive capacity of farmlands. These issues are not confined to any one region or country, but are global in nature, requiring multidisciplinary, multi-organizational, and multinational approaches and educational efforts. Conservation agriculture is a concept designed for optimizing crop yields, and reaping economic and environmental benefits. The key elements of conservation agriculture are minimum disturbance of soil, rational organic soil cover using crop residues or cover crops, and the adoption of innovative and economically viable cropping systems and measures to reduce soil compaction through controlled traffic. Conservation agriculture offers promise in using crop residues for improving soil health, increasing productivity, reducing pollution, and enhancing sustainability and resilience of agriculture.

This book, *Applied Agricultural Practices for Mitigating Climate Change*, is designed to provide a discussion of each of the important aspects of effective factors on crop production, such as climate change, soil management, farm machinery, and different methods for sustainable agriculture. It deals with the effects of climate change on agriculture and mitigation strategies, while focusing on conservation agriculture as an effective strategy for sustainable agriculture and food security. Selected case studies are also provided. The book provides current information which can be utilized when dealing with these issues.

This book is comprised of 18 chapters. In different chapters, the book highlights the theme of conservation agriculture to mitigate climate change. Chapter 1 presents applications of plastics in the postharvest management of crops. Major applications of plastics are seen in packaging, drying, storage, and transportation of crops. Such plastics would be a boon to the postharvest management field in the near future.

Potassium-solubilizing microorganisms for sustainable agriculture are discussed in Chapter 2. The potassium-solubilizing microorganisms are the microorganisms that solubilize the insoluble potassium (K) to soluble forms of K for plant growth and development. The potassium-solubilizing microorganisms are the most important microbes for the solubilizing of the fixed form of K. Chapter 3 describes weed control for conservation agriculture in the climate-change scenario. It is discussed that weeds possess a better ability to survive and perform under adverse environmental conditions, which make them sturdy and highly competitive with crop plants. Different strategies to control weeds can be beneficial to fight with the problem of climate change. *In-situ* soil moisture conservation with organic mulching under the mid-hills region is highlighted in Chapter 4. *In-situ* soil moisture is one of the constraints in crop production during the winter season where there is hardly any source of irrigation under a hilly terrain.

A field trial was taken up to assess the *in-situ* soil moisture depletion under two organic mulching materials, *viz.*, maize stover and weed mulch.

Chapter 5 describes high-altitude protected vegetable cultivation for sustainable agriculture. It describes how protected cultivation is the only approach that has the potential to cater to the food and nutritional security challenges of the higher Himalayas because of climatic uncertainties. Numerous protected structures are available that enable crop-climate management, but harvest of solar radiations and conservation of temperature during winter remain a challenge to be accepted. Modernization of the approach toward protected cultivation is needed and demands immediate address through R&D innovations and interventions. Protected cultivation of vegetables offers distinct advantages of quality, productivity, and favorable market prices to the growers. Chapter 6 discusses the applications of remote sensing in crop production and soil conservation. It describes the application of remote sensing and GIS tool of precision agriculture. Different techniques have been discussed for land and water management. The broad objective of sustainable agriculture is to balance the inherent land resource with crop requirements, paying special attention to optimization of resource use toward achievement of sustained productivity over a long period. The chapter highlights the applications of RS and GIS technology in crop production and soil conservation.

Chapter 7 describes the causes of flash floods and remedial measures for their control in hilly regions. Floods in the hilly regions are a major cause of loss of life, property, crops, infrastructure, etc. Flash floods carry a higher amount of debris than normal floods and, as a result, cause huge damage to buildings, roads, bridges, hydropower stations, and other infrastructure. This chapter highlights flood problems in India, specifically in the hilly region, along with some of the important management issues which require immediate attention. Also, various short-term and long-term strategies are suggested for the effective management of flash floods. Chapter 8 discusses the role of crop modeling in mitigating the effects of climate change on crop production. Physiology-based crop simulation models have become a key tool in extrapolating the impact of climate change from limited experimental evidence to broader climatic zones. Different models are a simplification of reality; they allow a first assessment of the complexity of climate-change impact in agriculture. They are playing a significantly important role in assisting agriculture to adapt to climate change. In order to meet the increasing demand for assessment of climate-change impact, crop models need to be further improved and tested with climate-change scenarios including various changes in ambient temperature and CO_2 concentration.

Chapter 9 describes forestry and climate change. It discusses the adaptation capabilities of agroforestry with regards to climate change. The chapter highlights climate change and variability in the environment which have an impact on forests. Agroforestry's role in the sustainable management of degraded watersheds is discussed in Chapter 10. The watershed is described as a hydrological unit and social and ecological entity, which plays a crucial role in determining food, social and economical security, and provides life-support services to rural people. The watershed is the product of the interactions between land and water, particularly underlying geology, rainfall patterns, slope, soils, vegetative cover, and land use.

Chapter 11 describes infiltration studies of major soils under selected land-use practices in Sikkim, India. Estimation of infiltration rates using a double-ring infiltrometer is time-consuming and is very difficult in the hilly topography of Sikkim. In this chapter, an attempt is made to analyze the infiltration rate of four types of soil—namely, fine loamy, coarse loamy, loamy skeletal, and fine silty—under two different land-use watersheds. The infiltration characteristic of soil was determined using the double-ring infiltrometer method. The observed infiltration characteristics were compared with the Horton's, Philip's, Kostiokov's, and Green-Ampt's infiltration models to determine the best-suited model for the watersheds. Conservation agriculture for improving soil biodiversity is described in Chapter 12. It also describes how conservation tillage and residue management help in improving soil properties and ultimately soil quality.

Chapter 13 describes the point-injection nitrogen application under rice residue wheat for resource conservation. The chapter shows that point-injected nitrogen application with a self-propelled nitrogen (liquid urea) applicator is a better alternative over the conventional practice of broadcasting of urea under straw mulched no-till wheat farming. The lower nitrogen accumulation in straw mulch in case of point-injected nitrogen application indicates the reduced nitrogen loss under this system of fertilizer urea application, particularly, under straw mulched conditions. Chapter 14 describes the importance of water in relation to plant growth. It describes how plant growth is affected by several factors such as seed variety, amount of water, soil type, amount of light, temperature, humidity, etc. Among these factors, water is essential for all living organisms, and plants are no exception. It also gives an understanding of the water budget and hydrological process which can be used for water resources management and environment planning and management.

Chapter 15 discusses the influence of deficit irrigation on various phenological stages of temperate fruits. It describes how deficit irrigation is an optimization strategy in which irrigation is applied at a reduced level in various growth stages of a crop; factors controlling stress conditions alter the normal equilibrium and lead to a series of morphological, physiological, biochemical, and molecular changes in plants which adversely affect their growth and productivity. High-pressure processing (HPP) of seafoods is discussed in Chapter 16. High-pressure processing is considered to be an alternative technology for food safety and preservation in the modern food industry. This chapter focuses on the application of high-pressure processing in seafoods. HPP is beneficial for the shucking or opening of shellfish, oysters, prawns and shrimp, as well as fish processing, etc. This chapter also emphasizes changes in seafoods due to HPP.

Chapter 17 discusses irrigation scheduling under deficit irrigation. It describes irrigation water management as a technique of irrigation scheduling and controlled water application in a manner that fulfills the crop-water requirement without any water loss. Water scarcity deficit irrigation is defined as the application of water lower than potential evapotranspiration, which is a key factor in utilizing sinking irrigation water use and is more appropriate under deficit-irrigation planning. Chapter 18 discusses the water resources scenario of the Indian Himalayan region. The chapter deals with the extensive review of the water resources of the Indian Himalayas considering the changing climate. It also deals with the influence of global warming on the water resources of the Indus Basin, and the harmful outcomes that pose a threat to the population as a whole, such as floods and droughts.

As editors, we realize that we have just begun to scratch the surface with some of the recent advances in conservation agriculture for climate change mitigation. We would like to take this opportunity to thank the chapter authors for their contributions. All of us (both the editors and authors) are thankful for the valuable and constructive comments that were received on each chapter. This book discusses real-world examples and is based on experiences at different agro-climatic regions throughout the world. It is hoped that this book will be useful for conservation agriculture, precision farming for climate change mitigation, and agriculture production systems.

Rohitashw Kumar
Vijay P Singh
Deepak Jhajharia
Rasoul Mirabbasi

Editors

Dr. Rohitashw Kumar (B.E., M.E., Ph. D.) is a Professor in the College of Agricultural Engineering and Technology, She-e-Kashmir University of Agricultural Sciences and Technology of Kashmir, Srinagar, India. He is also Professor Water Chair (Sheikkul Alam Shiekh Nuruddin Water Chair), Ministry of Water Resources, Govt. of India, at the National Institute of Technology, Srinagar (J&K). He obtained his Ph.D. degree in Water Resources Engineering from NIT, Hamirpur, and Master of Engineering Degree in Irrigation Water Management Engineering from MPUAT, Udaipur.
He received a Special Research Award in 2017 and Student Incentive Award-2015 (Ph.D. Research) from the Soil Conservation Society of India, New Delhi. He also got the first prize in India for best M. Tech thesis in Agricultural Engineering in year 2001. He graduated from Maharana Pratap University of Agricultural and Technology, Udaipur, India, in Agricultural Engineering. He has published over 80 papers in peer-reviewed journals, one book, four practical manuals, and 20 chapters in books. He has guided ten post-graduate students in Soil and Water Engineering. He has handled more than ten research projects as a principal or co-principal investigator. Since 2011, he has been Principal Investigator of the All India Coordinated Research Project on Plasticulture Engineering and Technology.

Prof. Vijay P Singh is a Distinguished Professor, a Regents Professor, and the inaugural holder of the Caroline and William N. Lehrer Distinguished Chair in Water Engineering at Texas A&M University. His research interests include Surface-Water Hydrology, Groundwater Hydrology, Hydraulics, Irrigation Engineering, Environmental Quality, Water Resources, entropy theory, copula theory, and mathematical modeling. He graduated with a B.Sc. in Engineering and Technology with an emphasis on Soil and Water Conservation Engineering in 1967 from U.P. Agricultural University, India. He earned an MS in Engineering with specialization in Hydrology in 1970 from the University Of Guelph, Canada, a Ph.D. in Civil Engineering with specialization in Hydrology and Water Resources in 1974 from the Colorado State University, Fort Collins, USA, and a D.Sc. in Environmental and Water Resources Engineering in 1998 from the University of the Witwatersrand, Johannesburg, South Africa. He has published extensively on a wide range of topics. His publications include more than 1,200 journal articles, 30 books, 70 edited books, 305 book chapters, and 315 conference proceedings papers. For seminar contributions, he has received more than 90 national and international awards, including three honorary doctorates. Currently, he serves as President-Elect of the American Academy of Water Resources Engineers, the American Society of Civil Engineers, and previously he served as President of the American Institute of Hydrology. He is Editor-in-Chief of two book series and one journal and serves on the editorial boards of more than 20 journals. He has served as Editor-in-Chief of three other journals. He is a Distinguished Member of the American Society of Civil Engineers, an Honorary Member of the American Water Resources Association, and a fellow of five professional societies. He is also a fellow or member of 11 national or international engineering or science academies.

Dr. Deepak Jhajharia (B. Tech., M. Tech. and Ph. D.) is currently working as a Professor in the Department of Soil & Water Conservation Engineering, College of Agricultural Engineering & Post Harvest Technology, Ranipool, Gangtok, Sikkim, India. He is also acting as Principal Investigator of the All India Coordinated Research Project on Plasticulture Engineering Technology CAEPHT (Gangtok) center, which is funded entirely by ICAR Central Institute of Post Harvest Engineering and Technology, Ludhiana, India, since 2016. He graduated from the College of Technology and Engineering (MPUAT), Udaipur, Rajasthan, India, in Agricultural Engineering in 1998, and did post-graduation from the Indian Institute of Technology Delhi, India, in Water Resources Engineering, the Department of Civil Engineering. He obtained his Ph.D. degree from the Department of Hydrology, Indian Institute of Technology Roorkee, Uttarakhand, India. He was awarded Young Talent Attraction – BJT of Science without Borders Program for International Scientists, National Council for Scientific and Technological Development (CNPq), Brazil, as Research Collaborator at Universidade Federal Rural de Pernambuco (UFRPE), Recife, PE, Brazil, in 2013. He has published over 50 papers in peer-reviewed journals, three extension bulletins, and seven chapters in books. He has guided eight M. Tech. theses in the field of Soil and Water Conservation Engineering along with many undergraduate theses in the field of Agricultural Engineering. He has handled nine research projects sponsored by governmental agencies as principal or co-principal investigator. He also conducted one 21-day summer school for scientists from ICAR and faculty members from different universities and one 90-day skill development training program on greenhouse technology for school drop-outs and unemployed rural youth from six states of northeast India. He was awarded the CSIRO Land and Water Publication Award 2013, CSIRO Australia for a global review paper published in the *Journal of Hydrology*. He is a Fellow of the Indian Association of Hydrologists, Roorkee (in 2015) and Society of Extension Education, Agra (in 2018). He is a recipient of the Distinguished Alumni Award (in 2016) from the College of Technology and Engineering Alumni Society, CTAE (MPUAT), Udaipur. He has also adjudged the Best Extension Scientist (2017–18) of the AICRP on PET in recognition of outstanding contribution to the extension and popularization of Plasticulture Technologies in Sikkim. He is also a life member of 14 different professional societies from India and abroad.

Dr. Rasoul Mirabbasi is an Associate Professor of Hydrology and Water Resources Engineering at Shahrekord University, Iran. His research focuses mainly on Statistical and Environmental Hydrology and Climate Change. In particular, he is working on Modeling Natural Hazards, including floods, droughts, winds, and pollution, toward a sustainable environment. Formerly, he was a Visiting Researcher at the University of Connecticut, United States. He has contributed to more than 150 publications in journals, books, or technical reports. He is the reviewer of about 20 Web of Science (ISI) journals. He is currently the Head of the Water Resources Center of Shahrekord University.

Contributors

Angrej Ali
Department of Horticulture, Faculty of
  Agriculture
Sher-e-Kashmir University of
  Agricultural Sciences and
  Technology of Kashmir
  (SKUAST-K)
Srinagar, Jammu and Kashmir, India

S. A. Banday
Division of Fruit Science, Faculty of
  Horticulture
Sher-e-Kashmir University of
  Agricultural Sciences and
  Technology of Kashmir
  (SKUAST-K)
Srinagar, Jammu and Kashmir, India

B. B. Basak
Directorate of Medicinal and Aromatic
  Plant Research
Anand Agricultural University
Anand, Gujarat, India

Zaffar Bashir
Division of Basic Sciences and
  Humanities
Sher-e-Kashmir University of
  Agricultural Sciences and
  Technology of Kashmir
  (SKUAST-K)

Sushmita M. Dadhich
Division of Agricultural Engineering
Sher-e-Kashmir University of
  Agricultural Sciences and
  Technology of Jammu (SKUAST-J)
Jammu, Jammu and Kashmir, India

Hemant Dadhich
Division of Agricultural Engineering
Sher-e-Kashmir University
  of Agricultural Sciences
  and Technology of Jammu
  (SKUAST-J)
Jammu, Jammu and Kashmir, India

Joy Kumar Dey
Department of Agronomy
Palli Siksha Bhavana (Institute of
  Agriculture), Visva-Bharati
Sriniketan, Birbhum, West
  Bengal, India

Jagvir Dixit
Division of Agricultural
  Engineering
Sher-e-Kashmir University of
  Agricultural Sciences and
  Technology of Kashmir
  (SKUAST-K)
Srinagar, India

Tsering Dolkar
Division of Fruit Science, Faculty of
  Horticulture
Sher-e-Kashmir University of
  Agricultural Sciences and
  Technology of Kashmir
  (SKUAST-K)
Srinagar, India

N. K. Garg
Department of Civil Engineering
Indian Institute of Technology
New Delhi, India

Mir Bintul Huda
Water Resources Chair, Water
  Resources Management Centre
National Institute of Technology
Srinagar, India

Juvaria Jeelani
Division of Soil Science
Sher-e-Kashmir University of
  Agricultural Sciences and Technology
  of Kashmir (SKUAST-K)
Srinagar, Jammu and Kashmir, India

D. Jhajharia
Department of Soil and Water
  Conservation Engineering
College of Agricultural Engineering
  and Post Harvest Technology, CAU
Gangtok, Sikkim, India

S. J. Kale
ICAR-Central Institute of Postharvest
  Engineering and Technology
Abohar, Punjab, India

M. S. Kanwar
Krishi Vigyan Kendra
Sher-e-Kashmir University of
  Agricultural Sciences and
  Technology of Kashmir (SKUAST-K)
Nyoma-Changthang, Ladakh, Jammu
  and Kashmir, India

Rajesh Kaushal
ICAR-Indian Institute of Soil and Water
  Conservation
Dehradun, Uttarakhand, India

Nayar Afaq Kirmani
Division of Soil Science
Sher-e-Kashmir University of
  Agricultural Sciences and
  Technology of Kashmir (SKUAST-K)
Srinagar, Jammu and Kashmir, India

Amit Kumar
Division of Fruit Science, Faculty of
  Horticulture
Sher-e-Kashmir University of
  Agricultural Sciences and Technology
  of Kashmir (SKUAST-K)
Srinagar, Jammu and Kashmir, India

Rohitashw Kumar
College of Agricultural Engineering
  and Technology
Sher-e-Kashmir University of
  Agricultural Sciences and
  Technology of Kashmir
  (SKUAST-K)
Srinagar, India

Nancy Loria
ICAR-Indian Institute of Soil and Water
  Conservation, Research Centre
Chandigarh, India

J. S. Mahal
Punjab Agricultural University
Ludhiana, Punjab

Y. Marwein
School of Natural Resource
  Management
College of Postgraduate Studies,
  (Central Agricultural University,
  Imphal)
Barapani, Ri-Bhoi, Meghalaya, India

Prerna Nath
ICAR-Central Institute of Postharvest
  Engineering and Technology
Abohar, Punjab, India

Sharmistha Pal
ICAR-Indian Institute of Soil and Water
  Conservation, Research Centre
Chandigarh, India

Kusum Pandey
Department of Soil and Water
   Engineering
Punjab Agricultural University
Ludhiana, Punjab, India

Pankaj Panwar
ICAR-Indian Institute of Soil and Water
   Conservation, Research Centre
Chandigarh, India

G. T. Patle
Department of Irrigation and Drainage
   Engineering, College of Agricultural
   Engineering and Post Harvest
   Technology
Central Agricultural University
Gangtok, Sikkim, India

K. N. Qaisar
Faculty of Forestry
Sher-e-Kashmir University of
   Agricultural Sciences and
   Technology of Kashmir
   (SKUAST-K)
Ganderbal, India

Nasir Ahmad Rather
Department of Civil Engineering
Baba Ghulam Shah Badshah
   University
Rajouri, India

Lala I. P. Ray
School of Natural Resource
   Management
College of Postgraduate Studies,
   (Central Agricultural University,
   Imphal)
Barapani, Ri-Bhoi, Meghalaya, India

Rishi Richa
College of Agricultural Engineering
   and Technology
Sher-e-Kashmir University of
   Agricultural Sciences and
   Technology of Kashmir
   (SKUAST-K)
Srinagar, India

S. Sarvade
College of Agriculture, Balaghat
Jawaharlal Nehru Krishi Vishwa
   Vidyalaya
Jabalpur, Madhya Pradesh, India

N. C. Shahi
Department of Post Harvest Process
   and Food Engineering, College of
   Technology
G. B. Pant University of Agriculture
   and Technology
Pantnagar, Uttarakhand, India

M. K. Sharma
Division of Fruit Science, Faculty of
   Horticulture
Sher-e-Kashmir University of
   Agricultural Sciences and
   Technology of Kashmir (SKUAST-K)
Srinagar, Jammu and Kashmir, India

Parmeet Singh
Division of Agronomy, Faculty of
   Agriculture
Sher-e-Kashmir University of
   Agricultural Sciences and
   Technology of Kashmir (SKUAST-K)
Sopore, India

Lal Singh
Division of Agronomy, Faculty of
    Agriculture
Sher-e-Kashmir University of
    Agricultural Sciences and
    Technology of Kashmir
    (SKUAST-K)
Sopore, India

Indra Singh
ICAR-Indian Institute of Soil and Water
    Conservation, Research Centre
Chandigarh, India

Sapam Raju Singh
Department of Water Resources
    Engineering
Indian Institute of Technology
Delhi, India

Anupama Singh
Department of Post Harvest Process
    and Food Engineering, College of
    Technology
G. B. Pant University of Agriculture
    and Technology
Pantnagar, Uttarakhand, India

Ranjna Sirohi
Department of Post Harvest Process
    and Food Engineering, College of
    Technology
G. B. Pant University of Agriculture
    and Technology
Pantnagar, Uttarakhand, India

J. A. Sofi
Division of Soil Science
Sher-e-Kashmir University of
    Agricultural Sciences and
    Technology of Kashmir (SKUAST-K)
Srinagar, Jammu and Kashmir, India

Dinesh Kumar Vishwakarma
College of Agricultural Engineering
    and Technology
Sher-e-Kashmir University of
    Agricultural Sciences and
    Technology of Kashmir (SKUAST-K)
Srinagar, India

M. Y. Zargar
Sher-e-Kashmir University of
    Agricultural Sciences and
    Technology of Kashmir (SKUAST-K)
Srinagar, India

1 Application of Plastics in Postharvest Management of Crops

S. J. Kale and Prerna Nath
ICAR-Central Institute of Postharvest
Engineering and Technology

CONTENTS

1.1 POSTHARVEST MANAGEMENT OF CROPS

Postharvest management involves a number of unit operations such as precooling, cleaning, sorting, grading, storing, packaging, etc. (Table 1.1). The moment a crop is harvested from ground or picked from its parent plant it starts to deteriorate. Hence, appropriate postharvest management strategies are indispensable to ensure better quality of crops, whether they are sold for fresh consumption or used to prepare

1

TABLE 1.1

Different Postharvest Management Operations

Perishable Crops		Durable Crops	
1. Field handling	9. Unitization	1. Field handling	9. Treatment
2. Receipt	10. Storage	2. Threshing	10. Storage
3. Conveying	11. Dispatch	3. Receipt	11. Processing
4. Cleaning	12. Transport	4. Pre-Cleaning	12. Retailing
5. Sorting	13. Wholesaling	5. Drying	13. Consumption
6. Grading	14. Retailing	6. Storage	
7. Treatment	15. Consumption	7. Transport	
8. Packaging		8. Cleaning	

processed products. Both quantitative and qualitative losses occur at all stages in postharvest handling system due to various extrinsic and intrinsic factors. Extrinsic factors include surrounding temperature, relative humidity (RH), gas composition, mechanical injury, etc., whereas intrinsic factors include crop variety, respiration rate, moisture content, microbial load, etc. By and large, both factors causing postharvest losses vary from place to place and crop to crop. Deterioration of fresh crops mainly causes for weight loss, bruising injury, physiological breakdown due to ripening processes, temperature injury, chilling injury, or attack by microorganisms. All the crops counting fruits, vegetables, cereals, oilseeds, pulses, root crops, flowers, etc. are living organisms having respiratory system. They respire even after harvest. During respiration, they take oxygen in and release carbon dioxide. Respiration causes significant weight loss. It also accumulates heat in the container, changes gas composition of surrounding, and thereby deteriorates the produce. Therefore, it is essential that crops must be kept in admirable conditions to acquire excellent quality with maximum shelf life. This constraint underlines the importance of postharvest management of crops.

The postharvest management is not new to the contemporary world. Rather, it is being practiced since ancient times. A number of crops are have been dried, stored, or processed since the beginning of the civilization. However, with time, the postharvest management practices became more sophisticated and technically sound. In recent times, advanced techniques of cleaning, grading, sorting, storing, packaging, transportation, processing, etc. have been well established to minimize the postharvest losses and acquire nutritional quality of crops. Postharvest operations are mainly dealt with food handling during which contamination as well as damage to the food may take place. Hence, care should be taken to select appropriate materials to avoid any possible damage. Various materials including metals, stainless steel, plastics, woods, fibers, glass, papers, etc. are used in postharvest management operations. However, plastics found to be the most used materials for the said purpose.

A report indicates that the packaging industry in India has seen a strong diffusion of plastics as compared with global standards (Anonymous, 2014). However, agriculture sector has still not explored the benefits of plastics to a considerable extent (Figure 1.1). The major applications of plastics in agriculture include mulching, irrigation, and protected cultivation.

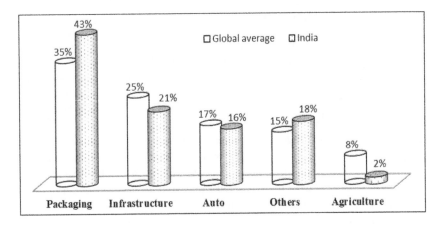

FIGURE 1.1   Use of plastics in different fields. (Anonymous, 2014.)

In postharvest management operation, plastics are mainly used in packaging systems, transportation containers, grading and sorting machines, cold store units, pack-houses, etc. Moreover, it is evident that plastics have a noteworthy role in preservation of longevity and quality of harvested crops. Plastics are mainly favored for their extraordinary characteristics and versatility of applications. They are light weight, corrosion resistant, moisture proof, highly versatile, adoptable, and can be moulded into attractive shapes. In many cases, food grade plastics are found as comparable as food grade stainless steels used in food processing industry. By considering the importance of plastics, an attempt has been made to discuss some of the important applications of plastics in postharvest management of various crops. These applications are discussed one by one in subsequent text.

1.2   FIELD HANDLING OF CROPS

Field handling is the first postharvest unit operation carried out immediately after harvesting. It involves three major operations namely collection, packaging, and transportation of crops. Crops are collected and subsequently transported, with or without packaging, to the desired destination. Harvested crops are collected in the field containers such as bags, plastic buckets, plastic crates, woven baskets, gunny bags, etc. Harvesting bags with shoulder or waist slings are used for fruits with firmer skins like citrus. They are easy to carry and leave both hands free. Plastic buckets or other containers are also suitable for fruits that are more easily crushed, such as tomatoes. However presently in India, plastic crates have replaced all other containers due to their strength, inertness, portability, corrosion resistance, etc. They are smooth with no sharp edges or projections to damage the produce. They are quite sturdy and hence do not bend out of shape when lifted or tipped. Packaging of harvested crops directly into packages in the field immediately after harvest reduces the damage caused by multiple handling. However, it is commonly used by commercial growers in developed countries and not so common practice in rural areas.

1.3 REMOVAL OF FIELD HEAT

Temperature management immediately after harvesting is the most essential step in extending shelf life of crops. Freshly harvested fruits and vegetables contain significant amount of field heat which needs to be removed prior to packaging, transportation or storage. Conventional cost-effective practice is to harvest these crops in the evening so that they are subjected to lower atmospheric temperatures during 9 h. Sometimes, these crops are immersed in cold water which cleans and cools the crops. Recently, farmers have started to use shadenets having 30%–50% shading intensity to provide cooler environment compared to ambient air temperatures. These shadenets are mainly made of UV-stabilized plastics. In most of the cases, shadenet house used for crop production is used for removing field heat from the crops up to certain extent. This operation also helps to achieve thermal equilibrium in the harvested crops.

1.4 FIELD CURING OF ROOT, TUBER, AND BULB CROPS

Curing of root, tuber, and bulb crops such as potatoes, sweet potatoes, and onions is an important practice if these crops are to be stored for substantial period of time. Curing is accomplished by holding the produce at high temperature and high relative humidity for several days after harvesting. The main objective of curing is to heal the wounds and form a new protective layer of cells. Although curing adds additional cost, the extended length of storage life makes the practice economically viable. Optimum conditions for curing vary crop to crop. Potato is cured at 15°C–20°C and 90%–95% RH for 5–10 days, whereas sweet potato is cured at 30°C–32°C and 85%–90% RH for 4–7 days (Calverley, 1998).

Onion and garlic bulbs are directly exposed to the sun in the field after harvest thereby allowing the external skin layers and neck tissue to dry prior to handling and storage. Onions are uprooted, windrowed and left in the soil to dry for 5–8 days, depending on the air temperature and moisture. The crop leaves are used to cover and shade the bulbs during the curing process to protect the crop from excessive heat and sunburn. However, curing becomes difficult in regions where harvesting coincides with rainy/moist season. Unexpected rains also damage the bulbs during curing. Under such circumstances, polytunnels, and polythene sheet covers found very useful in curing the crops. The crops can also be cured after packing into 15–25 kg net sacks.

1.5 GRADING AND SORTING

Grading and sorting are two important unit operations to increase the market value of a produce. Specially designed graders and sorters are used to accomplish these unit operations. Although various materials are used in the fabrication of these equipments, the plastics are also being used due to their low weight, sturdiness, corrosion resistance, etc.

1.6 CONVEYING

In some respects the most important and probably the most undervalued postharvest operation is the conveying of produce. Conveying involves physically moving or lifting the produce in bulk from one location to other. Belt conveyors, screw conveyors or elevators are used for this purpose. Fresh produce can be moved by solid or liquid systems. Metal chain-link, rubber, plastic or canvas belts, and wood, metal or rubber rollers are examples of the solid systems used to move either loose or packaged produce. Hydro-handling along flumes can be used for water tolerant commodities such as apples, carrots, and tomatoes. Elevator belts or buckets are often used in the packing line systems. Engineered plastics are extensively used in conveyors and bucket elevators.

1.7 PACKAGING OF FRESH AND PROCESSED CROPS

Packaging is one of the most important steps in long and complicated journey of fresh and processed crops. Different types of packaging systems such as flexible bags, crates, baskets, cartons, bulk bins, palletized containers, etc. are used for handling, transportation and marketing of fresh and processed crops. The principal roles of food packaging are to protect food products from outside influences and damages, contain the food, and provide consumers with ingredients and nutritional information (Coles, 2003). The goal of food packaging is to contain food in a cost-effective way that satisfies industry requirements and consumer desires, maintains food safety and minimizes environmental safety (Marsh and Bugusu, 2007; Yadav, 2017).

1.7.1 CLASSIFICATION OF PACKAGING SYSTEMS

Packaging systems can be classified as given below:

1. Flexible sacks: made of plastics and nets
2. Wooden crates
3. Cartons (fiberboard boxes)
4. Plastic crates
5. Pallet boxes and shipping containers
6. Baskets made of woven strips of leaves, bamboo, plastic, etc.

Plastic nets are suitable only for hard crops such as coconuts and root crops (potatoes, onions, etc.), whereas plastic films find applicability in various places. It can be easily observed that almost all the packaging systems involve plastics. In present times, plastics have become an integral part of packaging systems of fresh and processed crops. Almost all the foods are being packaged in different types of plastics. The food industry is also dependent on plastics in the form of packaging. Moreover, packaging standards have become stricter with introduction of newer Indian norms closer to global standards which are also driving the use plastics in packaging.

Different types of plastics are available in the market for packaging purposes. However, each material has its own appropriateness. A report indicates that polyethylene (PE) is the most used material followed by polypropylene (PP) and polyethylene terephthalate (PET) (Anonymous, 2014). Plastics used in food packaging are commonly grouped as flexible and rigid plastics. Flexible packaging consists of monolayer or multilayer films of plastics, whereas multi-layered laminated sheets of plastics include PE, PP, PET, and polyvinyl chloride (PVC). PE and PP account

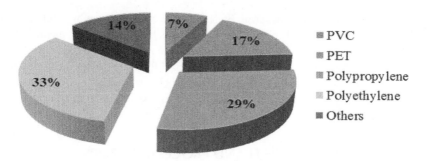

FIGURE 1.2   Share of different flexible packaging materials.

almost 62% of plastics usage in the packaging industry (Figure 1.2). With increase in income, consumer preferences for packaged foods are expected to increase further.

In food packaging systems, plastics are used as liners, flexible bags, wraps, boxes, etc. In most of the paper boxes as in case of tetra pack, plastic films are often used to line packing boxes in order to reduce water losses or prevent friction damage.

1.7.2 PLASTIC BAGS

Polythene bags are the chief containers for fresh fruits and vegetables packaging. Plastic bags are cheaper and involve very low costs in bagging and sealing. They are clear thereby allowing easy inspection of the stuffing and readily accept high-quality graphics. Plastic films are available in a wide range of thicknesses and grades and may be engineered to control the environmental gases inside the bag. The film material is enough permeable to maintain the correct mix of oxygen, carbon dioxide, and water vapors inside the bag. In addition to engineered plastic films, various patches and valves have been developed that affix to low-cost ordinary plastic film bags. These devices respond to temperature and control the mix of environmental gases.

1.7.3 SHRINK WRAP

Shrink wrapping is one of the most recent developments in packaging of freshly harvested fruits and vegetables. In this type of packaging, individual fruit or vegetable item is wrapped with polythene film. Shrink wrapping has been used successfully to package kinnow, capsicum, cabbage, cauliflower, potatoes, apples, cucumbers, bittergaurd, and various other fruits and vegetables. Shrink wrapping with plastic can reduce weight loss and shrinkage, protect the produce from disease, reduce mechanical damage and provide a good surface for stick-on labels (Dhall et al., 2012). Various studies have been conducted to determine the efficacy of shrink-wrap packaging in extending the shelf life of perishables. In a study, green cucumbers were shrink wrapped with Cryovac D955 (60 gauge) plastic film and stored at $12 \pm 1°C$, 90%–95% RH as well as ambient conditions (29°C–33°C, 65%–70% RH) (Dhall et al., 2012). It was noted that at 12°C and 90%–95% RH, shrink wrapped cucumber showed minimum physiological loss in weight (0.66%) as compared with unwrapped fruits (11.11%) at the end of refrigerated storage of 15 days.

1.7.4 Rigid Plastic Packages

Rigid plastic packages include clamshells. Packages having top and bottom with heat formed from one or two pieces of plastic are known as clamshells. Clamshells, which are inexpensive, versatile, provide excellent protection to the produce, and present a very pleasing consumer package. They are often used with smaller packs of high-value crops, such as small fruits, berries, mushrooms, etc., or food items that are easily damaged by crushing. They are also used extensively in packaging of minimally processed, pre-cut, and prepared salads. Moulded polystyrene containers have been found as a substitute for corrugated fiberboard. At present these containers are not cost-effective, but as environmental pressures grow, they may be more common. Similarly, heavy-moulded polystyrene pallet bins have been used as a substitute for wooden pallet bins. Although at present their cost is almost double that of wooden bins, they are durable, are easier to clean, recyclable, do not decay when wet, do not harbor disease, may be nested, and made collapsible.

1.7.5 Biodegradable Films

Environmental pressure is increasing continuously due to human interruptions. Under such circumstances, disposal and recyclability of packaging of all kinds of materials is becoming very important issue. It has been studied that common polyethylene takes about 200–400 years to breakdown in a landfill. However, addition of 6% starch reduces this time to about 20 years or less. Hence, nowadays, the packaging material companies are developing starch-based polyethylene substitutes that break down in a landfill as fast as ordinary paper. This move toward biodegradable or recyclable plastic packaging materials may be driven by cost in the long term, but by legislation in the near term. Some authorities have proposed a total ban on plastics.

1.7.6 Modified Atmospheric Packaging

Modified atmospheric packaging (MAP) is almost impracticable without using suitable plastics. MAP of fresh fruits and vegetables is based on modifying the concentrations of O_2 and CO_2 in the atmosphere that generated inside package (Mangaraj et al., 2009). It is desirable that the natural interaction occurring between respiration of the product and packaging generates an atmosphere with low levels of O_2 and/or a high concentration of CO_2. The growth of harmful microorganisms is thereby reduced and the shelf life of product is extended. In a modified atmosphere packaging, gases of the internal and the external ambient atmosphere try to equilibrate by permeation through the package walls at a rate dependent on the differential pressures between the gases of the headspace and those of the ambient atmosphere. In this context, the barrier to gases and water vapor provided by packaging material must be considered. Thus it can be stated that the success of the MAP largely depends upon the barrier (packaging) material used. These packages are made of plastic films with relatively high gas permeability.

Packaging films with a wide range of physical properties are used in MAP. There are several groupings in MAP films such as in the plural, vinyl polymers, styrene polymers, polyamides, polyesters, and other polymers. PP is part of the polyolefin group and used largely in MAP, in both forms: continuous and perforated. Although various types of plastic films for packaging are available, relatively few have been used to pack fresh fruits and vegetables and even fewer have a gas permeability that makes them suitable for MAP. It has been recommended that the permeability of CO_2 should be three to five times the permeability of O_2. Many polymers used to formulate packaging films are within this criterion.

1.8　STORAGE

Significant volume of fresh and processed food materials is stored before ultimate consumption. Very simple to the most modern storage structures are utilized for storage of agricultural crops. Clamps, root cellars, evaporative cooling structures, bunkers, cover and plinth (CAP), godowns, ventilated structures and such type of storage structures are being used conventionally in India (Kale et al., 2016; Kale and Nath, 2018). However, since recent times, with introduction of mechanical refrigeration systems, automatic conveyors and elevators, the storage structures have seen a paradigm shift. Presently, cold stores, modified and controlled atmospheric storage, hypobaric storage, vertical metal silos, hermetic storage, etc. are used for bulk storage of various crops. It can be seen that plastic is extensively used in almost all type of storage structures as construction material. It can be noticed that out of all these structures, CAP storage is impossible without use of suitable plastics. In India, CAP is still used to store very large volume of wheat and paddy. In fact, the CAP storage is the necessity in India as Indian grain production increased faster than storage capacity (Bhardwaj, 2015).

The storage of food grains under large polythene is being practiced in India as well as other countries since long period. In CAP storage, outdoor stacks of bagged grains are covered with a waterproof polythene sheet (Figure 1.3). The advantage of CAP storage is its low cost. It is considered that the cost of CAP storage is only one-fourth of the cost of godown storage. However, CAP storage is vulnerable to

FIGURE 1.3　CAP storage of food grains.

wind damage and needs to be inspected frequently to detect the damage. The system requires careful management if severe losses are to be avoided. Careful quality control is achieved with regular sampling. Security is also a problem and extra fencing is to be included in the cost calculation.

For CAP construction, a plinth with hooks for the ropes lashing the stack is constructed on a suitable site. Dunnage is provided and the covers are made of black polythene sheet of 250 μm thickness shaped to suit the stack. The covers are held down by nets and nylon lashing. Condensation is prevented by placing a layer of paddy husk-filled sacks on top of the stack under the polythene. For typical 150 ton CAP storage, the commonly constructed size is 8.55 m × 6.30 m for 3,000 bags each of 50 kg capacity. It is generally provided on a raised platform where grains are protected from rats and dampness of ground. The grain bags are stacked in a standard size wooden dunnage. The stacks are covered with 250–350 μm low-density polyethylene (LDPE) sheets from the top and all four sides. Wheat grains are generally stored in such CAP storage for 6–12 months. It is the most economical storage structure and widely used by the Food Corporation of India for bagged grains (Jain and Patil, 2012).

1.9 TRANSPORTATION OF CROPS

Transportation is an inevitable postharvest operation in the journey of crops from farm to fork. Crops are transported many times after harvesting till consumption. Transportation occurs within the field, field to pack house, field to cold storage, pack house to cold storage, field to warehouse, cold storage to distant markets, markets to consumers, etc. Freshly harvested grains are generally transported in bulk/gunny bags using trolleys, whereas fresh fruits and vegetables are transported using rigid crates, sacks, wooden boxes, etc. One of the most important containers used for transporting fruits and vegetables is the returnable/reusable moulded plastics crates (Figure 1.4). These reusable boxes are moulded from high-density polythene (HDPE) and widely used for transporting the crops in many countries.

FIGURE 1.4 Returnable plastic crates. (Kitinoja, 2013.)

Returnable plastic crates can be made to almost any specifications. They are strong, rigid, smooth, easily cleaned and made to stack and nest when empty in order to conserve space. In spite of the cost, their capacity for reuse can make them an economical investment. However, they are found economical only in large quantities. They are attractive, have many alternative uses, and are subject to high pilferage. These crates are easy to clean due to their smooth surface and protective for products because of their rigidity. They can be used many times, reducing the cost of transport. They are available in different sizes and colors and are resistant to adverse weather conditions. However, plastic crates can damage some soft produce due to their hard surfaces, thus liners are recommended when using such crates. One major drawback is that these crates deteriorate rapidly when exposed to the sunlight, unless treated with UV inhibitor.

1.10 DRYING OF CROPS

Drying is a classical method of food preservation, which involves moisture removal through application of heat. Applied heat raises the vapor pressure in the produce and thus removes moisture in the form of vapors. Drying extends shelf life of the product and also reduces weight and volume of the product which in turn facilitates transportation and reduces storage space. Drying is achieved through the simplest methods like sun drying to the most modern methods such as infrared drying, microwave drying, refractive window drying, etc. Application of plastics in some of the important drying methods is discussed below.

1.10.1 OPEN SUN DRYING

Open sun drying is one of the traditional methods of drying. It involves spreading the harvested crops under the sun in open yards (Figure 1.5). This is still one of the most common drying methods used in India and other developing countries. It is a continuous process where moisture content, air temperature and product temperature change simultaneously. In this process, the rate of drying is largely affected by

FIGURE 1.5 Open sun drying of grains and chilli spread on plastic sheets.

air temperature, relative humidity, sunshine hours, solar radiation, wind velocity, duration of rain showers during the drying period, etc.

Open sun drying is predominantly accomplished using plastics. Owing to easy availability, low cost and inertness of plastics, farmers use them for sun drying of harvested and threshed grains, fruits, vegetables, etc. Drying is completed even in the farms using plastics. There is no need of any furnished floor for drying of crops on polythene. Tarpaulins made of high-density polythene (300–500 g/m²) are used for drying purposes. Although, open sun drying is economical and simple, it has drawbacks such as no control over drying rate, non-uniform drying and chances of deterioration due to exposure of products to rain, birds, dust, storm, rodents, insects, and pests, which results in poor quality of dried products.

Direct exposure of products to the sun's UV radiations may also reduce the nutrients such as vitamins, carotenoids, etc. in the dried product. Likewise, UV radiations in the sun rays change the organoleptic properties such as texture, color, and flavor of food materials (Sangamithra et al., 2014). Therefore, indirect type sun drying is found advantageous over open sun drying method. Polyhouse dryers may be considered as indirect type solar dryers.

1.10.2 POLYHOUSE DRYING

The main objective of polyhouse dryers is to maximize the utilization of solar radiations. Polyhouse dryers mainly consist of drying chamber, exhaust fan, and chimney. Roof and walls of such dryers are made of transparent UV-stabilized polythene sheets (Figure 1.6). This sheet has a transmittivity of approximately 92% for visible radiation. It traps solar energy during daytime and maintains desired temperature for drying of produce (Shahi et al., 2011). UV-stabilized polythene sheet plays a vital role in preventing the deterioration of product in the form of loss in nutritional and organoleptic properties. It allows only short wavelength radiations to penetrate inside and converts them into long wavelength when they strike on the surface of the product or floor. Long wavelength radiations cannot move out and thus increase the temperature inside dryer. Polythene sheets used in the construction of polyhouse

FIGURE 1.6 Polyhouse dryers for drying of crops.

dryers have superior characteristics in terms of transparency, transmittivity, anti-corrosion property, self-adhesive, retraction ratio, tensile properties, tear-resistant, anti-puncture, water-proof, moisture-proof, dust-proof, etc. and hence outnumbered all other materials in polyhouse drying applications. In some studies, polycarbonate sheets have also been tested for polyhouse dryers (Janjai et al., 2011). Report indicates that a black surface inside polyhouse dryer increases the efficacy of converting light into heat (Shahi et al., 2011). Black polythene sheets of 25–100 μm thickness are used on the floor of the dryer for such purpose.

Polyhouse dryers are still not very popular among farmers. However, efforts are made to develop and popularize such dryers. A number of studies has been conducted to develop the polyhouse dryers. Ekechukwu and Norton (1999) designed a natural circulation solar dryer covered with a polythene sheet. It consisted of semi-cylindrical drying chamber with a cylindrical chimney at one end. The dimensions of drying chamber were 6.67 m long, 3.0 m wide and 2.3 m high. Kulanthaisami et al. (2009) developed and tested a semi-cylindrical solar tunnel dryer covered with UV-stabilized semi-transparent polythene of 200 μm for drying coconuts. Drying chamber of solar tunnel dryer was 18 m long and 3.75 m wide and drying capacity of 5,000 coconuts. The polythene sheet used in this dryer was opaque to long-wave radiations. Similarly, a polyhouse dryer suitable for hot and arid region of north western India was developed at ICAR-CIPHET, Abohar, Punjab, India (Kadam et al., 2011). Dryer was a Quonset shape low-cost polyhouse having dimensions of 6 m long, 4 m wide, 1.8 m ridge height and 32.75 m³ volume. It was oriented in east–west direction. The polyhouse frame was constructed using bamboo and covered with 200 μm thick UV-stabilized polythene sheet. The polyhouse dryer was provided a 9-in. diameter exhaust fan placed opposite the door to remove the moisture accumulated inside. Study reported that the performance of polyhouse dryer was decidedly dependent on solar radiation, ambient temperature and relative humidity. Increase in polyhouse temperature varied from 0.7°C to 19°C, whereas RH inside polyhouse dryer varied from 16% to 25.70% by air exchange at a flow rate of 6.1 m³/s.

In a study, Shahi et al. (2011) developed a solar polyhouse tunnel dryer having 5 m length, 4 m breath, 3.2 m central height and side heights of 2.5 m left and 1.5 m right. This dryer was consisted of drying chamber, small exhaust fan, and a metal duct. Transparent UV-stabilized polythene sheet of 200 μm thickness was used to construct the dryer. Top surface of dryer was curved to increase the area of radiation. The orientation of dryer was in north–south direction to achieve maximum penetration of solar radiations in the dryer. Polyhouse dryer was equipped with a fan of 1,000–1,200 m³/h airflow rate and 1 kW power to achieve forced ventilation. The concrete floor inside dryer was painted black for better absorption of solar radiations. A glass wool insulation of 2 in. thickness was also provided to floor in order to reduce heat loss through it. The capacity of solar tunnel ranged from 1 to 1.5 quintal of fresh fruits and vegetables depending upon the materials and thickness of the spreading layer. This dryer was found efficient in drying of different crops in Kashmir valley. Similar studies have also been conducted by various workers and demonstrated the efficacy of polyhouse dryers in drying of agricultural products efficiently (Janjai and Keawprasert, 2006; Janjai et al., 2011; Seveda, 2012; Arjoo and Yadav, 2017).

Attempts have also been made to develop polyhouse dryers and evaluate their performance at farmers' field. In this line, a simple semi-circular solar tunnel polyhouse

FIGURE 1.7 Polyhouse dryer in the field of a farmer.

dryer was developed in the field of a farmer at Pali, Rajasthan, India having dryer floor area as 5 m × 3.75 m (Figure 1.7). The orientation of dryer was in east–west direction. The framed structure was covered with UV-stabilized polythene sheet (200 μm) having ridge height as 2.25 m. The dryer was found suitable in drying various agricultural products on large scale under controlled environment. It was large enough to permit a person to enter inside and carry out operations such as loading and unloading the crops to be dried.

It can be stated that by using apposite polyhouse dryers, farmers can easily and properly manage their high-moisture perishable crops and thus reduce the postharvest losses noticeably.

1.10.3 Refractive Window Drying

Polyhouse drying is nearly impractical without application of suitable plastics. Similar is the case of refractive window drying. Refractive window drying is a novel drying technique that converts foods in the form of liquids or slurries into flakes or powders within a typical residence time of 3–5 min (Nindo and Tang, 2007; Raghavi et al., 2018). Generally, conventional drying methods involve high temperatures and/or longer durations which results in degradation of heat-sensitive bioactive compounds (Pavan et al., 2012). This limitation is addressed by refractive window drying. In this drying system, heat energy is transferred from hot water (≥ 95°C) circulated beneath a plastic conveyer belt (Mylar™) is used to dry a thin layer of liquid or slurried food product spread on the belt surface. On the end portions of this belt, cold water is circulated to assist the detachment of dried material by scraping (Nindo et al., 2003). The most important component of refractive window drying system is the conveyor belt made of food grade plastic sheet (Figure 1.8). In general, drying is completed only when the heat transfer takes place from hot medium (air or hot surface/liquid) to the product to be dried. In this system, heat is refracted from hot water to the product through plastic conveyor belt. Hence, characteristics of plastic belt decide the drying efficiency as well as end product quality.

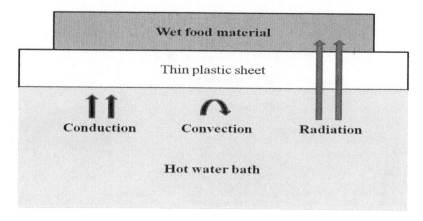

FIGURE 1.8   Representation of refractive window drying system.

In the refractive window drying system, product is dried as a thin film and cold air is circulated over product layer to achieve its cooling which results in dried products exhibiting excellent color, vitamins and antioxidant retention in comparison to other conventional drying methods (Caparino et al., 2012). Reports indicate that products dried using refractive window drying system is comparable in quality to freeze-dried products. However, the refractive window drying is advantageous over freeze-drying as being faster, energy efficient, and cheaper process (Nindo and Tang, 2007).

1.11   CONCLUSION

Postharvest management is essential to minimize the postharvest losses of agricultural crops. Various postharvest unit operations are practiced to achieve safe handling, shelf-life enhancement, and value addition of the agricultural crops. Numerous indigenous and engineered materials are used in different postharvest management operations. However, recently, plastics have outnumbered all other materials due to their bountiful advantages. Simple to the most complicated postharvest operations are carried out using various types of plastics. Major applications of plastics are seen in packaging, drying, storing, and transportation of crops. It appears that postharvest operations like packaging are almost impossible without plastics. However, major drawback in application of plastics in postharvest management of crops is their adverse effect on environment. Efforts are being implemented to develop and commercialize the biodegradable plastics. Such plastics would be boon to the postharvest management field in near future.

REFERENCES

Anonymous, 2014. Potential of plastics industry in northern India with special focus on plasticulture and food processing. A report on Plastics Industry, FICCI.
Arjoo, Y., Yadav, Y.K. 2017. Performance evaluation of a solar tunnel dryer for around the year use. *Current Agriculture Research Journal*, 5(3). doi: 10.12944/CARJ.5.3.22.
Bhardwaj, S., 2015. Recent advances in cover and plinth (CAP) and on-farm storage. *International Journal of Farm Sciences*, 5(2), pp. 259–264.

Calverley, D.J.B., 1998. Principles of storage for roots and tubers. In Calverley, D.J.B. (ed) *Storage and Processing of Roots and Tubers in the Tropics*. Food and Agriculture Organization of the United Nations, Agro-industries and Post-Harvest Management Service Agricultural Support Systems Division, Rome.

Caparino, O.A., Tang, J., Nindo, C.I., Sablani, S.S., Powers, J.R. and Fellman, J.K., 2012. Effect of drying methods on the physical properties and microstructures of mango (Philippine 'Carabao'var.) powder. *Journal of Food Engineering*, 111(1), pp. 135–148.

Coles, R., 2003. Plastic in food packaging. In Coles, R., Mcdowell, D. and Kirwan, M.I. (eds) *Food Packaging Technology*. Blackwell Publishing, CRC Press, London, UK, pp. 1–31.

Dhall, R.K., Sharma, S.R. and Mahajan, B.V.C., 2012. Effect of shrink wrap packaging for maintaining quality of cucumber during storage. *Journal of food science and technology*, 49(4), pp. 495–499.

Ekechukwu, O.V. and Norton, B., 1999. Review of solar-energy drying systems III: low temperature air-heating solar collectors for crop drying applications. *Energy Conversion and Management*, 40(6), pp. 657–667.

Jain, D. and Patil, R.T., 2012. Modelling of thermal environment in covered and plinth storage of wheat as effect of colour of plastic sheet. *Journal of Agricultural Engineering*, 49(1), pp. 36–42.

Janjai, S., Intawee, P., Kaewkiew, J., Sritus, C. and Khamvongsa, V., 2011. A large-scale solar greenhouse dryer using polycarbonate cover: modeling and testing in a tropical environment of Lao People's Democratic Republic. *Renewable Energy*, 36(3), pp. 1053–1062.

Janjai, S. and Keawprasert, T.K., 2006. Design and performance of a solar tunnel dryer with. *International Energy Journal*, 7(3), pp. 187–194.

Kadam, D.M., Nangare, D.D., Singh, R. and Kumar, S., 2011. Low-cost greenhouse technology for drying onion (*Allium Cepa* l.) slices. *Journal of Food Process Engineering*, 34(1), pp. 67–82.

Kale, S.J. and Nath, P., 2018. Kinetics of quality changes in tomatoes stored in evaporative cooled room in hot region. *International Journal of Current Microbiology and Applied Sciences*, 7(6), pp. 1104–1112.

Kale, S.J., Nath, P., Jalgaonkar, K.R. and Mahawar, M.K., 2016. Low cost storage structures for fruits and vegetables handling in Indian conditions. *Indian Horticulture Journal*, 6(3), pp. 376–379.

Kitinoja, L., 2013. Returnable Plastic Crate (RPC) systems can reduce postharvest losses and improve earnings for fresh produce operations. PEF White Paper, (13–01).

Kulanthaisami, S., Subramanian, P., Mahendiran, R., Venkatachalam, P. and Sampathrajan, A., 2009. Drying characteristics of coconut in solar tunnel dryer. *Madras Agricultural Journal*, 96(1/6), pp. 265–269.

Mangaraj, S., Goswami, T.K. and Mahajan, P.V., 2009. Applications of plastic films for modified atmosphere packaging of fruits and vegetables: a review. *Food Engineering Reviews*, 1(2), p. 133.

Marsh, K. and Bugusu, B., 2007. Food packaging-roles, materials, and environmental issues. *Journal of Food Science*, 72(3), pp. 39–55.

Nindo, C.I., Feng, H., Shen, G.Q., Tang, J. and Kang, D.H., 2003. Energy utilization and microbial reduction in a new film drying system. *Journal of Food Processing and Preservation*, 27(2), pp. 117–136.

Nindo, C.I. and Tang, J., 2007. Refractance window dehydration technology: a novel contact drying method. *Drying Technology*, 25(1), pp. 37–48.

Pavan, M.A., Schmidt, S.J. and Feng, H., 2012. Water sorption behavior and thermal analysis of freeze-dried, refractance window-dried and hot-air dried açaí (Euterpe oleracea Martius) juice. *LWT-Food Science and Technology*, 48(1), pp. 75–81.

Raghavi, L.M., Moses, J.A. and Anandharamakrishnan, C., 2018. Refractance window drying of foods: a review. *Journal of Food Engineering*, 222, pp. 267–275.

Sangamithra, A., Swamy, G.J., Prema, R.S., Priyavarshini, R., Chandrasekar, V. and Sasikala, S., 2014. An overview of a polyhouse dryer. *Renewable and Sustainable Energy Reviews*, 40, pp. 902–910.

Seveda, M.S., 2012. Design and development of walk-in type hemicylindrical solar tunnel dryer for industrial use. *ISRN Renewable Energy*, 2012, pp. 1–9.

Shahi, N.C., Khan, J.N., Lohani, U.C., Singh, A. and Kumar, A., 2011. Development of polyhouse type solar dryer for Kashmir valley. *Journal of Food Science and Technology*, 48(3), pp. 290–295.

Yadav, Y.K., 2017. Performance evaluation of solar tunnel dryer for drying of garlic. *Current Agriculture Research Journal*, 5(2), pp. 220–226.

2 Potassium-Solubilizing Microorganisms for Sustainable Agriculture

Zaffar Bashir, M. Y. Zargar, and
Dinesh Kumar Vishwakarma
Sher-e-Kashmir University of Agricultural Sciences
and Technology of Kashmir (SKUAST-K)

CONTENTS

2.1 INTRODUCTION

As per United Nations estimates, the global human population is projected to reach 8.9 billion by 2050, with developing country of the Asia and Africa to absorb the vast majority of the increase (Wood, 2001). The populations of the developing countries in the world continue to increase at an increased rate; so the demands of foods will be one of the greatest challenges faced by the human population. To meet this challenge, the soil biological system is needed to be focused with great effort. Soil is the natural body on the earth's crust. There are several minerals containing essential elements in the soil, but most important are nitrogen (N), phosphorus (P), and potassium (K). After N and P, K is the most important plant nutrient forgrowth, metabolism, and development of plants. In addition to increasing plant resistance to diseases, pests, and abiotic stresses, K is required to activate over 80 different enzymes responsible for plant and animal processes such as energy metabolism, starch synthesis, nitrate reduction, photosynthesis, and sugar degradation (Almeida et al., 2015; Cecílio Filho et al., 2015; Gallegos-Cedillo et al., 2016; Hussain et al.,

2016; White and Karley, 2010; Yang et al., 2015). For inadequate K intake, the plants will have poorly developed roots, slow growth, lower yields, (White and Karley, 2010) and prone to diseases (Amtmann et al., 2008; Armengaud et al., 2009) and pest (Amtmann and Armengaud, 2009; Troufflard et al., 2010). Sometime K requirement increases in the plant where agricultural soils lack sufficient phyto-available K for crop production (Mengel and Kirkby, 2001; Mikhailouskaya and Tcherhysh, 2005; Rashid et al., 2004; Rengel and Damon, 2008; Sindhu et al., 2012). It is generally supplied as K-fertilizers in both intensive and extensive agricultural systems (Zhang et al., 2013). K-solubilizing microorganisms (KSMs) are able to release potassium from insoluble minerals (Gundala et al., 2013; Archana et al., 2012, 2013; Sindhu et al., 2012). In addition, researchers have discovered that the K-solubilizing bacteria (KSB) which can provide beneficial effects on plant growth through suppressing pathogens and improving soil nutrients and structure.

The KSMs can promote K-solubilization from silicate mineral that can enhance the fertility status of soils. Rhizospheric microorganisms contribute directly and indirectly to the physical, chemical, and biological parameters of soil through their beneficial or detrimental activities. Rhizospheric bacteria helps in soil processes such as mobilization and mineralization of nutrients, soil organic matter decomposition and solubilization of K (Verma et al., 2012; Abhilash et al., 2013), and phosphate solubilization, nitrogen fixation, nitrification, denitrification, and sulphur reduction (Khan et al., 2007). The bacteria that possess potassium-solubilizing ability are called potassium-solubilizing bacteria or KSB and they can convert the insoluble potassium compounds into soluble forms and make them available to the plants (Zeng et al., 2012). An overdose use of chemical fertilizers have the negative environmental impacts and can increase the costs of crop production; therefore, there is an urgent need to implicit eco-friendly and cost-effective agro-technologies to hike crop production. So, the utilization of potassium solubilizers is considered to be a sound strategy in improving the productivity of agricultural lands. This new technique is also claimed to show the ability to restore the productivity of degraded, marginally productive, and unproductive agricultural soils (Rajawat et al., 2016). However, utilization of KSMs is found to be limited because of inexperienced farmers and practitioners (World Bank, 2007).

2.2 POTASSIUM IN SOIL

Potassium, an essential plant nutrient, has a major role in crop production. Its amount in a soil depends on the parent material as well as degree of weathering. Potassium is known to exist in structural, non-exchangeable, and water soluble forms. Potassium content of Indian soils has traditionally been considered as adequate, but in the recent years, the importance of K and the need for its continuous optimal availability for the better crop production has been observed as deficient due to the hidden hunger of K (Leaungvutiviroj et al., 2010). Among the major plant nutrients, potassium is one of the most abundant elements in soil. It is also one of the seven most common elements in the earth crust. On average, the surface layer (lithosphere) contains 2.6% potassium. The present conceptual understanding of soil potassium availability is the existence of four distinct K pools differing in accessibility to plant roots with reversible transfer of K between the pools (Syers, 2003). The K content of Indian soils varies from less than 0.5% to 3.00%. The average total K content of those soils is 1.52% (Mengel and Kirkby, 1987). However, total K is rather poorly correlated with available K and is rarely used to describe K fertility status of soil. The immediate source of K for plants is the small amount that present in the soil solution its average concentration is ranged from 1% to 2%. As this is removed, the equilibrium is distributed and K in the non-exchangeable and soil mineral fraction will be drawn upon. The availability of K to the plants depends directly on the concentration of K in soil solution and indirectly on soil (Goldstein, 1994). Soil solution K plays a vital role in providing the pathway for K uptake from the soil by plant roots (Öborn et al., 2005).

2.2.1 POTASSIUM FIXATION IN SOIL

In addition to releasing potassium, soil minerals can also fix potassium significantly affecting its availability. This involves the adsorption of K ions onto sites in the interlayers of weathered sheet silicates, such as vermiculite. The degree of K fixation in soils depends on the type of clay mineral and its charge density, moisture content, competing ions, and soil pH. Mont morillonite, vermiculite, and weathered micas are the major clay minerals that tend to fix K (Sparks, 1987). Soil wetting and drying also significantly affects the K fixation. The fixation process of K is relatively fast, whereas the release of fixed K is very slow due to the strong binding force between K and clay minerals (Öborn et al., 2005). Whether a soil fixes or releases K highly depends on the K concentration in the soil solution. As mentioned above, in addition to organic acids, the H^+ concentration in soil solution seems to play a key role in K release from clay minerals. Therefore, optimization of soil pH may be a means of enhancing K release. For optimized K fertilizer management practices, it is crucial to understand the factors that regulate K release from soil non-exchangeable pool.

2.3 K-SOLUBILIZING MICROORGANISM

A group of microorganisms was reported to be involved in the solubilization of fixed forms of K into available forms, which is absorbed by plants (Gundala et al., 2013). Microbial inoculants that can dissolve fixed K from rocks and mineral as well as enhance plant growth and yield are also eco-friendly. The first evidence of microbial involvement in solubilization of rock potassium had shown by Muentz (1890). A wide range of KSMs namely *Bacillus sp*, *Pseudomonas sp*, (Sheng et al., 2008; Lian et al., 2002; Liu et al., 2012; Basak and Biswas, 2012; Singh et al., 2010) have been reported to release potassium in accessible form from K-bearing minerals in soils. Several fungal and bacterial species, popularly called as potassium-solubilizing microorganisms or KSMs that assist plants growth by solubilization of insoluble forms of K. The KSMs are ubiquitous whose numbers vary from soil to soil. Rhizospheric microorganisms contribute significantly in the solubilization of locked form of soil minerals in the soil (Sindhu et al., 2009). Many microorganisms like bacteria, fungi and actinomycetes were colonized even on the surface of mountain rocks (Groudev,

1987; Gundala et al., 2013) have been reported that the silicate-solubilizing bacteria *B. mucilaginosus* sub spp. Siliceus liberates K from feldspar and alumino silicates. It has also been reported as silicate-solubilizing bacteria present in rhizosphere as well as non-rhizosphere soil (Lian et al., 2002; Liu et al., 2001). The K-solubilizing rhizobacteria were isolated from the roots of cereal crops which grown in potassium- and silicate-amended soil. A wide range of the rhizospheric microorganisms are reported as the K-solubilizers include *B. mucilaginosus* (Sugumaran and Janarthanam, 2007; Zarjani et al., 2013), *B. edaphicus* (Sheng et al., 2002), *B. circulanscan* (Lian et al., 2002), *Burkholderia, A. ferrooxidans* (Sheng and Huang, 2002; Sheng and He, 2006) *Arthrobacter sp.* (Zarjani et al., 2013), *Enterobacter hormaechei* (KSB-8) (Prajapati et al., 2013); *Paenibacillus mucilaginosus* (Liu et al., 2012), *P. frequentans, Cladosporium* (Argelis et al., 1993); *Aminobacter, Sphingomonas, Burkholderia* (Uroz et al., 2007); *Paenibacillus glucanolyticus* (Sangeeth et al., 2012).

2.4 MECHANISM OF POTASSIUM SOLUBILIZATION

Indigenous rhizospheric microorganisms have the potential to solubilize the fixed or locked form of nutrients (potassium) from trace mineral sources. Mechanism of K-solubilization could be mainly attributed to excrete organic acids which either directly dissolves rock K or chelate silicon ions to extract K into solution (Prajapati et al., 2013). Many researchers have quantitatively investigated the ability of KSMs to solubilize insoluble potassium in liquid Aleksandrov broth medium (Archana et al., 2013; Maurya et al., 2014). By the mechanism of potassium solubilization, insoluble potassium and structural unavailable forms of potassium compounds are solubilized due to the production of various types of organic acids (Table 2.1). These acids are

---

TABLE 2.1

Microorganisms Produce Various Organic Acids Which Solubilize Insoluble Potassium to Soluble Potassium

S. No.	Microorganisms: Bacteria and Fungi	Organic Acid Produced	References
1.	*Aspergillus flavus*	Citric, Oxalic, Gluconic, Succinic	Maliha et al., 2004
2.	*Aspergilluscandidus, Aspergillusflavus*	Malic, Gluconic, Oxalic	Shin et al., 2006
3.	*Serratia marcescens (CC-BC14)*	Citric, Lactic	Chen et al., 2006
4.	*Chryseobacterium (CC-BC05)*	Citric	Chen et al., 2006
5.	*Trichoderma sp, A.terreus, A.wenti, Pencillium sp, Fusarium oxysporium*	Lactic, malic, acetic, Tartaric, Fumaric, Citric, Gluconic	Akintokun et al., 2007
6.	*Aspergillus niger, Pencillium sp,*	Oxalic, Citric	Arwidsson et al., 2010
7.	*Pseudomonas Trivalis (BIHB 769)*	Gluconic, Lactic, Succinic, Formic, and Malic	
8.	*Aspergillus awamori S19*	Malic, Citric, and Fumaric	Jain et al., 2012
9.	*Enterobacter sp.FS-11*	Malic, Gluconic	Shahid et al., 2012
10.	*Aspergillus niger FSI*	Citric, Gluconic, and Oxalic	Mendes et al., 2013

accompanied by acidolysis, complexolysis exchange reactions, and these are key processes attributed to their conversion in soluble form (Uroz et al., 2009). The organic and inorganic acids convert insoluble K (muscovite, mica, biotite, and feldspar) to the soluble form of K that result increasing the availability of nutrients to plant. Sheng and He (2006) reported that solubilization of illite and feldspar by microorganisms is due to the production of organic acids like oxalic acid and tartaric acids, gluconic acid and 2-ketogluconic acid, oxalic acid, citric acid, malic acid, and succinic acid. Tartaric acid seems to be the most frequent agent of mineral K-solubilization (Zarjani et al., 2013). Other organic acids, such as acetic, citric, lactic, propionic, glycolic, oxalic, succinic acid, fumaric, tartaric, etc. have also been identified among K-solubilizers (Wu et al., 2005). Sheng and Huang, 2002 found that K release from the minerals was affected by pH, O_2, and the bacterial strain used. The efficiency of the K-solubilization by different microorganisms was found to vary with the nature of potassium-bearing minerals and aerobic conditions (Uroz et al., 2009).

Thus, the synthesis and discharge of organic acids by the microorganisms into the surrounding environment acidify the microbe's cells and their surrounding environment that ultimately lead to the release of K ions from the mineral K by protonation and acidification (Goldstein, 1994).

2.5 POTENTIAL ROLE OF POTASSIC BIOFERTILIZER

Use of chemical fertilizer has a negative impact on the environment. Potassium-solubilizing microbes (KSM) could serve as inoculants. They convert fixed form of potassium in the soil into available form that plants can use. This is a promising strategy for the improvement of plant absorption of potassium and so reducing the use of chemical fertilizer (Zhang and Kong, 2014). Potassic bio-fertilizers in agriculture can improve soil fertility, yield-attributing characters, and thereby final yield has been reported by many workers. In addition, their application in soil improves soil biota and minimizes the sole use of chemical fertilizers. It is an established fact that the Indian soil is rich source of potassium-containing secondary minerals which is not available to plant but can be available to plant using KSB. So the inoculations with KSM in the soil become necessary to restore and maintain the effective microbial populations for solubilization of fixed K and availability of other macro- and micronutrients to harvest good sustainable yield of various crops.

2.6 EFFECT OF KSMS ON PLANT GROWTH AND YIELD

Potassium solubilization is done by a wide range of bacteria, fungal strains, and actinomycetes (Ahmad et al., 2016; Bakhshandeh et al., 2017; Gundala et al., 2013; Meena et al., 2014).

There are strong evidences that soil bacteria are capable of transforming soil K to the forms available to plant (Meena et al., 2015; Meena et al., 2014; Meena et al., 2016). Inoculation of seedling and seeds with the KSMs generally showed significant enhancement of germination percentage, plant growth, and yield and K uptake by plants under glass house and field conditions (Zhang et al., 2013). The application of organo minerals with combination of siliate bacteria for enhancing plant growth and yield of maize and wheat was first reported by Aleksandrov (1985). More importantly, research investigation conducted under field level test crops such as wheat, forage crop, maize, and sudan grass crops have revealed that KSMs could drastically reduce the usage of chemical (Xie, 1998).

As reported by previous researchers (Sindhu et al., 2012; Zeng et al., 2012), the enhancement of plant K nutrition might be due to the stimulation of root growth or the elongation of root hairs by specific microorganisms. Thus, no direct increase in the availability of soil solution K is expected. The KSMs have been isolated from rhizospheric soil of various plants and from K-bearing minerals such as wheat (Parmar and Sindhu, 2013; Zhang et al., 2013), feldspar (Sheng et al., 2008), potato-soybean cropping sequence (Biswas, 2011), Iranian soils (Zarjani et al., 2013), Ceramic industry soil, mica core of Andhra Pradesh (Gundala et al., 2013), common bean (Kumar et al., 2012), biofertilizers (Zakaria, 2009), sorghum, maize, bajra, chilli (Archana et al., 2013), cotton, tomato, soybean, ground nut, and banana (Archana et al., 2012), soil of Tianmu Mountain, Zhejiang province (China) (Hu et al., 2006), rice (Muralikannan and Anthomiraj, 1998), tea, valencia orange (Shaaban et al., 2012), black pepper (Sangeeth et al., 2012), potato (Abdel-Salam and Shams, 2012), growth by improving by N-fixer, P-, and K-solubilizers are another beneficial effect of microorganisms with K-solubilizing potential (Basak and Biswas, 2012). A hydroponics study was carried out by Singh et al. (2010) to evaluate the effect of *B. mucilaginosus, Azotobacter chroococcum*, and *Rhizobium spp.* on their ability to mobilize K from waste mica using maize and wheat as the test crops under a phytotron growth chamber. Thus, the application of KSB as biofertilizers for agriculture improvement can reduce the use of agrochemicals and support eco-friendly crop production (Archana et al., 2013).

Therefore, it is imperative to isolate more species of mineral-solubilizing bacteria to enrich the pool of microbial species and genes as microbial fertilizers, which will be a great benefit to the ecological development of agriculture (Liu et al., 2012).

2.7   FUTURE PROSPECTS

Potassium solubilizers are the basic components of soil microbial community and play an important role in making K available to plants. These solubilizers have good potential for making use of fixed K and release K slowly under soil systems with low K availability in tropical and subtropical developing countries (Figure 2.1).

The mechanism of K-solubilization by microorganisms have been studied in detail but the K-solubilization is a complex phenomenon affected by many factors, such as solubilizers used, nutritional status of soil, mineral type and environmental factors. Moreover, the stability of the KSMs after inoculation in soil is also important for K-solubilization to benefit crop growth and development. Therefore, further study is needed to understand the problem of development of efficient and indigenous potassium-solubilizing microbial consortium for growth and yield of crops.

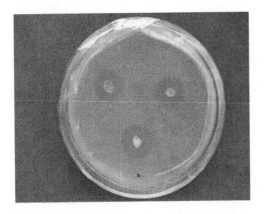

FIGURE 2.1 KSM on Aleksandrov medium and direct and indirect ways to boost growth of plants.

2.8 CONCLUSION

Rhizosphere microorganisms contribute significantly in solubilization of fixed forms of K in the soil inoculation of KSMs in soil, which has been shown to improve solubilization of insoluble mineral K resulting in higher crop performances. Although adequate amount of K is present in the soil, its availability to the plants is limited and thus becomes limiting factor for crop production. Applied chemical K fertilizers are ineffective and their prices increases at increased rate. KSMs bring out the locked K into plant utilizable form through the process of solubilization. The insoluble K is solubilized with the secretion of low molecular weight organic acids.

REFERENCES

Abdel-Salam, M.A. and Shams, A.S., 2012. Feldspar-K fertilization of potato (Solanum tuberosum L.) augmented by biofertilizer. *Journal of Agriculture and Environmental Sciences, 12*(6), pp. 694–699.

Abhilash, P.C., Dubey, R.K., Tripathi, V., Srivastava, P., Verma, J.P. and Singh, H.B., 2013. Remediation and management of POPs-contaminated soils in a warming climate: challenges and perspectives. *Environmental Science and Pollution Research, 20*(8), pp. 5879–5885.

Ahmad, M., Nadeem, S.M., Naveed, M. and Zahir, Z.A., 2016. Potassium-solubilizing bacteria and their application in agriculture. In: Meena, V.S., Maurya, B.R., Verma, J.P. and Meena, R.S. (Eds.), *Potassium Solubilizing Microorganisms for Sustainable Agriculture* (pp. 293–313). Springer, New Delhi.

Akintokun, A.K., Akande, G.A., Akintokun, P.O., Popoola, T.O.S. and Babalola, A.O., 2007. Solubilization of insoluble phosphate by organic acid-producing fungi isolated from Nigerian soil. *International Journal of Soil Science, 2*(4), pp. 301–307.

Aleksandrov, V.G., 1985. Organo-mineral fertilizers and silicate bacteria. *Dokl Akad Nauk, 7*, pp. 43–48.

Almeida, H.J., Pancelli, M.A., Prado, R.M., Cavalcante, V.S. and Cruz, F.J.R., 2015. Effect of potassium on nutritional status and productivity of peanuts in succession with sugar cane. *Journal of Soil Science and Plant Nutrition, 15*(1), pp. 1–10.

Amtmann, A., Troufflard, S. and Armengaud, P., 2008. The effect of potassium nutrition on pest and disease resistance in plants. Physiologia *Plantarum,* 133(4), pp. 682–691.

Amtmann, A. and Armengaud, P., 2009. Effects of N, P, K and S on metabolism: new knowledge gained from multi-level analysis. Current *Opinion* in *Plant Biology,* 12(3), pp. 275–283.

Armengaud, P., Sulpice, R., Miller, A.J., Stitt, M., Amtmann, A. and Gibon, Y., 2009. Multilevel analysis of primary metabolism provides new insights into the role of potassium nutrition for glycolysis and nitrogen assimilation in Arabidopsis roots. Plant Physiology, 150(2), pp. 772–785.

Arwidsson, Z., Johansson, E., von Kronhelm, T., Allard, B. and van Hees, P., 2010. Remediation of metal contaminated soil by organic metabolites from fungi I—production of organic acids. Water, Air, and Soil Pollution, 205(1–4), p. 215.

Archana, D.S., Nandish, M.S., Savalagi, V.P. and Alagawadi, A.R., 2012. Screening of potassium solubilizing bacteria (KSB) for plant growth promotionalactivity. *BIOINFOLET—A Quarterly Journal of Life Sciences*, 9(4), pp. 627–630.

Archana, D.S., Nandish, M.S., Savalagi, V.P. and Alagawadi, A.R., 2013. Characterization of potassium solubilizing bacteria (KSB) from rhizosphere soil. *BIOINFOLET—A Quarterly Journal of Life Sciences,* 10(1b), pp. 248–257.

Argelis, D.T., Gonzala, D.A., Vizcaino, C. and Gartia, M.T., 1993. Biochemical mechanisms of stone alteration carried out by filamentous fungi living in monuments. *Biogeochemistry, 19*(3), pp. 129–147.

Bakhshandeh, E., Pirdashti, H. and Lendeh, K.S., 2017. Phosphate and potassium-solubilizing bacteria effect on the growth of rice. *Ecological Engineering, 103*, pp. 164–169.

Basak, B. and Biswas, D., 2012. *Modification of Waste Mica for Alternative Source of Potassium: Evaluation of Potassium Release in Soil from Waste Mica Treated with Potassium Solubilizing Bacteria (KSB).* LAP—LAMBERT Academic Publishing, Riga, pp. 2–4.

Biswas, D.R., 2011. Nutrient recycling potential of rock phosphate and waste mica enriched compost on crop productivity and changes in soil fertility under potato–soybean cropping sequence in an Inceptisol of Indo-Gangetic plains of India. *Nutrient Cycling in Agroecosystems, 89*(1), pp. 15–30.

Cecílio Filho, A.B., Feltrim, A.L., Mendoza Cortez, J.W., Gonsalves, M.V., Pavani, L.C. and Barbosa, J.C., 2015. Nitrogen and potassium application by fertigation at different watermelon planting densities. *Journal of Soil Science and Plant Nutrition, 15*(4), pp. 928–937.

Chen, Y.P., Rekha, P.D., Arun, A.B., Shen, F.T., Lai, W.A. and Young, C.C., 2006. Phosphate solubilizing bacteria from subtropical soil and their tricalcium phosphate solubilizing abilities. *Applied Soil Ecology, 34*(1), pp. 33–41.

Gallegos-Cedillo, V.M., Urrestarazu, M. and Álvaro, J.E., 2016. Influence of salinity on transport of Nitrates and Potassium by means of the xylem sap content between roots and shoots in young tomato plants. *Journal of Soil Science and Plant Nutrition, 16*(4), pp. 991–998.

Goldstein, A.H., 1994. Involvement of the quinoprotein glucose dehydrogenase in the solubilization of exogenous phosphates by gram-negative bacteria. In: Torriani-Gorini, A., Yagil, E. and Silver, S. (Eds.), *Phosphate in Microorganisms: Cellular and Molecular Biology* (pp. 197–203). ASM Press, Washington, DC.

Groudev, S.N., 1987. Use of heterotrophic microorganisms in mineral biotechnology. *Acta Biotechnologica, 7*(4), pp. 299–306.

Gundala, P.B., Chinthala, P. and Sreenivasulu, B., 2013. A new facultative alkaliphilic, potassium solubilizing, Bacillus Sp. SVUNM9 isolated from mica cores of Nellore District, Andhra Pradesh, India. Research and reviews. *Journal of Microbiology and Biotechnology, 2*(1), pp. 1–7.

Hu, X., Chen, J. and Guo, J., 2006. Two phosphate-and potassium-solubilizing bacteria isolated from Tianmu Mountain, Zhejiang, China. *World Journal of Microbiology and Biotechnology, 22*(9), pp. 983–990.

Hussain, Z., Khattak, R.A., Irshad, M., Mahmood, Q. and An, P., 2016. Effect of saline irrigation water on the leachability of salts, growth and chemical composition of wheat (Triticum aestivum L.) in saline-sodic soil supplemented with phosphorus and potassium. *Journal of Soil Science and Plant Nutrition, 16*(3), pp. 604–620.

Jain, R., Saxena, J. and Sharma, V., 2012. Solubilization of inorganic phosphates by Aspergillus awamori S19 isolated from rhizosphere soil of a semi-arid region. *Annals of Microbiology, 62*(2), pp. 725–735.

Khan, M.S., Zaidi, A. and Wani, P.A., 2007. Role of phosphate-solubilizing microorganisms in sustainable agriculture—a review. *Agronomy for Sustainable Development, 27*(1), pp. 29–43.

Kumar, P., Dubey, R.C. and Maheshwari, D.K., 2012. Bacillus strains isolated from rhizosphere showed plant growth promoting and antagonistic activity against phytopathogens. *Microbiological Research, 167*(8), pp. 493–499.

Leaungvutiviroj, C., Ruangphisarn, P., Hansanimitkul, P., Shinkawa, H. and Sasaki, K., 2010. Development of a new biofertilizer with a high capacity for N2 fixation, phosphate and potassium solubilization and auxin production. *Bioscience, Biotechnology, and Biochemistry, 74*(5), pp. 1098–1101.

Lian, B., Fu, P.Q., Mo, D.M. and Liu, C.Q., 2002. A comprehensive review of the mechanism of potassium releasing by silicate bacteria. *Acta Mineralogica Sinica, 22*(2), pp. 179–183.

Liu, D., Lian, B. and Dong, H., 2012. Isolation of Paenibacillus sp. and assessment of its potential for enhancing mineral weathering. *Geomicrobiology Journal, 29*(5), pp. 413–421.

Liu, G., Lin, Y. and Huang, Z., 2001. Screening of silicate bacteria with potassium-releasing and antagonistical activity. *Chinese Journal of Applied & Environmental Biology, 7*(1), pp. 66–68.

Maliha, R., Khalil, S., Ayub, N., Alam, S. and Latif, F. 2004. Organic Acids Production and Phosphate Solubilization by Phosphate Solubilizing Microorganisms (PSM) Under in vitro Conditions. *Pakistan Journal of Biological Science, 7*(2), pp. 187–196.

Maurya, B.R., Meena, V.S. and Meena, O.P., 2014. Influence of Inceptisol and Alfisol's potassium solubilizing bacteria (KSB) isolates on release of K from waste mica. *Vegetos, 27*(1), pp. 181–187.

Meena, V.S., Maurya, B.R. and Bahadur, I., 2015. Potassium solubilization by bacterial strain in waste mica. *Bangladesh Journal of Botany, 43*(2), pp. 235–237.

Meena, V.S., Maurya, B.R. and Verma, J.P., 2014. Does a rhizospheric microorganism enhance K$^+$ availability in agricultural soils? *Microbiological Research, 169*(5–6), pp. 337–347.

Meena, V.S., Maurya, B.R., Verma, J.P. and Meena, R.S. eds., 2016. *Potassium Solubilizing Microorganisms for Sustainable Agriculture.* Springer, New Delhi.

Mendes, G.O., Dias, C.S., Silva, I.R., Júnior, J.I.R., Pereira, O.L. and Costa, M.D., 2013. Fungal rock phosphate solubilization using sugarcane bagasse. *World Journal of Microbiology and Biotechnology, 29*(1), pp. 43–50.

Mengel, K. and Kirkby, E.A., 1987. Potassium in crop production. In *Advances in* Agronomy. (4th ed., Vol. 33, pp. 59–110). International Potash Institute and Academic Press, Bern.

Mengel, K. and Kirkby, E.A. 2001. *Principles of Plant Nutrition* (5th ed.). Kluwer Academic Publishers, Dordrecht, p. 849.

Mikhailouskaya, N. and Tcherhysh, A. 2005. K-mobilizing bacteria and their effect on wheatyield. *Latvian Journal of Agronomy, 8*, pp. 154–157.

Muentz, A., 1890. Sur la decomposition des roches et la formation de la terre arable. *Comptes Rendus Academic Science, 110*, pp. 1370–1372.

Muralikannan, N. and Anthomiraj, S. 1998. Occurrence of silicate solubilizing bacteria in riceecosystem. *Madras Agriculture Journal, 85*, pp. 47–50.

Öborn, I., Andrist-Rangel, Y., Askekaard, M., Grant, C.A., Watson, C.A. and Edwards, A.C., 2005. Critical aspects of potassium management in agricultural systems. *Soil Use and Management, 21*(1), pp. 102–112.

Parmar, P. and Sindhu, S.S., 2013. Potassium solubilization by rhizosphere bacteria: influence of nutritional and environmental conditions. *Journal of Microbiology Research, 3*(1), pp. 25–31.

Prajapati, K., Sharma, M.C. and Modi, H.A., 2013. Growth promoting effect of potassium solubilizing microorganisms on Abelmoscus esculantus. *International Journal of Agriculture Sciences, 3*(1), pp. 181–188.

Rajawat, M.V.S., Singh, S., Tyagi, S.P. and Saxena, A.K., 2016. A modified plate assay for rapid screening of potassium-solubilizing bacteria. *Pedosphere, 26*(5), pp. 768–773.

Rashid, M., Khalil, S., Ayub, N., Alam, S. and Latif, F., 2004. Organic acids production and phosphate solubilization by phosphate solubilizing microorganisms (PSM) under in vitro conditions. *Pakistan Journal of Biological Sciences, 7*(2), pp. 187–196.

Rengel, Z. and Damon, P.M., 2008. Crops and genotypes differ in efficiency of potassium uptake and use. *Physiologia Plantarum, 133*(4), pp. 624–636.

Sangeeth, K.P., Bhai, R.S. and Srinivasan, V., 2012. Paenibacillus glucanolyticus, a promising potassium solubilizing bacterium isolated from black pepper (Piper nigrum L.) rhizosphere. *Journal of Spices and Aromatic Crops, 21*(2), pp. 118–124.

Shaaban, E.A., El-Shamma, I.M.S., El Shazly, S., El-Gazzar, A. and Abdel-Hak, R.E., 2012. Efficiency of rock-feldspar combined with silicate dissolving bacteria on yield and fruit quality of valencia orange fruits in reclaimed soils. *Journal of Applied Sciences Research, 8*, pp. 4504–4510.

Shahid, M., Hameed, S., Imran, A., Ali, S. and van Elsas, J.D., 2012. Root colonization and growth promotion of sunflower (Helianthus annuus L.) by phosphate solubilizing Enterobacter sp. Fs-11. *World Journal of Microbiology and Biotechnology, 28*(8), pp. 2749–2758.

Sheng, X.F. and He, L.Y., 2006. Solubilization of potassium-bearing minerals by a wild-type strain of Bacillus edaphicus and its mutants and increased potassium uptake by wheat. *Canadian Journal of Microbiology, 52*(1), pp. 66–72.

Sheng, X.F., He, L.Y. and Huang, W.Y., 2002. The conditions of releasing potassium by a silicate-dissolving bacterial strain NBT. Agricultural Sciences in China, 1(6), pp. 662–666.

Sheng, X. and Huang, W., 2002. Mechanism of potassium release from feldspar affected by the sprain Nbt of silicate bacterium. *Acta Pedologica Sinica, 39*(6), pp. 863–871.

Sheng, X.F., Zhao, F., He, L.Y., Qiu, G. and Chen, L., 2008. Isolation and characterization of silicate mineral-solubilizing Bacillus globisporus Q12 from the surfaces of weathered feldspar. *Canadian Journal of Microbiology, 54*(12), pp. 1064–1068.

Shin, W., Ryu, J., Kim, Y., Yang, J., Madhaiyan, M. and Sa, T., 2006, July. Phosphate solubilization and growth promotion of maize (Zea mays L.) by the rhizosphere soil fungus Penicillium oxalicum. *18th world conference of soil science, Philadelphia, Pennsylvania.*

Sindhu, S.S., Parmar, P. and Phour, M., 2012. Nutrient cycling: potassium solubilization by microorganisms and improvement of crop growth. In *Geomicrobiology and Biogeochemistry* (pp. 175–198). Springer, Berlin and Heidelberg.

Sindhu, S.S., Verma, M.K. and Mor, S., 2009. Molecular genetics of phosphate solubilization in rhizosphere bacteria and its role in plant growth promotion. In *Phosphate Solubilizing Microbes and Crop Productivity* (pp. 199–228). Nova Science Publishers, New York.

Singh, G., Biswas, D.R. and Marwaha, T.S., 2010. Mobilization of potassium from waste mica by plant growth promoting rhizobacteria and its assimilation by maize (Zea mays) and wheat (Triticum aestivum L.): a hydroponics study under phytotron growth chamber. *Journal of Plant Nutrition*, *33*(8), pp. 1236–1251.

Sparks, D.L., 1987. Potassium dynamics in soils. Advances in Soil Science pp. (1–63). Springer, New York.

Sugumaran, P. and Janarthanam, B., 2007. Solubilization of potassium containing minerals by bacteria and their effect on plant growth. *World Journal of Agricultural Sciences*, *3*(3), pp. 350–355.

Syers, J.K. 2003. Potassium in soils: current concepts. Feed the soil to feed the people the role of potash in susta-inable agriculture. *Proceedings of IPI Golden Jubilee Congress 1952–2002 held at Basel*, Switzerland, 8–10 October 2002, p. 301.

Troufflard, S., Mullen, W., Larson, T.R., Graham, I.A., Crozier, A., Amtmann, A. and Armengaud, P., 2010. Potassium deficiency induces the biosynthesis of oxylipins and glucosinolates in Arabidopsis thaliana. *BMC Plant Biology*, *10*(1), p. 172.

Uroz, S., Calvaruso, C., Turpault, M.P. and Frey-Klett, P., 2009. Mineral weathering by bacteria: ecology, actors and mechanisms. *Trends in Microbiology*, *17*(8), pp. 378–387.

Uroz, S., Calvaruso, C., Turpault, M.P., Pierrat, J.C., Mustin, C. and Frey-Klett, P., 2007. Effect of the mycorrhizosphere on the genotypic and metabolic diversity of the bacterial communities involved in mineral weathering in a forest soil. *Applied and Environmental Microbiology*, *73*(9), pp. 3019–3027.

Verma, J.P., Yadav, J. and Tiwari, K.N., 2012. Enhancement of nodulation and yield of chickpea by co-inoculation of indigenous mesorhizobium spp. and plant growth–promoting Rhizobacteria in Eastern Uttar Pradesh. *Communications in Soil Science and Plant Analysis*, *43*(3), pp. 605–621.

White, P.J. and Karley, A.J., 2010. Potassium. In: Hell, R. and Mendel, R.R. (Eds.), *Cell Biology of Metals and Nutrients, Plant Cell Monographs* (Vol. 17, pp. 199–224). Springer, Berlin and Heidelberg.

Wood, N.T., 2001. Nodulation by numbers: the role of ethylene in symbiotic nitrogen fixation. *Trends in Plant Science*, *6*(11), pp. 501–502.

World Bank. 2007. *World Development Report 2008: Agriculture for Development*. The World Bank, Washington, DC.

Wu, S.C., Cao, Z.H., Li, Z.G., Cheung, K.C. and Wong, M.H., 2005. Effects of biofertilizer containing N-fixer, P and K solubilizers and AM fungi on maize growth: a greenhouse trial. *Geoderma*, *125*(1–2), pp. 155–166.

Xie, J.C., 1998. Present situation and prospects for the world's fertilizer use. *Plant Nutrition and Fertilizer Science*, *4*, pp. 321–330.

Yang, B.M., Yao, L.X., Li, G.L., He, Z.H. and Zhou, C.M., 2015. Dynamic changes of nutrition in litchi foliar and effects of potassium-nitrogen fertilization ratio. *Journal of Soil Science and Plant Nutrition*, *15*(1), pp. 98–110.

Zakaria, A.A.B., 2009. *Growth Optimization of Potassium Solubilizing Bacteria Isolated from Biofertilizer*. Faculty of Chemical & Natural Resources Engineering, University Malaysia Pahang, Gambang, p. 40.

Zarjani, J.K., Aliasgharzad, N., Oustan, S., Emadi, M. and Ahmadi, A., 2013. Isolation and characterization of potassium solubilizing bacteria in some Iranian soils. *Archives of Agronomy and Soil Science*, *59*(12), pp. 1713–1723.

Zeng, X., Liu, X., Tang, J., Hu, S., Jiang, P., Li, W. and Xu, L., 2012. Characterization and potassium-solubilizing ability of Bacillus Circulans Z 1–3. *Advanced Science Letters*, *10*(1), pp. 173–176.

Zhang, C. and Kong, F., 2014. Isolation and identification of potassium-solubilizing bacteria from tobacco rhizospheric soil and their effect on tobacco plants. *Applied Soil Ecology*, *82*, pp. 18–25.

Zhang, A.M., Zhao, G.Y., Gao, T.G., Wang, W., Li, J., Zhang, S.F. and Zhu, B.C., 2013. Solubilization of insoluble potassium and phosphate by Paenibacillus kribensis CX-7: a soil microorganism with biological control potential. *African Journal of Microbiology Research*, *7*(1), pp. 41–47.

3 Weed Control for Conservation Agriculture in Climate Change Scenario

Parmeet Singh and Lal Singh
Sher-e-Kashmir University of Agricultural Sciences
and Technology of Kashmir (SKUAST-K)

CONTENTS

3.1 INTRODUCTION

As the population increases, food demands placed on the agricultural production systems will test the capabilities of current agriculture practices. Moreover, adequate food production in the future can only be achieved through the implementation of sustainable growing practices that minimize environmental degradation and preserve resources while maintaining high-yielding, profitable systems. To this end, conservation agriculture (CA) is a system designed to achieve sustainability by improving the biological functions of the agro-ecosystem with limited mechanical practices and judicious use of chemical inputs. CA is characterized by three linked principles, viz. (i) continuous minimum mechanical soil disturbance, (ii) permanent organic soil cover, and (iii) diversification of crop species grown in sequences and/or associations. While sometimes mistakenly used synonymously, it is the less intensive conservation tillage system that has become more recognized and adopted within the agricultural community. A host of benefits can be achieved through employing components of CA or conservation tillage, including reduced soil erosion and water runoff, increased productivity through improved soil quality, increased water availability, increased biotic diversity, and reduced labor demands. CA systems require a total paradigm shift from conventional agriculture r to management of crops, soil, water, nutrients, weeds, and farm machinery.

3.2   PROSPECTS OF CA

Globally, about 125 million ha area is practiced following the concepts and technologies for CA (Table 3.1). The major countries being USA (26.5 million ha), Brazil (25.5 million ha), Argentina (25.5 ha), Canada (13.5 million ha), and Australia (17.0 million ha), India has also started practicing CA technology, and about 3 million ha area of wheat grown under the rice–wheat system in the Indo-Gangetic plain is believed to be under resourced conservation technologies. The tillage system is gradually undergoing a paradigm shift from frequent tillage operations before sowing crops called as conventional tillage (CT), to no-tillage operation before sowing a crop, called as zero tillage (ZT). The ZT technology in rice–wheat cropping system is now foreshadowing nothing less than the end of an age-old concept, popularly known as more you till, more you eat. The need of the hour now is to infuse new technologies for further enhancing and sustaining the productivity as well as to tap new sources of growth in agricultural productivity. In this context, the role of CA in improving efficiency, equity, and environment is well recognized. The adoption of CA offers avenues for much needed diversification of agriculture, thus expanding the opportunities for cultivation of different crops during different seasons in the year. The prospects for introduction of sugarcane, pulses, vegetables, etc. as intercrop with wheat and winter maize provide good avenues for intensification and diversification

TABLE 3.1
Global Adoption of Conservation Agriculture Systems

Country	Area (M ha)	% of Global Total
USA	26.5	21.2
Brazil	25.5	20.4
Argentina	25.5	20.4
Australia	17.0	13.6
Canada	13.5	10.8
Russian Federation	4.5	3.6
China	3.1	2.5
Paraguay	2.4	1.9
Kazakhstan	1.6	1.3
Others	5.3	4.2
Total	124.8	100.0

of rice–wheat system. Resource conserving technologies help integrate crop, live-stock, land, and water management research in agro-ecological intensification of both low- and high-potential environments. Such technologies need to be developed and popularized extensively.

3.3 WEED MANAGEMENT

Weeds are one of the biggest constraints to the adoption of CA. Tillage affects weeds by uprooting, dismembering, and burying them deep enough to prevent emergence, by moving their seeds both vertically and horizontally, and by changing the soil environment and so promoting or inhibiting weed seed germination and emergence. Any reduction in tillage intensity or frequency may, therefore, have an influence on weed management. As the density of certain annual and perennial weeds can increase under CA, effective weed control techniques are required to manage weeds successfully. Crop yield losses in CA due to weeds may vary, depending on weed community and intensity. Weed species shifts and losses in crop yield as a result of increased weed density have been cited as major hurdles to the widespread adoption of CA. Implementation of CA has often caused yield reduction because reduced tillage failed to control weed interference. However, the recent development of post-emergence broad-spectrum herbicides provides an opportunity to control weeds in CA. Crop yields can be similar for conventional and conservation tillage systems if weeds are controlled and crop stands are uniform (Mahajan et al., 2002). The crop-ping system also plays an important role to influence weed flora in CA. There is also evidence of allelopathic properties of cereal residues in inhibiting weed germination (Jung et al., 2004). Weeds under CA may also be controlled when the cover crop is harvested or killed by herbicides. Farming practices that maintain soil microorgan-isms and microbial activity can also lead to weed suppression. Various approaches including the use of preventive measures, crop residue as mulches, intercropping, competitive crop cultivars, herbicide-tolerant cultivars, and herbicides are needed to manage weeds effectively in a CA system.

3.3.1 Preventive Measures

Seeds of most crops are contaminated with weeds, especially where weed seeds resemble the shape, size of crop seeds, and have similar life cycles. To obtain weed-free crop seeds, cultural, and mechanical measures need to be adopted. In undis-turbed or no-till systems, seeds of weeds and volunteer crops are deposited in the topsoil. There is no weed seed burial by tillage operations under CA. Therefore, an appropriate strategy is needed to avoid high weed infestations and prevent unaccept-able competition with the emerging crop.

3.3.2 Cultural Practices

In CA systems, stale seedbed practice is a valuable way of reducing weed pressure. This practice has been found very effective in zero-till wheat in the north-western Indo-Gangetic Plains. The main advantage of the stale seedbed practice is that the crop emerges in weed-free environments, and acquires a competitive advantage over late-emerging weed seedlings. Studies have suggested a small difference in weed populations between conventional and zero-till fields (Derksen et al., 1993), and in some cases, fewer weeds have been observed in zero-till conditions (Hobbs and Gupta, 2001; Singh et al., 2001; Malik et al., 2002). In CA system, time of sowing can be manipulated in such a way that ecological conditions for the germination of weed seeds are not met. In the north-western part of the Indo-Gangetic Plains, farmers advanced wheat seeding by 2 weeks to get a head start over noxious weed, Phalaris minor (Singh et al., 1999). One of the pillars of CA is ground cover with dead or live mulch, which leaves less time for weeds to establish during fallow or a turnaround period. Some other common problems under CA include emergence from recently produced weed seeds that remain near the soil surface, lack of disruption of perennial weed roots, interception of herbicides by thick surface residues, and change in timing of weed emergence. Shrestha et al. (2002) concluded that long-term changes in weed flora are driven by an interaction of several factors, such as tillage, environment, crop rotation, crop type, and the timing and type of weed management practice. Continuous ZT increased the population density of *Echinochloa colona* and *Cyperus iria* (*C. iria*) in rice but reduced the population of *Avena ludoviciana* and *Chenopodium album* in subsequent wheat (Mishra and Singh, 2012). Rotational tillage systems significantly reduced the seed density of *C. iria*, *Artemisia ludoviciana*, and *Montipora hispida* compared to continuous ZT or CT. It was concluded that continuous ZT with effective weed management using recommended herbicide + hand weeding was more remunerative and energy efficient in rice-wheat cropping system.

3.3.3 CROP RESIDUES

In CA, crop residues present on the soil surface improve soil and moisture conservation, and soil tilth. The germination response of weeds to residue depends on the quantity, position vertical or flat, and below- or above-weed seeds, allelopathic potential of the residue, and weed biology (Chauhan et al., 2006). Crop residues, when uniformly and densely present under CA, could suppress weed seedling emergence, delay the time of emergence, and allow the crop to gain an initial advantage in terms of early vigor over weeds. Cover crops, such as *Sesbania* can produce a green biomass of up to 30 t/ha within 60 days, and control most of the weeds, leaving fields almost weed free. In addition to reducing weed emergence, high amount of residue may prolong or delay emergence, which may have implications for weed management in CA. Delayed weed emergence allows the crop to take competitive advantage over weeds, and these weed seedlings are likely to have less impact on crop yield loss and weed seed production. Plants emerging earlier produce a greater number of seeds than the later emerging ones (Chauhan and Johnson, 2010).

3.3.4 INTERCROPPING

Intercrops can be more effective than sole crops in pre-empting resources used by weeds and suppressing weed growth. Intercropping of short-duration, quick-growing, and early-maturing legume crops with long-duration and wide-spaced crops leads to covering ground quickly and suppressing emerging weeds effectively. Maize–legume intercropping results in higher canopy cover and decreased light availability for weeds, leading to reduction in weed density and dry matter compared with sole crops. Brown manuring involving growing of *Sesbania* along with direct-seeded rice or maize as inter- or mixed-crop for 25–30 days and then killing *Sesbania* by 2,4-D spray or mechanical means has been found to be a highly beneficial resource-conserving technology for soil and water conservation, weed control, and nutrient supplementation (Sharma et al., 2010).

3.3.5 CROP DIVERSIFICATION

Continuous cultivation of a single crop or crops having similar management practices allow certain weed species to become dominant in the system, and over time, these weed species become hard to control. Therefore, it is very important to rotate crops having a different growing period. Different crops require different management practices, which may help in disturbing the growing cycle of weeds and prevent selection of the weed flora toward increased abundance of problem species. Crop rotation is an effective practice for management of *Phalaris minor* because selection pressure is diversified by changing patterns of disturbances (Bhan and Kumar, 1997; Chhokar and Malik, 2002). Crop rotation also allows farmers to use new herbicides, and this practice may control problematic weeds.

3.3.6 CHEMICAL WEED MANAGEMENT

Weed management using herbicides has become an integral part of modern agriculture. In CT systems, crop residues generally are not present at the time of pre-emergence herbicide application. However, in CA systems, residues are present at the time of herbicide application, and may decrease the herbicide's effectiveness as the residues intercept the herbicide, thus reducing the amount of herbicide that can reach the soil surface and kill-germinating seeds. The efficacy of herbicides may also depend on the herbicide formulations under CA systems. For example, pre-emergence herbicides applied as granules may provide better weed control than liquid formations in no-till systems. Depending on the herbicide chemical properties and formulations, some herbicides intercepted by crop residues in CA systems are prone to volatilization, photo degradation, and other losses. As the effect of no-till systems on weed control varies with weed species and herbicides used, choosing an appropriate herbicide and appropriate timing is very critical in CA systems (Chauhan et al., 2006). Nevertheless, injudicious and continuous use of a single herbicide over a long period of time may result in the development of resistant biotypes, shifts in weed flora, and negative effects on the succeeding crop and environment. Therefore, for the sustenance of CA systems, herbicide rotation and/or integration of weed management practices are needed. Any single method of weed control cannot provide season-long and effective weed control under CA systems. Therefore, a combination of different weed management strategies should be evaluated for widening the weed control spectrum and efficacy for sustainable crop production. The use of clean crop seeds and seeders and field sanitation (irrigation canals and bunds free from weeds) should be integrated for effective weed management. Combining good agronomic practices, timeliness of operations, fertilizer, water management, and retaining crop residues on the soil surface improve the weed control efficiency of applied herbicides and competitiveness against weeds. Approaches such as stale seedbed practice, uniform and dense crop establishment, use of cover crops and crop residues as mulch, crop rotations, and practices for enhanced crop competitiveness with a combination of pre- and post-emergence herbicides could be integrated to develop sustainable and effective weed management strategies under CA systems. Based on extensive field experiments on conservation agriculture systems in diversified cropping systems at the IARI, New Delhi during the last decade, the following broad conclusions have been made:

- It is possible to achieve the same or even higher yield with ZT as with CT.
- Retention of crop residues on soil surface is essential for success of ZT in the long-run.
- ZT along with residue has beneficial effects on soil moisture, temperature moderation, and weed control.
- Zero-till systems cause shift in weed flora and may result in emergence of perennial weeds like *Cyperus* and *Cynodon.*
- Restricting tillage reduces weed control options and increases reliance on herbicides.
- Altering tillage practices change weed seed depth in the soil, which play a role in weed species shifts and affect the efficacy of control practices.
- CA is a machine-, herbicide- and management-driven agriculture for its successful adoption.
- Integrated weed management involving chemical and non-chemical methods (residue, cover crops, varieties, etc.) is essential for success of CA systems in the long run.

3.4   LIMITATIONS IN ADOPTION OF CA SYSTEMS

CA has problems both for scientists and farmers to overcome the past mindset and explore the opportunities. Spread of conservation agriculture is constrained due to non-availability of suitable machinery, competing use of crop residues, weed management problems, particularly of perennial species, localized insect and disease infestation, and more importantly likelihood of lower crop productivity, at least in the short term. Biophysical, economic, social, and cultural constrains limit the adoption of this promising innovation of the 20th century by the resource-poor small land farmers of south and south-east Asia (Lal, 2007). Despite several payoffs, there are also many trade-offs to adoption of CA systems (Table 3.2).

3.5   CLIMATE CHANGE AND WEED MANAGEMENT

Global climate change will alter many elements of the future crop production. Atmospheric CO_2 concentration, average temperature, and tropospheric ozone (O_3) concentration will be higher, droughts will be more frequent and severe, more intense precipitation events will lead to increased flooding, some soils will degrade,

TABLE 3.2

Two Sides of No-Till Conservation Agriculture

Payoffs	Trade-Offs
Reduces soil erosion	Transition from conventional farming to no-till farming is difficult
Conserves water	Necessary equipment is costly
Improves soil health	Heavier reliance on herbicides
Reduces fuel and labor costs	Prevalence of weeds, disease, and other pests may shift in unexpected ways
Reduces sediment and fertilizer pollution of lakes and streams	May initially require more N fertilizers
Sequesters carbon	Can slow germination and reduce yields

Source: Huggins and Reganold (2008).

and climatic extremes will be more likely to occur (IPCC, 2007). Changes in climate influence not only the performance of individual organism but also impact interactions with other organisms at various stages in their life histories via changes in morphology, physiology, and chemistry. Weeds have better adaptability to the changing environment by virtue of greater genetic diversity. Evidences indicate that few weeds species respond strongly to recent increases in atmospheric CO_2. To our belief, weeds may be a better source of genetic materials for genetic engineering of crop plants on account of hardiness to stress factors, relatedness, and their co-existence with crop plants and may offer a better chance of introgression and interaction at macro-molecular level (i.e., up and downstream components which otherwise may be absent in unrelated and non-co-existed species).

Numerous studies can be found in literature regarding the crops–weeds interaction in field conditions suggesting that weeds pose a very serious and potential threat to agricultural production resulting approximately one-third yield loss by virtue of their competitiveness with special mention to obnoxious or invasive weeds like *Euphorbia geniculata* which outplay the crop plants in almost every aspect and led to big loss to crop production in India. Recently, agriculture scientists have started paying more attention toward crop–weed competitiveness in a high CO_2 environment (Ziska and George, 2004) and it has been suggested that possibly recent increases in atmospheric CO_2 during the 20th century may have been a factor in the selection of weed species and a contributing factor of invasiveness of weed species. Very recently, available literature on crop–weed interaction under climate change pertaining to Indian farming system was reviewed by Mahajan et al. (2012). They opined that productivity of rice and wheat would decline due to the increased climate variability, particularly due to gap in water supply and demand, and weed incidence (2,050 and beyond). Plants with C_4 carbon fixation pathway (mostly weeds) have a competitive advantage over crop plants possessing the more common C_3 pathway at elevated temperature. For a C_3 crop such as rice and wheat, elevated CO_2 may have positive effects on crop competitiveness with C_4 weeds (Mahajan et al., 2012), however, this may not be so always. All crop–weed competition studies, where the photosynthetic pathway is the same, weed growth is favored as CO_2 is increased. Therefore, the problems of *Phalaris minor* and *Avena ludoviciana* in wheat would aggravate with increase in CO_2 due to climate change. Under water stress conditions, *P. minor* had advantage over wheat with CO_2 enrichment (Naidu and Varshney, 2011). Studies on the effect of CO_2 enrichment on weed species at the Directorate of Weed Science, Jabalpur revealed that a few weed species such as *Dactyloctenium aegyptium* and *E. colona* responded to elevated CO_2, but *Cyperus rotundus* and *Eleusine indica* did not respond to CO_2 enrichment. In addition, efficacy of several herbicides reduced under high CO_2 environment (In house study, DWSR). Many of these weeds reproduce by vegetative means, for example, *Cynodon dactylon* in rice and *Convolvulus arvensis* and *C. arvense* in wheat.

These weeds may show a strong response in growth with increase in atmospheric CO_2. India is a country where erratic rainfall is not uncommon. A shift in weed species composition can also be expected under erratic rainfall because of climate change. Owing to a sudden change in climate, environmental stresses on a crop may increase and as a result the crop may become less competitive with weeds. The aberrations in weather conditions not only affect crop–weed competition, but also trigger weed seed germination in several flushes causing serious weed management issues. Three flushes of *P. minor* are not uncommon in the wheat fields in northwest India, which are not controlled by a single application of herbicide (Mahajan et al., 2012).

3.5.1 CHALLENGES

Many questions are to be answered in context of weeds and weed management under the regime of climate change. Most important include:

- How an individual factor will affect crop, weeds, and associated microorganisms?
- How multi-factor climate change (i.e., CO_2, O_3, UV radiation, other greenhouse gases, temperature, etc.) will affect the relative competitiveness of crop, weeds, and microbes? Who will dominate whom?
- Weed dynamics under climate conditions.
- How will a change in precipitation (seems to be almost certain) effect weed growth?
- What are the physiological, biochemical, and molecular basis and mechanism of dominance?
- What are ways to sustain/increase the productivity of crops in changing climate?
- How we can predict the possible losses of crop yields in futuristic climate change conditions?

3.6 CONCLUSION

From the above mentioned studies, it can be inferred that weeds possess better ability to survive and perform under adverse environmental conditions which make them sturdy and highly competitive with crop plants. Now a big question arises in this context, can we exploit these attributes of weeds for the crop improvement? If yes, then there is no other alternate better than weeds simply because of the co-existence of weed and crop plants. An advantage using weeds as a source of gene(s) may be other co-ordinated regulatory aspects of the transgene(s). As both weeds and crops grow in the same environment, it is expected that internal machinery (at least partly) which is required for the functioning of transgene(s) might be present already in crop plants. Development and availability of the sophisticated molecular tools provide us liberty to play at molecular level and transfer the genetic material into crop plants, thus breaking the reproductive barriers for inter-specific and inter-generic transfer of the genetic material. However, success of such approaches requires integration and collaborative efforts from all the corners of scientists to bring together expertise in weed science, molecular biology, and plant physiology. Following strategies can be beneficial to fight with the problem of climate change which seems to be certain in years to come:

- A thorough study on weed dynamics under climate conditions.
- Identification of crop cultivars resilient to climate changes
- Preventive measures: Early planting of crops can be effective by means of avoiding the high temperature; however, scope of such strategy is limited as it depends on the maturity of the preceding crop also.
- Return to CT practices: Looks difficult as it required again lots of labor work, fuel, and feed.
- Engineering of crops for climate resilience: Most viable and dynamic strategy is to engineer.
- The crop plants which can perform better under futuristic climate change conditions.
- For this purpose, weeds can be a good source of genetic materials for raising transgenic crops.

REFERENCES

Bhan, V.M. and Kumar, S., 1997, Integrated management of Phalaris minor in rice–wheat ecosystems in India. In *Proceedings of International Conference on Ecological Agriculture: Towards Sustainable Development*, November 15–17, Chandigarh, India, 2, pp. 400–415.

Chauhan, B.S., Gill, G.S. and Preston, C., 2006. Tillage system effects on weed ecology, herbicide activity and persistence: a review. *Australian Journal of Experimental Agriculture*, 46(12), pp. 1557–1570.

Chauhan, B.S. and Johnson, D.E., 2010. Implications of narrow crop row spacing and delayed Echinochloa colona and Echinochloa crus-galli emergence for weed growth and crop yield loss in aerobic rice. *Field Crops Research*, 117(2–3), pp. 177–182.

Chhokar, R.S. and Malik, R.K., 2002. Isoproturon-resistant littleseed canarygrass (Phalaris minor) and its response to alternate herbicides. *Weed Technology*, 16(1), pp. 116–123.

Derksen, D.A., Lafond, G.P., Thomas, A.G., Loeppky, H.A. and Swanton, C.J., 1993. Impact of agronomic practices on weed communities: tillage systems. *Weed Science*, 41(3), pp. 409–417.

Hobbs, P.R. and Gupta, R.K., 2003. Resource-conserving technologies for wheat in the rice–wheat system. In Ladha, J.K., Hill, J., Gupta, R.K., Duxbury, J. and Buresh, R.J. (Eds.), Improving the productivity and sustainability of rice–wheat systems: Issues and impacts. Special publication 65, Madison, WIS: ASA, pp. 149–171.

Huggins, D.R. and Reganold, J.P., 2008. No-till: the quiet revolution. *Scientific American*, *299*(1), pp. 70–77.

IPCC (Intergovernmental Panel on Climate Change). 2007. The Fourth IPCC Assessment Report. Cambridge University Press, Cambridge, UK. www.ipcc.ch/pdf/assessment-report/ar4/syr/ar4_syr_spm.pdf.

Jung, W.S., Kim, K.H., Ahn, J.K., Hahn, S.J. and Chung, I.M., 2004. Allelopathic potential of rice (Oryza sativa L.) residues against Echinochloa crus-galli. *Crop Protection*, *23*(3), pp. 211–218.

Lal, R. 2007. Constraints to adopting no-till farming in developing countries. *Soil and Tillage Research*, *94*, pp. 1–3.

Mahajan, G., Brar, L.S. and Walia, U.S., 2002. Phalaris minor response in wheat in relation to planting dates, tillage and herbicides. *Indian Journal of Weed Science*, *34*(1–2), pp. 114–115.

Malik, R.K., Yadav, A., Singh, S., Malik, R.S., Balyan, R.S., Banga, R.S., Sardana, P.K., Jaipal, S., Hobbs, P.R., Gill, G. and Singh, S., 2002. Herbicide resistance management and evolution of zero tillage-A success story. *Research Bulletin. CCS Haryana Agricultural University, Hisar*, pp. 1–43.

Mishra, J.S. and Singh, V.P., 2012. Tillage and weed control effects on productivity of a dry seeded rice–wheat system on a Vertisol in Central India. *Soil and Tillage Research*, *123*, pp. 11–20.

Naidu, V.S.G.R. and Varshney, J.A.Y.G., 2011. Interactive effect of elevated CO. *Indian Journal of Agricultural Sciences*, *81*(11), pp. 1026–1029.

Sharma, A.R., Ratan, S., Dhyani, S.K. and Dube, R.K., 2010. Effect of live mulching with annual legumes on performance of maize (Zea mays) and residual effect on following wheat (Triticum aestivum). *Indian Journal of Agronomy*, *55*(3), pp. 177–184.

Shrestha, A., Knezevic, S.Z., Roy, R.C., Ball-Coelho, B.R. and Swanton, C.J., 2002. Effect of tillage, cover crop and crop rotation on the composition of weed flora in a sandy soil. *Weed Research*, *42*(1), pp. 76–87.

Singh, S., Kirkwood, R.C. and Marshall, G., 1999. Biology and control of Phalaris minor Retz (littleseed canarygrass) in wheat. *Crop Protection*, *18*(1), pp. 1–16.

Singh, R.G., Singh, V.P., Singh, G. and Yadav, S.K., 2001. Weed management studies in zero-till wheat in rice-wheat cropping system. *Indian Journal of Weed Science*, *33*(3–4), pp. 95–99.

Ziska, L.H. and George, K., 2004. Rising carbon dioxide and invasive, noxious plants: potential threats and consequences. *World Resource Review*, *16*(4), pp. 427–447.

4 In-situ Soil Moisture Conservation with Organic Mulching under Mid Hills of Meghalaya, India

Y. Marwein and Lala I. P. Ray
Central Agricultural University

Joy Kumar Dey
Palli Siksha Bhavana (Institute of Agriculture)

CONTENTS

4.1   INTRODUCTION

Increase in agricultural productivity demands optimum utilization of natural resources like soil and water. Mulching is one of the important practices for restoring water; among various mechanical and agronomic measures, it reduces soil erosion, increases *in-situ* soil moisture storage and improves the productivity of crops (Bhatt and Rao, 2005). The practice of mulching has been widely used as management tool in many parts of the world. However, the effect varies with soils, climate and kind of mulch materials used, and the rate of application. The surface mulch favorably influences the soil moisture regime by controlling evaporation (Ramakrishna et al., 2006; Montenegro et al., 2013).

To increase water availability to crops, it is necessary to adopt the *in-situ* moisture conservation techniques including large-scale soil and water conservation practices and various water harvesting measures (Lannotti, 2007; Chavan et al., 2009). Mulching can help to improve crop yield and optimize water use (Lamont, 1999; Parmar et al., 2013; Prasad et al., 2014; Saikia et al., 2014). Mulches which are derived from plant material are called organic mulch, viz., grass, straws, leaves, etc. (Lannotti, 2007; Chavan et al., 2009; Rashid and Hossain, 2014); it can be fruitfully utilized to retain moisture. A field trial was taken up to assess the *in-situ* soil moisture depletion under two organic mulching materials, viz., maize stover and weed mulch with rajma as a test crop at the mid hill of Meghalaya.

4.2 MATERIALS AND METHODS

A field trial was carried out during winter season (2015–2016) at the experimental farm of the College of Postgraduate Studies, Umiam, which is located at Ri-Bhoi district, Meghalaya to study the soil moisture depletion pattern under organic mulch. The performance of a legume crop, rajma cultivar "selection-9" was studied under the mulching practices. The experimental soil is sandy clay loam with pH of 4.83 and organic carbon (1.96%). The weekly rainfall (mm), average maximum and minimum temperature (°C) and relative humidity (%) are shown in Figure 4.1 and the weekly rainy day (day), pan evaporation (mm) and the wind speed (km/h) are shown in Figure 4.2. During the experimentation period, maximum weekly rainfall of 129.2 mm was received during 42nd standard week (October); the total amount of 325.2 mm was received during the crop-growing season. The total number of rainy days recorded was 14; the highest number of rainy day occurred during the 39th standard meteorological week (5 days).

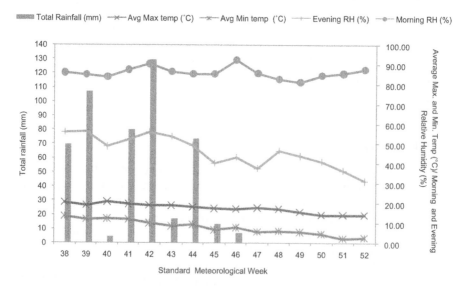

FIGURE 4.1 Weekly variations of rainfall, temperature, and humidity.

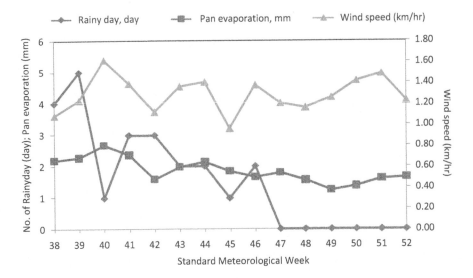

FIGURE 4.2 Weekly variations of rainy days, pan evaporation, and wind speed.

To moisture stress developed at the 0–15 and 16–30 cm soil depth, tensiometers were also installed at the respective mulch and un-mulch plots. A calibration curve was prepared for the tensiometer prior to the installation and shown in Figure 4.3. A soil moisture characteristics curve was also prepared to know the moisture holding capacity of the soil. According to the prepared soil moisture characteristic, curve soil moisture contents in Field capacity (FC) and Permanent wilting point (PWP) are 29.34% and 8.66%, respectively. The prepared curve is shown in Figure 4.4. Mulching was applied @ 5 t/ha 1 day after sowing of the seed. The maize stover was chopped first and then spread over the plot. Standard agronomic practices were followed during crop growing period and the crop was harvested at maturity.

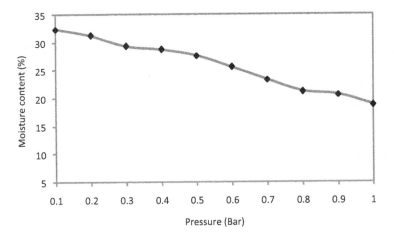

FIGURE 4.3 Calibration curve of a tensiometer.

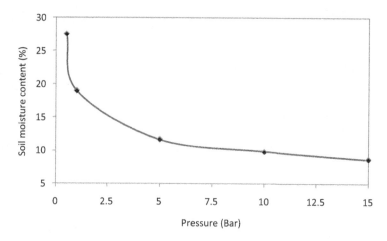

FIGURE 4.4   Soil moisture characteristics curve.

4.3   RESULTS AND DISCUSSION

4.3.1   SEED EMERGENCE

The emergence percentage for the rajma varieties under different organic mulching is presented in Table 4.1. Weed mulch (79.75%) gave higher germination than un-mulched (75.60%) indicating a better *in-situ* soil moisture holding capacity by mulching practices. The results were well in agreements with the findings of Bonanno and Lamont (1987), Sharma and Acharya (2000), Roy et al. (2010), Sharma et al. (2010), and Nwokwu and Aniekwe (2014).

4.3.2   SOIL MOISTURE DEPLETION

The soil moisture values were recorded to determine the depletion pattern of soil moisture at two different depths: 0–15 and 16–30 cm for un-mulched and organic mulch treatment plots, respectively as shown in Figures 4.5 and 4.6. The weekly recorded data of *in-situ* soil moisture is shown in the figure and the depletion status was found steady, however, the value of *in-situ* moisture content recorded was found more at lower depth than the upper layer. The same trend was observed for maize stover mulch and un-mulch treatment. Weed mulch showed better soil moisture retention than maize stover mulch.

TABLE 4.1
Effect of Mulching on Emergence of Rajma

Treatments	Emergence Percentage (%)
Un-mulch	75.60
Maize stover mulch	77.50
Weed mulch	79.75

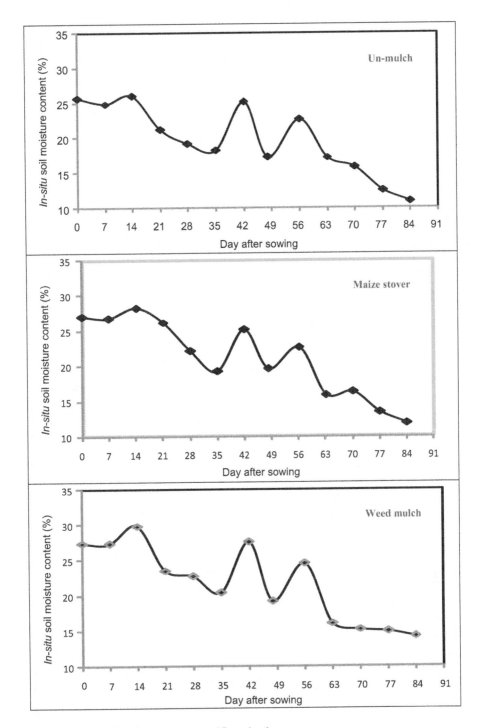

FIGURE 4.5   Soil moisture content at 15 cm depth.

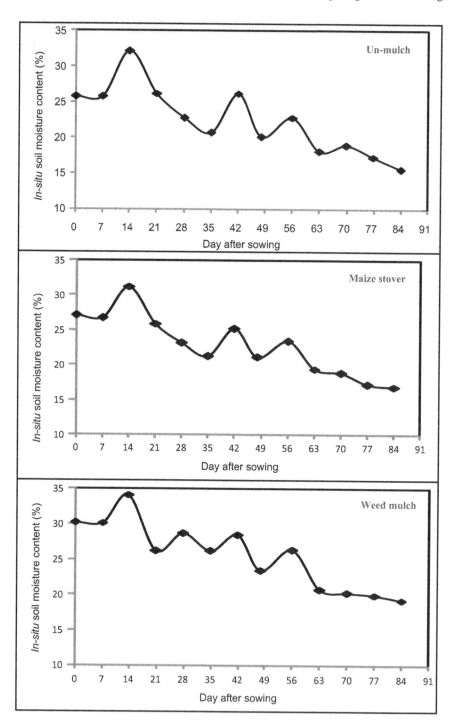

FIGURE 4.6   Soil moisture content at 30 cm depth.

At the 15 cm depth, the soil moisture fluctuation was recorded, where the lowest was 10.98% at 84 days after sowing (DAS) and the highest value was 26% at 14 DAS. While at the 30 cm depth, the soil moisture status showed a highest value at 14 DAS as 32.15% and the lowest at 84 DAS as 15.64%. The weekly *in-situ* soil moisture status for the maize stover mulch at 0–15 cm depth was found highest at 14 DAS (28.23%) and latter decreased; the lowest recorded value was at 84 DAS (11.91%); during the final stage of the crop. During later stage of vegetative and pod formation, the *in-situ* moisture was found higher and again it starts decreasing. While at 15–30 cm depth the *in-situ* soil moisture status of the maize stover mulch was found to be ranged between the lowest at 84 DAS (16.89%) and the highest at 14 DAS (31.23%). It may be noted that the *in-situ* soil moisture depletion pattern was not steady compared to un-mulch condition and the value of soil moisture was found higher at lower depth. It may be noted that the *in-situ* soil moisture content was found more compared to the other mulch treatments. The recorded soil moisture status for weed mulch at 15 cm depth was found to be lowest at 84 DAS (14.25%) and the highest value was observed at 14 DAS (29.87%); whereas at the 30 cm depth of weed mulching, the soil moisture status were found to be ranged between the lowest of 19.24% at 84 DAS and the highest of 34.14% at 14 DAS. Similar findings were reported by Ahmed et al. (2007), Chavan et al. (2009), Parmar et al. (2013), Saikia et al. (2014).

4.3.3 TENSIOMETER OBSERVATIONS

The tensiometer readings of the pressure gauge were recorded by the tensiometer installed in the un-mulched and organic mulched treatment fields at two different depths: 15 and 30 cm as shown in Figure 4.7.

The variation of tensiometer reading from 0 to 87 day of culture shows a steady increasing of soil moisture tension value compared to the other two organic mulch treatments condition. The minimum value of soil moisture tension at 15 and 30 cm depth was recorded as 2.6 and 1.8 kPa, respectively. For maize stover mulch treatment, the soil moisture tension at 15 and 30 cm depths were 2.2 and 2 kPa, respectively. However, for weed mulch treatment, the soil moisture tension at 15 and 30 cm depths were 1.9 and 1 kPa, respectively. The soil moisture tension values were found higher for the upper layer of soil depth compared to the bottom, which indicates the availability of relatively more soil moisture at the bottom layer of soil for all organic mulch and un-mulch treatment plots. Similar trend in organic mulching treatments was also reported by Sinkevičienė et al. (2009), Chavan et al. (2009). The difference between the high and lower tensions is more at the top of 15 cm depth (8 kPa) at the end of the growing season, whereas 5 kPa was recorded at the 30 cm depth. Mulches reduce soil deterioration by preventing runoff and soil loss, minimize the weed infestation, increase the *in-situ* soil moisture availability and reduce water evaporation (Sarangi et al., 2010). The *in-situ* soil moisture depletion pattern was not found steady compared to un-mulching condition, and the value of soil moisture was found higher at lower depth, under organic mulching and un-mulch condition. Similar findings were reported by Ahmed et al. (2007), Chavan et al. (2009), Parmar et al. (2013), Saikia et al. (2014). Weed mulch showed better soil moisture retention compared to maize stover mulch.

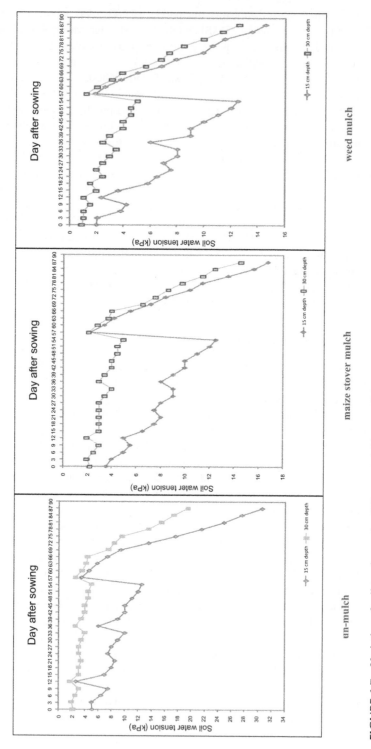

FIGURE 4.7   Variation of soil tension under different mulching conditions.

4.3.4 CROP YIELD PARAMETERS

The organic mulching and the varieties selected for the field trial gave positive results on the yield parameters. The reductions in number of pods per plant may also be attributed due to the abscission of flowers and pods (Malik et al., 2006) and/or by the failure of fertilization due to the production of unviable pollens under moisture stress conditions under un-mulched plots (Ahmed and Suliman, 2010).

Increasing the length of pod may be related with the age of the plant and its genetic characters as reported by Singh et al. (1994), Rashid and Hossain (2014). Weed mulching treatment registered an increase in seed yield of rajma due to the availability of optimum soil moisture content at the seeds development stage, which enabled higher nutrient uptake, greater dry matter accumulation, more grains per pod and increased 100 seed weight. Better control of weeds under mulch which could have also favored to increase the yield as reported by Barman et al. (2005), Chawla (2006), and Chinnathurai et al. (2012).

4.4 CONCLUSION

Weed mulch is more effective in maintaining the optimum soil moisture content at the seeds development stage, which enabled higher nutrient uptake, greater dry matter accumulation, more grains per pod, and increased seed index. Better control of weeds under mulch which could have also favored to increase the yield per plant. The soil moisture depletion was found rapid under un-mulch compared to organic mulching. Hence, organic mulch can play a major role in maintaining *in-situ* soil moisture and be used by the farmers of the hilly regions to cultivate winter crop.

ACKNOWLEDGMENT

The authors would like to thank the Research Advisory Committee Members, Dean, College of Postgraduate Studies, Barapani for the logistic help and support rendered during the tenure of the experiment. The financial support received from CAU, Imphal during the tenure of Master Degree is also thankfully acknowledged.

REFERENCES

Ahmed, Z.I., Ansar, M., Iqbal, M. and Minhas, N.M., 2007. Effect of planting geometry and mulching on moisture conservation, weed control and wheat growth under rainfed conditions. *Pakistan Journal of Botany, 39*(4), p. 1189.

Ahmed, F.E. and Suliman, A.S.H., 2010. Effect of water stress applied at different stages of growth on seed yield and water-use efficiency of cowpea. *Agriculture and Biology Journal of North America, 1*(4), pp. 534–540.

Barman, D., Rajni, K., Pal, R. and Upadhyaya, R.C., 2005. Effect of mulching on cut flower production and corm multiplication in gladiolus. *Journal of Ornamental Horticulture, 8*(2), pp. 152–154.

Bhatt, R.M. and Rao, N.K.S., 2005. Influence of pod load on response of okra to water stress. *Indian Journal of Plant Physiology, 10*(1), p. 54.

Bonanno, A.R. and Lamont Jr, W.J., 1987. Effect of polyethylene mulches, irrigation method, and row covers on soil and air temperature and yield of muskmelon. *Journal of the American Society for Horticultural Science, 112*(5), pp. 735–738.

Chavan, M.L., Phad, P.R., Khodke, U.M. and Jadhav, S.B., 2009. Effect of organic mulches on soil moisture conservation and yield of rabi sorghum (M-35-1). *International Journal of Agricultural Engineering, 2*(2), pp. 322–328.

Chawla, S.L., 2006. Effect of irrigation regimes and mulching on vegetative growth, quality and yield of flowers of African marigold (Tagetes erecta L.) cv. "Doubie Mix" (Doctoral dissertation, Department of Horticulture Rajasthan College of Agriculture, Udaipur).

Chinnathurai, S.J., Veeramani, A. and Prema, P., 2012. Weed dynamics, yield and economics of pigeonpea influenced by growth promoters and mulching. *Indian Journal of Weed Science, 44*(3), pp. 186–190.

Lamont, W.J., 1999. The use of different colored mulches for yield and earliness. Proceedings of the New England Vegetable and Berry Growers Conference and Trade Show, Sturbridge, MA (pp. 299–302).

Lannotti, M., 2007. What is mulch and which mulch should you use where? http://garden. Love to know.com. (Accessed on: September 28/2016).

Malik, A., Waheed, A., Qadir, G. and Asghar, R., 2006. Interactive effects of irrigation and phosphorus on green gram (Vigna radiata L.). *Pakistan Journal of Botany, 38*(4), p. 1119.

Montenegro, A.D.A., Abrantes, J.R.C.B., De Lima, J.L.M.P., Singh, V.P. and Santos, T.E.M., 2013. Impact of mulching on soil and water dynamics under intermittent simulated rainfall. *Catena, 109*, pp. 139–149.

Nwokwu, G. and Aniekwe, L., 2014. Impact of different mulching materials on the growth and yield of watermelon (Citrullus lanatus) in Abakaliki, Southeastern Nigeria. *Journal of Biolgy Agriculture and Healthcare, 4*(23), pp. 22–30.

Parmar, H.N., Polara, N.D. and Viradiya, R.R., 2013. Effect of mulching material on growth, yield and quality of watermelon (Citrullus lanatus Thunb) Cv. Kiran. *Universal Journal of Agricultural Research, 1*(2), pp. 30–37.

Prasad, B.V.G., Chakravorty, S., Saren, B.K. and Panda, D., 2014. Effect of mulching on phys-iological growth determinants of productivity in French bean (Phaseolus vulgaris L.). *HortFlora Research Spectrum, 3*(2), pp. 162–165.

Ramakrishna, A., Tam, H.M., Wani, S.P. and Long, T.D., 2006. Effect of mulch on soil temperature, moisture, weed infestation and yield of groundnut in northern Vietnam. *Field Crops Research, 95*(2–3), pp. 115–125.

Rashid, A.S.M.H. and Hossain, M.M., 2014. Yield and quality of green pod production of bush bean (Phaseolus vulgaris L.) as influenced by harvesting time. *American-Eurasian Journal of Agricultural & Environmental Sciences, 14*(11), pp. 1221–1227.

Roy, S., Arunachalam, K., Dutta, B.K. and Arunachalam, A., 2010. Effect of organic amendments of soil on growth and productivity of three common crops viz. Zea mays, Phaseolus vulgaris and Abelmoschus esculentus. *Applied Soil Ecology, 45*(2), pp. 78–84.

Saikia, U.S., Kumar, A., Das, S., Pradhan, R., Goswami, B., Wungleng, V.C., Rajkhowa, D.J. and Ngachan, S.V., 2014. Effect of mulching on microclimate, growth and yield of mustard (Brassica juncea) under mid-hill condition of Meghalaya. *Journal of Agrometeorology, 16*(1), pp. 144–145.

Sarangi, S.K., Saikia, U.S. and Lama, T.D., 2010. Effect of rice (Oryza sativa) straw mulching on the performance of rapeseed (Brassica campestris) varieties in rice-rapeseed crop-ping system. *Indian Journal of Agricultural Sciences, 80*(7), pp. 603–605.

Sharma, P.K. and Acharya, C.L., 2000. Carry-over of residual soil moisture with mulching and conservation tillage practices for sowing of rainfed wheat (Triticum aestivum L.) in north-west India. *Soil and Tillage Research, 57*(1–2), pp. 43–52.

Sharma, A.R., Singh, R., Dhyani, S.K. and Dube, R.K., 2010. Moisture conservation and nitrogen recycling through legume mulching in rainfed maize (Zea mays)–wheat (Triticum aestivum) cropping system. *Nutrient Cycling in Agroecosystems, 87*(2), pp. 187–197.

Singh, D.N., Nandi, A. and Tripathy, P., 1994. Genetic variability and character association in French bean (Phaseolus vulgaris). *Indian Journal of Agricultural Sciences, 64,* pp. 114–116.

Sinkevičienė, A., Jodaugienė, D., Pupalienė, R. and Urbonienė, M., 2009. The influence of organic mulches on soil properties and crop yield. *Agronomy Research, 7*(1), pp. 485–491.

5 High-Altitude Protected Vegetable Cultivation – A Way for Sustainable Agriculture

M. S. Kanwar

Sher-e-Kashmir University of Agricultural Sciences and Technology of Kashmir (SKUAST-K)

CONTENTS

5.1   INTRODUCTION

High-altitude environment can be defined as environment above forest line at an altitude of 2,500–3,000 m above the mean sea level. It is only above this altitude range that significant physiological, structural, and ecological adaptations are observed in plant – also in humans – which can be distinctly correlated with altitude. Extremely low temperature and meagre precipitation are characteristic signatures of high-altitude climate. Besides Ladakh, similar high-altitude cold and arid climate also prevails in parts of Uttar Kashi, Chamoli, Pithoragarh districts in Uttarakhand, Spiti area of Lahaul & Spiti district and Sumdo area of Kinnaur district and Pangi of Chamba district in Himachal Pradesh, and in some part of north Sikkim. The area of arid temperate region in India is 79,278 km², which is nearly 4.72% of the total area of country. Nonetheless, the major portion of Indian cold arid (68,321 km²) constitutes Ladakh. In Himachal Pradesh it comes out to be about 20% of the state area.

Protected cultivation is the modification of the natural environment to achieve optimum plant growth. Modifications can be made to both the aerial and root environments to increase crop yields, which extend the growing season and permit plant growth during periods of the year not commonly used to grow crops in an open field. It involves protection of different production stages of crop development from adverse environmental conditions, such as extreme temperature, hailstorm, scorching sun, and heavy rain/snow with ultimate objective of more economic yield. Protected structures in the form of tunnels, trenches, and polyhouses/greenhouses permit enrichment of greenhouse air with CO_2, humidification, effective insect pest, disease control, and photoperiodic control. Modified environments so created enable the people to raise crops and vegetables in off-season, enhance the period of crop duration, increase the crop productivity per unit area, and inspire the growers for establishment of entrepreneurship. A protected structure could be a structure formed of glass, with a heating (and usually cooling) system that was used year-round, but especially in winter. Then came houses built of thermoplastic (Plexiglas and others), followed by Quonsets covered with plastic, which may or may not be heated, have one or two layers, and be used year-round or for only a few months every year. The kinds of protective structures used by growers in higher altitudes range from simple structures such as rain shelters, shade houses, mulches, row covers, and plastic tunnels, to permanent structures covered in plastic or glass with computerized environment controls. The type of protected structure one needs will be determined by the crops to be grown and investment competency and, to a lesser extent, by management intensity and market strategy. High value crop production under controlled modified environments is now an accepted reality throughout the world with an estimated 405,000 ha of protected structures/greenhouses spread over all the continents (Reddy, 2016).

The degree of sophistication and technology depend on local climatic conditions and the socio-economic environment. Protected cultivation of vegetables offers distinct advantages of quality, productivity, and favorable market prices to the growers. Vegetable growers can substantially increase their income by protected cultivation of vegetables in off season, as the vegetables produced in the normal season generally do not fetch good returns due to large availability of these vegetables in the market. Protected cultivation is a familiar technology in all regions of the country but has gained momentum in hill states of India because of the immediate benefits realized and wider applications to bring socio-economic transformation in higher Himalayas. It allows precision farming and overcomes limitations of space and disadvantages of climate change. Better carbon, nutrient, and water assimilation, low photo-respiration and higher photo-synthetic activity under protected conditions offer outstanding scope for production of organic vegetables.

5.2   CLIMATE AND FARMING

Such a geographical setting manifests itself into some peculiar agro-climatic conditions that prevail in high Himalayas. Extreme cold and aridity coupled with large seasonal as well as diurnal variation in temperature are limiting factors affecting agricultural productivity adversely. Winter temperature plummets to as low as 30°C below freezing point (Stobdan et al., 2018). In summers, nonetheless, temperature can surge up to 39°C (Akbar et al., 2013). Precipitation is mainly in the form of snow during winter months. The glaciers and melting snow are the main sources of water for agriculture. The area receives clear and bright sunshine even in winters. Wind blows with high speed; sometimes it blows up to 50–60 km/h, especially in the afternoon in some lofty parts of the region, with the result fierce dust storms rage over the valleys and the sandy plateaus.

Harsh climate coupled with poorly developed soil and limited source of irrigation water results in low farm productivity. Cropping is generally restricted to the period of May–September during which wheat and barley is grown besides peas and mustard. Double cropping is practiced only in the pockets lying below 3,000 m altitude. Wheat is grown up to 3,600 m above mean sea level and barley up to 4,400 m. Though food security is taken care of by an efficient public distribution system, nutritional and income insecurities are the principal concerns for the policy makers. It is the later which is driving farmers of the region, especially younger generation out of farming and toward other opportunity outside their villages, area, and home land.

Fresh vegetables and fruits are always in short supply in the region. Only 11,867 tonnes of fresh vegetables (Stobdan et al., 2018) are produced locally during the growing season whereas demands of only the army stationed in Ladakh is around 37,200 tonnes annually (Mishra et al., 2010). Rest is imported through long distances and even air lifted to a limited extent – during winters when roads are closed. Obviously these "fresh vegetables" do not remain fresh when it ends its journey on the chopper board in kitchen.

5.3   HIGH ALTITUDES AND PROTECTED CULTIVATION

Higher Himalayas are known for their cold arid climate which has several characteristic features as given below:

1. Extreme tmperatures (–38°C in winter and +37°C in summer)
2. Soils coarse textured and highly permeable
3. Low plant population
4. Precipitation in valleys is low (less than 100 mm) mostly in the form of snow
5. Relative humidity quite low (between 20% and 40%)
6. Oxygen concentration in air is low enough to affect physiology of living organisms
7. Sharp temperature fluctuations. There may be experience of frostbite and sunstroke in the same season
8. Thin atmosphere due to high altitude so intense solar radiation impregnated with more ultraviolet and infra-red rays that affect biological material
9. Cropping season is of short duration (120–150 days)
10. Only irrigated agriculture is possible
11. Winter is characteristically devoid of any vegetational greenery
12. Atmospheric pressure low
13. Altitude of the region ranges between 9,000 and 15,000 ft, above mean sea level (AMSL) with mountain peaks high up to 26,000 ft AMSL
14. High wind velocity
15. High number of clear sky days

The climates prevalent in higher Himalayas are cold arid, dry temperate and temperate which offer both opportunities and challenges to agricultural scholars to manipulate the climatic parameters to usher the noble era of food and nutritional security in these remote areas. Interdisciplinary research and development projects in the areas of designing of protected structures for climate management, integrated crop management, crop selection, and varietal development and natural resources management are the prime fields of immediate considerations.

5.3.1   PROBLEMS

There are numerous problems in high altitudes hindering the cultivation of vegetable crops in open environment. Low temperature period is considerably longer enabling the necessity of protected cultivation of vegetables. As already mentioned, temperature may dip as low as −38°C in cold arid arresting plant life in open conditions. Strong winds are most common in valley areas causing soil erosion. Low humidity coupled with high solar intensity is another problem which demands frequent irrigation. High-altitude conditions have certain characteristics.

5.3.2 OBJECTIVES

Though main objective remains the high economic yield along with better quality in case of protected cultivation, but specifically in high altitudes objectives are somewhat different. During winters and autumn, growing vegetables are major objectives because in open no cultivation is possible in sub-zero temperature. Cool season vegetables are preferred in winters and autumn under protected conditions. In this case good yield may be attributed to a particular vegetable or genotype of a particular vegetable. During summer, growing warm season vegetables for longer duration is major objective in most parts of high altitudes. In colder parts of high altitudes, it is difficult to grow even cool season vegetables during summer season. Therefore, protected cultivation offers growing of cool season vegetables with high yield and earliness.

5.3.3 PROTECTED STRUCTURES

The common structures that are being used in high altitudes are naturally ventilated polyhouse. In some cases only shade nets are used to protect the crop from hot sun and winds during summer. Very simple and reasonably cheaper grass thawing, stone pillars/walls or wooden or bamboo poles with clear polyethylene, or insect proof nylon nets as cladding material are also being used for vegetable production. Normally poles of 10–12 ft height are suitable for most of the vegetables. For net house cultivation ultraviolet radiation (UV) stabilized nylon nets of 40–50 mesh size are used. For longer life and prevention of sagging good support with crisscross wiring of GI wire is essential. Greenhouse structures require constant maintenance and repair. Many of the selected greenhouse covers must be replaced on a regular basis. Cooling and watering systems must be maintained and routinely serviced. In addition, contingency plans and backup systems must be in place in case any of these major systems should break down. Even a 1-day loss of cooling during summer or water during a critical period can result in drastic reduction in crop yields. Along with the essential skills, capital and labor to build, maintain and grow a crop, producers must develop markets willing to pay the relatively high prices necessary to make the enterprise economically viable. Greenhouse-grown vegetables cannot compete with comparable field-grown crops based on price; therefore, greenhouse-grown vegetables are often marketed to buyers based on superior quality and off-season availability. Common types of greenhouses used in India are tunnels, low cost greenhouse/polyhouse, and medium cost greenhouse.

5.4   POINTS TO BE CONSIDERED FOR COOL CLIMATE

As the polyhouse technology involves technical aspects, the type of designs and materials should be according to the local climate. All the points are general in nature and irrespective of the crop.

1. The height should be less than hot and humid areas (3–4.5 m), and top ventilation should not be there.
2. It should be closed completely during nights to maintain higher temperature using day time solar radiation.
3. The structural strength shall be more and roof shall be strong enough to withstand the probable snow load.
4. If possible, heating arrangement must be made.
5. In dry temperate areas of Himachal Pradesh and Ladakh of Jammu and Kashmir, "lean-to," gothic and "Z" type polyhouses are successful (Spehia, 2016). Here the covering material can be double layered to reduce the heat loss from inside.
6. Water sup must be ensured on regular basis along with rain water harvesting structures. Drainage system should be good.
7. Use plastic mulches to reduce evaporation and weed control.

A polyhouse is a controllable dynamic system, managed for intensive production of high quality, fresh market produce. Polyhouse production allows for crop production under very diverse conditions. However, there are a number of variables that polyhouse growers have to manage in order to obtain maximum sustainable production from their crops. These variables include; air temperature, root zone temperature, vapor pressure deficit, fertilizer feed, carbon dioxide enrichment, selection of growing media, and plant maintenance. The task of managing these inter-related variables simultaneously can appear over whelming; however, there are successful strategies that a grower can use. The main approach is to try to optimize these variables to obtain maximum performance from the crop over the production season. The goal of optimization can be used to determine how to control these variables in the polyhouse for maximum yield and profit, taking into account the costs of operation and the increased value under the modified environment. The polyhouse system is complex; to simplify the decision-making process growers use indicators. An indicator can be thought of as a small window to a bigger world, you do not get the entire picture, but you do gain an understanding of what is happening. Another way to look at it is to understand the basic rules of thumb which can be used to obtain insights on the direction and dynamics of the crop-environment interaction. Indicators provide information concerning complex systems in order to make them more easily understandable. They quickly reveal changes in the polyhouse which may require alterations in management strategies. Indicators also help identify the specific changes in crop management that need to be made. Over time, and with experience, growers will be able to build on these basic indicators to improve their ability to respond to changes in the crop and to anticipate the needs of the crop.

5.5 GREENHOUSE DESIGNING

Three main factors should be considered namely, load limitations, light penetration, and cost when designing greenhouse. The primary load considerations include snow and wind. Roof slopes of at least 28° and heated air in the greenhouse should prevent snow accumulation on the roof. Bracing along sides of the greenhouse and roof should be sufficient to withstand wind, particularly in the spring. Bracing along the roof also should be sufficient to withstand crop loads if tomato or cucumber vines are to be supported by twine attached to the bracing. A concrete footing is preferred for a permanent greenhouse. A wide door at one end of the greenhouse will ensure easy access for equipment. Without sacrificing strength, support structures should be kept to a minimum to maximize light penetration. Glazing materials should be highly transparent. Overhead electrical lines, irrigation systems, and heating ducts should be kept to a minimum. Support structures should be painted with a reflective, light-colored material for maximum light reflection. Most greenhouse crops grow best in light whose wavelengths range from 400 to 700 nm. This range of wavelengths is called photo synthetically active radiation. Most greenhouse coverings will accommodate these short waves of visible light. Polyethylene and fiberglass tend to scatter light, while acrylic and polycarbonate tend to allow radiation to pass through directly. Scattered or diffused light tends to benefit plants by reducing excess light on upper leaves and increasing reflected light to lower leaves. Plastic glazed greenhouses have several advantages over glass greenhouses, the main advantage being cost. Plastic is also adapted to various greenhouse designs, generally resistant to breakage, lightweight, and relatively easy to apply.

5.5.1 TYPES OF PLASTIC COVERINGS

1. Acrylic is resistant to weathering and breakage and is very transparent. Its ultra-violet radiation absorption rate is higher than glass. Double-layer acrylic transmits about 83% of light and reduces heat loss 20%–40% over single layer. This material does not yellow easily. Its disadvantages are that it is flammable, very expensive, and easily scratched.

2. Polycarbonate resists impact better and is more flexible, thinner, and less expensive than acrylic. Double-layer polycarbonate transmits about 75%–80% of light and reduces heat loss 40% over single-layer. This material scratches easily, has a high expansion/contraction rate, and starts turning yellow and losing transparency within a year (although new varieties with UV inhibitors do not yellow as quickly).

3. Fiberglass reinforced polyester panels are durable, attractive, and moderately priced. Compared to glass, Fiberglass reinforced polyester panels are more resistant to impact, transmit slightly less light, and weathering over time reduces light transmission. This plastic is easy to cut and comes in corrugated or flat panels. It provides superior weather ability only when coated with Tedlar. Fiberglass has a high expansion/contraction rate.

4. Polyethylene film (PE) is inexpensive but temporary, less attractive, and requires more maintenance than other plastics. It is easily destroyed by UV radiations from the sun although film treated with UV inhibitors will last 12–24 months longer than untreated. Because it comes in wider sheets it requires fewer structural framing members for support, resulting in greater light transmission. Using a double layer of six mil polyethylene on the outside and two mil as an inner barrier will help conserve heat; this inner layer also will help reduce water condensation. The inner layer should be 1–4 in. from the outside layer with layers kept separated by a small fan (creating an insulating dead air space) or wood spacers. Two layers reduce heat loss 30%–40% and transmit 75%–87% of available light when new.

5. Polyvinyl chloride film has very high emissivity for long-wave radiation, which creates slightly higher air temperatures in the greenhouse at night. UV inhibitors can increase the life of the film. It is more expensive than polyethylene film and tends to accumulate dirt, which must be washed off in winter for better light transmission.

5.6 ENVIRONMENTAL CONSIDERATIONS IN GREENHOUSE CULTIVATION

5.6.1 TEMPERATURE

Regulating air temperature in the greenhouse is important for both vegetative growth and fruiting. To determine heating requirements, it is essential to know the minimum temperature requirements for the crop, the lowest outdoor temperature that might be expected, and the surface area of the greenhouse. Heat loss will also be affected by wind and site exposure. Greenhouse cooling is also important. Evaporative cooling is the most efficient and economical way to reduce greenhouse temperatures. Proper ventilation is important not only for temperature control, but also to replenish carbon dioxide and control relative humidity. Relative humidity above 90% will encourage disease problems. Roof ventilators are seldom used on plastic houses, which instead use side vents to provide both ventilation and cooling. Vents should be installed as high on the wall as possible. Heating, cooling, and ventilation should be automated to save labor and to ensure proper temperature control.

Temperature is the most important environmental factor for growth and development. Different metabolic processes like photosynthesis, respiration, etc. and different metabolic responses like enzymatic activity, membrane permeability, substrate concentration, and cumulative reactions are all regulated by temperature. Higher temperature in the early stages of crop growth promote leaf expansion which in turn helps in more light interception, flowering, and fruit development for optimum temperature ranges and beyond which the crop growth declines. The biochemical reactions will not take place at temperatures below the freezing point because of the loss of catalytic action and other properties of water, which is the main constituent of a cell. Similarly, at temperatures of 35°C or above, the molecular structure of the enzyme and protein systems is irreversibly altered leading to stunting of the growth. However, in most of the vegetable crops the diurnal temperature variation effects have not been sufficiently investigated. However, in general, diurnal differential is beneficial for growth and development. At comparatively low night temperatures, quality of growth, earliness, intensity of flowering, translocation of carbohydrates, and fruit development are benefited. At comparatively low night temperature condition, absorption of water is relatively high, rate of respiration is relatively low, less quantities of carbohydrates are used and greater quantities are stored, turgor pressure of the new cells is high, causing cell elongation, and new cells are produced and at high night temperature condition the rate of gross photosynthesis remains at a high level but a rate of respiration increases markedly. Highest net photosynthetic rate at 30°C has been observed when plants were at an early growth stage and at 35°C during mid late growth stages have been observed with decline of physiological efficiency below 25°C.

Soil temperature influences seed germination and water uptake by the plants. Plants extract water from warm soils more quickly and easily than from cold soils. Minimum soil temperature for seed germination in majority of vegetable crops ranges between 2°C and 15°C (Sharma, 2016). However, optimum temperature for seed germination in most of the vegetable crops lies between 20°C and 30°C (Kanwar, 2011). Soil temperature below 10°C retards the growth and development of chilli plants, while 17°C is optimum. Raising the temperature increased the shoot growth but root growth was retarded above 30°C. Days to flowering was reduced from 87 to 65 days under high temperature conditions (Sharma, 2016).

5.6.2 LIGHT

Solar radiant energy entering the greenhouse is reflected from various surfaces within the structure and passes out through the cover; the rest is absorbed by the plants, soil benches, and is converted into heat energy. The thermal energy dissipated as latent heat in transpiration, warms greenhouse air by conduction, and convection and it is emitted as long wave radiation. Thus sunlight enters the greenhouse as a short wave radiation and its wavelength is changed when it is absorbed and converted into heat and portion of long wave radiation that is trapped within the structure. The retention of long wave radiation in the structure is called as greenhouse effect. Higher yield is generally realized in the crops grown under glass, poly house, or poly tunnel because glass or polyethylene used to develop these structures has the property of absorbing ultraviolet and infrared rays. Polythene sheet will not allow the radiation from the soil surface once after reaching the earth. Tomato plants grown in red or blue light produce a greater dry weight over the same period of time than plants grown in green and white light. Gimenez et al. (2002) observed increased over all growth parameters of lettuce grown under tunnels, and they also observed that 30% reduction in solar radiation gave this highest plant growth rate. The envelop of the greenhouse must be transparent to the visible range of radiation in the solar spectrum (between 400 and 700 nm) and opaque to long wave and low energy infrared radiation. The effect of wavelength of light and heat transmission is very important in greenhouse technology. The classical material glass has a high transmittance throughout the visible portion of light spectrum, in the wave length 400–700 nm range important for photosynthesis.

5.6.3 RELATIVE HUMIDITY

The optimum relative humidity plays an important role in determining the yield and quality of glasshouse crops. Humidity levels in the area surrounding the plants affect all important processes of plant growth viz., transpiration, water balance, cooling of plants, ion transport, etc. High humidity is reported to be detrimental for plant growth either by increasing disease and pest incidence and causing physiological disorders or it may enhance vegetative growth. In general the humidity in a greenhouse is higher than the ambient. Therefore potential evaporation is less in greenhouses resulting in less water consumption per unit produce. Relative humidity can be regulated according to needs of the crop under greenhouse. It can be raised by spraying water and can be lowered by ventilation, and plant dry weight increased significantly in several species as relative humidity increased to 60%–70%.

5.6.4 CARBON DIOXIDE

The addition of supplementary carbon dioxide dose into the greenhouse has been found to significantly increase the yields of greenhouse tomatoes and other vegetables. Supplementary carbon dioxide is most effective on days when the greenhouse has been shut up for several days with no ventilation. Maximum results can be achieved by injecting 1,000–1,500 ppm CO_2 into the greenhouse using propane burners or other CO_2 generators (Boseland and Votava, 2012).

5.7 PLANNING OF PROTECTED STRUCTURES

5.7.1 SITE SELECTION AND PREPARATION

Selecting site for polyhouse should be based on water availability, natural temperature conditions (heat zone maps), frost free days, light, wind, access to market, etc. For a given farm, the best place to build a greenhouse is also based on a number of factors. Some serious thoughts about the location of the greenhouse can help minimize costs, maximize efficiency, and in the case of light maximize productivity. Ready access in all types of weather can help keep things simple.

5.7.2 ORIENTATION

The polyhouse design has to be according to the climatic zone and the lengthy direction of polyhouse (gutter direction) should be North-South in multi span polyhouses with top vents falling west/north west in North Indian condition. In single span polyhouses the orientation should be East-West.

5.7.3 ROADS

For easy marketing and other uses, polyhouse should be near to roads. It helps in taking the produce to markets well in time without any botheration and with minimum expenditure.

5.7.4 ACCESS TO UTILITIES – WATER

Irrigation water is needed year-round in the polyhouse. Trenching and installation of water lines can be done well in advance of construction once the site is identified. If additional greenhouses may be added in the future, consider not putting the end of the water line at a hydrant that will be in a greenhouse. Extend the line and place the end of the line at a hydrant outside – where it could be dug up if the line needed to be extended for more polyhouses. Remember to put several cubic feets of gravel around the drain at the base of the hydrant so that it can quickly drain.

5.7.5 LIGHT

Since polyhouse involves the business of harvesting light, maximizing exposure to the sun is very important. Start with an awareness of the path of the sun over the day and the seasons and if there are buildings or trees that cast shadows. The length of the shadow is longest in late December and can reach over twice the height of the object making a shadow. A building 15 ft tall can cast over a 30 ft shadow in December but perhaps less than a 5 ft shadow in June. If the soil slopes to the south or toward the sun, it may be a few degrees warmer than level ground, which is good, but it may make building the polyhouse more difficult. To maximize light in poly-houses, an east-west orientation is preferred at northern latitudes for maximum light during the winter. This means that the ends of the polyhouse face east and west and during the day the sun enters mostly from the side of the greenhouse. With multispan structures, the orientation should be north-west.

5.7.6 WIND

It helps to know the direction of the prevailing wind. Some wind can help with ventilation and cooling. Excessive wind can possibly damage the structure. With heated polyhouses, 15–20 mph winds can double heat loss. If blowing snow is a major factor, the orientation can be selected to help remove snow so it blows between the polyhouses and out rather than up against the walls of the greenhouse.

5.8 CLIMATE REGULATION, EQUIPMENT, AND MANAGEMENT

The resulting climate within the polyhouses has been along the years the product of the construction simplicity, the natural climate, and the limited tools available to regulate or modify it. Since the vegetable production calendar has been nowadays much more extended than years ago and out-of-season cropping is not yet the only goal as it was, growers have to face a higher number of limiting conditions that can be summarized as follows:

- Sub-optimal minimum temperatures along the cold season, including the risk of frost
- Excessive air humidity levels, condensation of water on the greenhouse cover and water dropping on the crop

- Over-optimal maximum temperatures during the clear days
- High water vapors pressure deficits
- Depletion of CO_2 concentration in the air for some hours in the day

5.8.1 TEMPERATURE

Temperature is the most important factor that influences the crop growth and production in the protected structures. When selecting a site for a polyhouse conditions that can create a type of bowl that traps cold air should be selected for warmer zones whiles same considerations should be taken into account when selecting the site for the polyhouse and the warmest possible site selected for colder zones. The type of polyhouse technology to be adopted for a specific area can be specified with the data on extreme minimum and maximum temperatures for whole year for few years. The fluctuation of outside temperature is important to plan the climatic control mechanism of the polyhouse. Normally, temperature inside the polyhouse is 20% more than outside. Two basic type of temperature control mechanisms are temperature reducing (tropical and sub-tropical areas between 40°N and 40°S) and heating mechanism (areas above and below 40°N and 40°S).

A. *Temperature reducing*: for this, basically, three types of mechanisms are being used: ventilation, shading, and evaporative cooling.
- *Ventilation*: ventilation serves two main functions: exchange air between interior and exterior and cooling by ventilating the hot air created inside polyhouse. The rate of air removal from polyhouse must increase as the altitude of the polyhouse site increases. The rate of increase of air exchange rate may be determined by multiplying with 0.04 for every increase of 304.8 m or 1,000 ft elevation up to 5,000 ft elevation and then factor will increase at 0.05 for every increase of 1,000 ft.
- *Shading*: This is very much specific for the tropical and sub-tropical areas. 40%–50% reduction of solar radiation in highest light intensity periods can reduce temperature below the shade to the extent of 20% of outside. Shading in polyhouse can be done by
  - *Application of shading compounds*: Spraying shading compounds on the outer side of the covering material of the polyhouse at the onset of warm season. The shading compounds used are common slaked lime (used for whitewash) or calcium carbonate. In case of lime, gum or sticker should be used to stick the coating properly on the plastic. Lime coating reduces the inside temperature by 3°C–4°C.
  - *Application of shade net*: The reduction in inside temperature by 6°C by using aluminium plated mesh as compared to polyhouse without shading at 33°C. shading can be done by
    i. Shade net above the polyhouse: shade net of desired specification is applied over and above the roof of polyhouse either with a gap or along the PE film. In hot dry climate this type of shading coupled with any evaporation cooling system or alone in naturally ventilated polyhouses, can reduce the temperature

very effectively. Still, it has not gained popularity probably due to problems associated with operational aspects.

ii.   Shade net below the roof: It is applied inside the polyhouse above head weight but its capacity to reduce the inside temperature is lesser compared to first method.

- *Evaporative cooling*: These are the systems practiced to cool down the temperature of the polyhouses by conversion of heat to latent heat. With more than 80% humidity, the effect of such cooling system turns to be less effective. I that case, air circulating fan will be more effective as compared to exhaust fan. Another method is misting or fogging. In misting, water is dispersed in the form of fine droplets measuring 50–100 μm. In general, misters are placed above the crop in such a manner that the system can uniformly cover the cropped area of the polyhouse. An intermittent misting for a very short duration, ideally, 5–15 s is generally recommended. In fogging, fog is generated artificially by ultra-fine water droplets (2–40 μm size) to modify the micro environment. The fog in the polyhouse maximizes the vapor pressure of the air by raising the ambient humidity and lowers the air to leaf vapor pressure gradient reducing transpiration. Fan-pad cooling system can be of positive pressure systems and negative pressure systems. In positive system, both fan and pad are in the same side of polyhouse. The fan sucks the air from outside and it blows through the wet pad into the polyhouse and is suitable for smaller polyhouses. In negative system, fan is at one side and wet pad on the opposite side. Fan drives out the inside air and due to pressure difference the hot and dry outside air enters into the polyhouse through wet pad which gets cooled down after passing through the pad and moves toward the fan reducing the temperature.

B. *Heating of polyhouse*: boiler and heater are the two main heating equipments used in commercial polyhouses throughout the world.

- *Boiler*: this system is also called "control heating system." Here the steam (102°C) or hot water (85°C) produced by the boiler is piped to the polyhouse. Hot water is used for smaller ranges (less than 700 mt) and steam is often used for longer ranges.
- *Heater*: this is called "localized heating system" as individual heater heats a specific area of the polyhouse. For this, unit or force heaters, convection heaters and radiant or infra-red heaters are used. Out of these, infra-red heaters of solar heating systems are best as they are low energy heaters and reduces fuel bill up to 50%.

5.8.2   Drainage of Rain Water

A half inch of rain displaced from the roof of a 30' × 96' polyhouse is approximately 900 gallons of water. If all the water is absorbed along the sides of the greenhouse, the soil inside the greenhouse wall can become very wet, particularly in the winter and after the spring thaw. This may not be a problem on sandy, well- drained soil, but if flooding is experienced inside the polyhouse, add drainage pipe (4" round plastic) in trenches along the walls and cover them with gravel. If the polyhouse is built on a slightly raised area, water moves away from the structure, additional drainage is probably not required. Natural slope of the land can also be used to drain rainwater away.

5.8.3 LIGHTING

Light conditions are also very important for plant growth. The problem of radiation deficit during the winter North Indian conditions is affecting the polyhouse crops and there is the need to improve the growing facilities toward optimizing the light balance in favour of the plants. Reduction of the greenhouse structure is a generalized step. Growers should pay more attention to the quality of the film in relation to the mechanical or thermal properties but with the optical ones too. The shape and height of the cover should also be according to the climate in the area. Moreover, lighting arrangements can also be made for increasing the lighting hours for long day plants during winters.

5.8.4 WATER AND NUTRIENTS

The use of the fertigation techniques, by means of localized irrigation supplying the water and minerals properly dosified and mixed from a unit provided with injecting, mixing, and measuring devices and control elements as well as a performant distribution system (Spehia, 2016), is generalized in polyhouses. The growers should be familiar with automatic installations, programs and the concepts of pH, and electrical conductivity of the nutrient solutions. In hi-tech polyhouses, automation of irrigation through equipment with computer and software facilities that can help in the watering decision by means of climatic measurements, namely the amount of solar radiation that is related to the volume of water transpired, is used. Computer modelling is used to calculate water and fertilizer requirements, irrigation scheduling, diseases and insect's prediction, and planting dates. Moreover, proper disease prediction is conducted by built-in software utilizing agroclimatic data for each region.

5.8.5 POLLINATION

When vegetables are to be grown under polyhouses, certain vegetables require pollination which can be achieved by honey bees or bumble bees. A full chapter in this book on rearing, domestication, and pollination is dedicated on the subject. Rearing and domestication can also be optional money making business for the grower.

5.9 INSTRUMENTS REQUIRED FOR POLYHOUSE

Temperature
- Thermometer: highly accurate instrument to measure the temperature (mostly air) in the range of 0°C–110°C.

- Digital thermometer: directly displays the temperature measured by thermocouple but cannot withstand high humidity conditions.

Humidity
- Dry and wet bulb thermometer: most common instrument to measure humidity.

pH
- Electro chemical pH meter: measures the voltage to determine the pH of the tested sample solution.

5.9.1 COVERING MATERIAL

A. Now-a-days basically UV stabilized transparent and water impermeably materials are being used as covering material which further can be divided into two groups:
- **Rigid transparent:** It mainly includes glass, polyvinyl chloride, and fiberglass reinforced plastic. The advantages of these types are long life and low heat loss but being costly make them uneconomical for individual grower and moreover is unsuitable for warm areas.
- **Flexible transparent:** it includes flexible plastic films like, vinyl etc. advantages include low cost, good transparency, light weight, etc. but heat loss and low durability are the main disadvantages. Out of these mainly, PE films of about 200 μm thickness, UV stabilized, PE-IR film (for sub-tropical type of climates), anti-drop, multi-layer, light diffusing, anti-dust, etc.

B. **Insect-proof (IP) nets:** these are fixed in the openings for ventilation in the polyhouse. Plastic or polyethylene nets of generally 40 mesh size are recommended. As temperature and humidity are main concerns in a polyhouse, care should be taken while selecting IP nets as decrease in mesh size increases the inside temperature and humidity and vice versa.

5.9.2 DESIGN OF LOAD

While designing a polyhouse, calculation of total load is very important and on the basis of total load only the type of foundation and roof structure should be determined. Total load includes dead load, crop load, wind load, and snow load. On an average, a polyhouse may be designed to bear a minimum of 25 kg/m^2 load in the non-snow fall zone with wind up to 150 km/h.

- **Dead load:** it is the weight of all structural materials used for construction and weight of equipments attached to the structure. In case of steel structure frame, the MS steel should be about 4.5–5.5 kg/m^2.
- **Crop load:** this is generally used for vegetables like tomato, capsicum, cucumber, etc. This generally includes load of trellising plants, load of

work, etc. and an average estimate of crop load may be considered as 20 kg/m^2.

- **Wind load:** it is created by wind blowing horizontally and transferred to the polyhouse structure. To estimate the average peak area, data of 25 years mean recurrent interval may be used to calculate the wind load.
- **Snow load:** this is applied vertically from the horizontal projection of roof and can be as high as 75 kg/m^2.

5.9.3 POTENTIAL CROPS

In high altitudes, every vegetable has its own potential. High-altitude environment gives possibility to grow variety of vegetables under protected structures. In summer season, number of vegetables can be grown with good yield and market potential. However, winter season offers limited choice to the growers despite of good market potential. Therefore, under protected cultivation, we can divide potential vegetables for summer and winter seasons.

Potential vegetables during summer season are as follows:

- Tomato
- Capsicum
- Brinjal
- Okra
- Cucumber
- Muskmelon
- Watermelon
- Sardamelon
- Summer squash
- Bottle gourd

Potential vegetables during winter season are as follows:

- Spinach
- Lettuce
- Swiss Chard
- Chinese cabbage
- Cabbage
- Turnip
- Kale
- Collards

5.9.4 LIMITATIONS IN PROTECTED CULTIVATION

Appropriate designing of protected structures with respect to problem areas, identification of suitable crops, non-availability effective package of varieties, and inputs are the factors limiting the adoption and area expansion in protected cultivation. Some more hurdles in accepting protected cultivation as entrepreneurship are enlisted below:

1. Lack of technical knowledge to both growers and manufacturers
2. Farmers do not know the objectives and scientific norms behind the greenhouse technology.
3. Manufactures of fabrication materials are not providing the materials according to standards and norms for greenhouse technology.
4. Only few farmers have taken the training for greenhouses technology and they are growing their crops productively with little drawbacks.
5. Most of the farmers do not know about package of practices.
6. Farmers are not aware about the standards or objectives behind the protected structures for creating environments for growing the better crops and for long duration
7. Nets used for protecting the crops from insects and viruses are neither insect proof nets nor shade nets,
8. Nets used for shading purpose in greenhouses are not standardized.
9. Farmers are not using the nets for shading the crops according to crops needs.
10. Mostly ventilators of greenhouses are found open or no proper nets are used by farmers for ventilation which give passage to the insects and virus vectors.
11. As farmers are not using proper nets for protecting the insects and virus, high infestation of insects and disease are found during the survey.
12. The threads used for staking the crops inside the greenhouse are not UV stabilized.
13. Farmers are not growing the crops according to seasons and times specified for particulars crops.
14. Farmers do not have the nursery growing techniques for disease free and quality seedlings on their farms as it is a prerequisite for protected cultivation.
15. Farmers do not have the provision of double door for greenhouses so they are making passage to insects with them during entry of greenhouses.
16. Manufactures and fabricators of greenhouses are not aware about the specifications for greenhouses for different crops so they are fabricating the greenhouses with uniformity for all crops.
17. Manufactures are not providing the technical knowledge how to grow crops inside the greenhouse.
18. Farmers are growing more than one crop in single greenhouse and there is no separation between the crops so insects and diseases are transmitting from one crop to another crops.
19. Some farmers have small size greenhouses which are not scientifically good for crops cultivation because required environments cannot be created inside them.

5.10   CONCLUSIONS

Protected cultivation is the only approach that has the potential to cater to the food and nutritional security challenges of higher Himalayas because of climatic uncertainties. Many more protected structures are available that enable crop climate management but harvest of solar radiations and conservation of temperature during winter remain the challenge to be accepted. Modernization of the approach toward protected cultivation is need of the hour and demand immediate redressal through R & D innovations and interventions. Protected cultivation of vegetables offers distinct advantages of quality, productivity, and favorable market prices to the growers. Vegetable growers can substantially increase their income by protected cultivation of vegetables in off season as the vegetables produced in the normal season generally do not fetch good returns due to large availability of these vegetables in the market. There is scope of entrepreneurship in off-season vegetable production, cultivation of rare exotic vegetable for export, commercial vegetable seedling nursery, and seed production as well as quality planting generation of high value crops provided suitable protected structures are evolved subsequently by followed integrated management strategies. Protected cultivation is a precision crop management technology with high input efficiency and when applied with organic crop production principles there will be complementary effects adding to the socio-economic benefits of farming commun.

REFERENCES

Akbar, P.I., Kanwar, M.S., Mir, M.S., and Hussain, A., 2013. Protected vegetable cultivation technology for cold arid agro-ecosystem of Ladakh. *International Journal of Horticulture 3*(19), pp. 109–113.

Gimenez, C., Otto, R.F., and Castilla, N., 2002. Productivity of leaf and root vegetable crops under direct covers. *Scentia Horticulturae 94*(1), pp. 1–11.

Kanwar, M.S., 2011. Temperature: decisive factor in vegetable production in cold arids. *Agribios Newsletter 9*(8), pp. 29–30.

Mishra, G.P., Singh, N., Kumar, H., and Singh, S.B., 2010. Protected cultivation for food and nutritional security at Ladakh. *Defense Science Journal 61*(2), pp. 219–225.

Reddy, P.P., 2016. *Sustainable Crop Production under Protected Cultivation.* Springer, Singapore, p. 434.

Sharma, J.P., 2016. Protected structures and organic vegetable production in higher Himalayas. In: Raj, A., Kanwar, M.S., Gupta, V., Ali, L., and Masoodi, T.H. (Eds.). *Protected Structures for High Himalayas: Design and Application.* PFDC, HMAARI (SKUAST-K), Leh, pp. 1–35.

Spehia, R.S., 2016. Green-house designs for high hills of Himachal Pradesh. In: Raj, A., Kanwar, M.S., Gupta, V., Ali, L., and Masoodi, T.H. (Eds.). *Protected Structures for High Himalayas: Design and Application.* PFDC, HMAARI (SKUAST-K), Leh, pp. 45–62.

Stobdan, T., Angmo, S., Angchok, D., Paljor, E., Dawa, T., Tsetan, T., and Chaurasia, O.P. 2018. Vegetable production scenario in trans-Himalayan Leh Ladakh region, India. *Defence Life Science Journal 3*(1), pp. 85–92.

6 Applications of Remote Sensing in Crop Production and Soil Conservation

Nayar Afaq Kirmani, J. A. Sofi, and Juvaria Jeelani
Sher-e-Kashmir University of Agricultural Sciences
and Technology of Kashmir (SKUAST-K)

CONTENTS

6.1 INTRODUCTION

Remote sensing techniques are widely used in agriculture and agronomy. Nowadays, use of remote sensing is necessary as the monitoring of agricultural activities faces special problems not common to other economic sectors. First, agricultural production follows strong seasonal patterns related to the biological lifecycle of crops. Second, the production depends on the physical landscape (e.g., soil type) as well as climatic driving variables and agricultural management practices. All variables are highly unpredictable with respect to space and time. Moreover, as productivity can change within short-time periods, due to unfavorable growing conditions, the agricultural monitoring systems should be well-timed. This is even more important, as many items are perishable. The solution for providing food security to all people of the world without affecting the agro-ecological balance lies in the adoption of new research tools, particularly from aerospace remote sensing, and combining them with conventional as well as frontier technologies like Geographic Information Systems (GIS). Sustainable agricultural development is one of the prime objectives in all countries in the world, whether developed or developing. The broad objective

69

of sustainable agriculture is to balance the inherent land resource with crop require-ments, paying special attention to optimization of resource use toward the achieve-ment of sustained productivity over a long period (Lal and Pierce, 1991). Sustainable agricultural development/sustainable increase in the crop production could be achieved by adopting a variety of agricultural technologies, which are summed up as follows:

1. Improved crop management technology through the use of high yield-ing, input responsive and soil, climatic, and biotic stresses – tolerant crop varieties;
2. Suitable cropping systems for different agro-ecological regions based on soil, terrain, and climatic suitability;
3. Integrated nutrient management for improving soil productivity and mini-mization of the risk of pollution of soil, water, and environment;
4. Integrated pest management for effective pests control as well as to reduce the adverse effects of pesticides on environment;
5. Soil and water conservation for controlling the soil degradation and improv-ing the moisture availability.
6. Input use efficiency maximization in terms of the economic return with minimal input.

Remote sensing can significantly contribute to providing a timely and accurate pic-ture of the agricultural sectors, as it is very suitable for gathering information over large areas with high revisit frequency. The present paper summarizes the main remote sensing applications, with a focus on regional to global applications. It pro-vides arguments for enhancing the investments in agricultural monitoring systems. It follows the strong conviction that a close monitoring of agricultural production systems is necessary, as agriculture must strongly increase its production for feeding

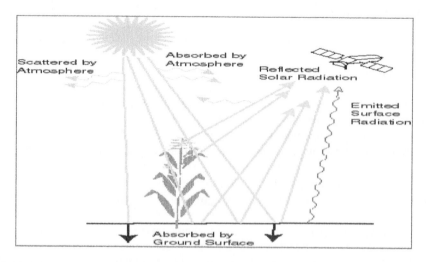

FIGURE 6.1 Basics of remote sensing and interaction of EMR with atmosphere.

to 9 billion people predicted by mid-century. Increasing in production must be achieved through minimizing the environmental impact on agriculture. Achieving this goal is difficult, as agriculture must cope with climate change and compete with land users not involved in food production (e.g., biofuel production, urban expansion, etc.). The necessary changes and transitions have to be monitored closely to provide decision makers with feedback on their policies and investments. Interactions vary with incident EM energy (wave length/frequency) and physical/chemical properties of surface material (Figure 6.1).

6.2 APPLICATIONS OF REMOTE SENSING AND GIS TECHNOLOGY IN CROP PRODUCTION AND SOIL CONSERVATION

Remote sensing and GIS technology are being effectively utilized in India in several areas for sustainable agricultural development and management. The areas of sustainable agricultural development/management include cropping system analysis, agro-ecological zonation, quantitative assessment of soil carbon dynamics and land productivity, soil erosion inventory, and integrated agricultural drought assessment and management and Integrated Mission for Sustainable Development (IMSD). The use of remote sensing and GIS technology in these areas of sustainable agricultural management and development are discussed below.

6.2.1 CROPPING SYSTEM ANALYSIS

Information on existing cropping systems in a region with respect to areal extent of crops, crop vigors/yield and yearly crop rotation/sequence practices is important for finding out the agricultural areas with low to medium crop productivity where sustainable increase in crop production can be achieved by adoption of suitable agronomic management packages including introduction of new crops, etc. GIS technology can play a vital role in the cropping system analysis of an area by spatially integrating temporal crop inventory information of various crop seasons of that area (Figure 6.2).

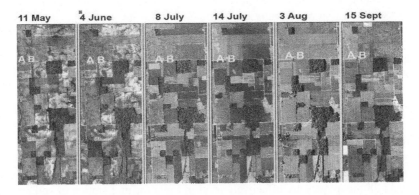

FIGURE 6.2 Crop condition analysis over the seasons.

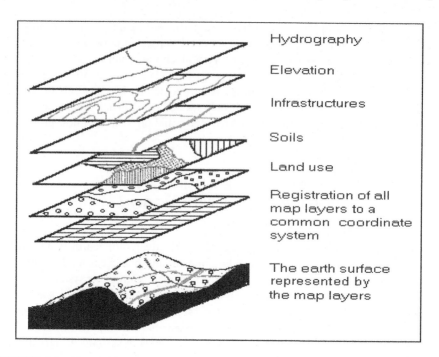

FIGURE 6.3   Use of different layers in GIS for land use planning and AEZ.

a. The rectangular field that is red in the first image goes through several changes over the course of the growing season. What do you think about those changes are?
b. The square field with a hole in the lower left, changes progressively as well. How does it change? What happens to the hole?

6.2.2   AGRO-ECOLOGICAL ZONE-BASED LAND USE PLANNING

Agro-ecological zoning (AEZ) is an important basis of the sustainable agricultural land use planning of a region. AEZ encompasses the delineation of landscapes into regions or zones that are broadly homogeneous with respect to agro-climate, soils, and terrain characteristics and also relatively uniform with respect to crop production possibilities. GIS technology is a very useful tool for automated logical integration of bioclimate, terrain, and soil resource information which are required for delineating AEZ in a region (Figure 6.3).

6.3   SOIL EROSION INVENTORY

The information on soil erosion such as quantification of erosional soil loss and soil conservation prioritization of watersheds/sub-watersheds provides vital inputs for sustainable agricultural management with respect to soil conservation. Remote sensing and GIS techniques are effectively used in India for preparation of soil erosion inventories by integration of physiography, soils, land use/land cover, slope map

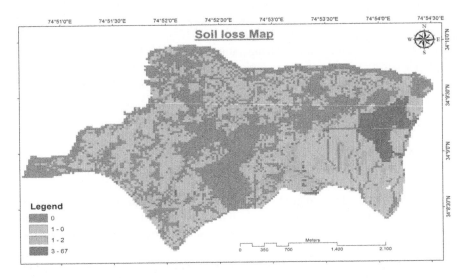

FIGURE 6.4   Soil erosion map of telbal micro watershed. (Khusboo, 2017.)

layers and use of ancillary data of agro-met and soil physico-chemical properties (Saha and Pande, 1993, 1994; Saha, 1996). In one such case study of a watershed in Northern India (Song Watershed, Dehradun District, Uttar Pradesh), satellite (IRS-1B, LISS-II) analyzed soil, land use/land cover, drainage, digital elevation model (DEM). It is a 3D representation of a terrain's surface commonly of a planet, moon, or asteroid created from a terrain's elevation data which is generated slope and slope length, and ancillary agro-met and soil characteristic data were used to assess erosional soil loss according to Universal Soil Loss Equation (USLE) in a GIS environment (Figure 6.4). The basic equation of the USLE model is as given below:

$$A = R \cdot K \cdot LS \cdot C \cdot P \tag{6.1}$$

where A is estimated soil loss (t/ha/year); R is rainfall erosivity factor; K is soil erodibility factor; LS is slope and slope length factor; C is land cover factor; and P is conservation practice factor.

6.4   SOIL CARBON DYNAMICS AND LAND PRODUCTIVITY ASSESSMENT

By studying the carbon dynamics of agro-ecosystems, it is possible to quantify the fixation and release of carbon in the soil-crop-plant system. This knowledge is essential for assessing the organic matter depletion of soil, long-term soil fertility, and sustained productivity of agro-ecosystems. An integrated remote Sensing and GIS-based methodology was developed for studying carbon dynamics such as annual crop Net Primary Productivity (NPP), soil organic matter decomposition and the annual soil carbon balance using models such as the Osnabruck – Biosphere and Century models (Saha and Pande, 1994). According to Osnabruck-Biosphere model,

$$NPP = f\left[NPP_i, F_s, F(CO_2)\right] \tag{6.2}$$

where NPP_i is Climatic (rainfall, temperature) potential NPP; F_s is Soil factor, and $F(CO_2)$ is nutrition factor of atmospheric CO_2 content to NPP.

The century model estimates decomposition loss (DL) of soil humus carbon and it is expressed as follows:

$$DL = f\left(K_i, T, M_d, T_d, C_i\right) \tag{6.3}$$

where, K_i is Maximum decomposition rate, T is Soil (Silt+Clay) content; M_d and T_d are rainfall and temperature factors, respectively, and C_i is initial soil humus carbon content.

Multi-temporal satellite data (IRS-1B: LISS-I) derived from annual crop biomass, soil map and soil characteristics and ancillary agro-met data were the major inputs used for the soil carbon dynamics study in a case study of Doon Valley (Dehradun district, Uttar Pradesh).

Quantitative assessment of land productivity is one of the important components of integrated nutrient management for sustainable agricultural development. The land productivity data provide information about the inherent fertility status of soils-capes, which is a useful guideline for supplementing soil nutrients from external sources, such as fertilizers/manure. In a case study from Jainti Watershed (Bihar), land productivity was assessed and mapped following the Storie Index model by GIS-aided integration of soil and slope maps and laboratory-measured soil physio-chemical data (Kudrat and Saha, 1993). According to the Storie Index model, land productivity (LP) is expressed as follows:

$$LP = f\left(A, B, C, X\right) \tag{6.4}$$

where A is a rating based on soil development; B is a rating based on soil texture; C is a rating based on terrain slope and X is a composite rating based on soil fertility, pH, drainage, erosion etc.

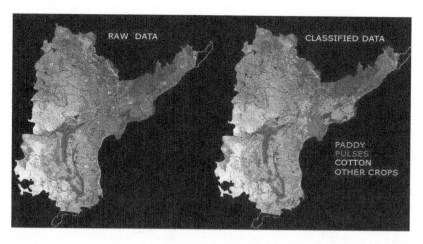

FIGURE 6.5   IRS-1C WiFS DATA (25-1-96) showing major crops in Andhra Pradesh, India.

Various studies carried out in several areas of sustainable agricultural management/development in India by integrated use of aerospace data and GIS have clearly indicated that remote sensing and GIS technology are very effective tools for suggesting action plans/management strategies for agricultural sustainability of any region (Figure 6.5).

REFERENCES

Khusboo, F., 2017. Characterization of Telbal micro-watershed using RS and GIS. Thesis, submitted in partial fulfillment for award of M.Sc. (Soil Sciences) to Sher-e-Kashmir University of Agricultural Sciences and Technology of Kashmir Shalimar, Srinagar, p. 93.

Kudrat, M. and Saha, S.K. 1993. Land productivity assessment and mapping through integration of satellite and terrain slope data. Journal of the Indian Society of Remote Sensing, *21*, p. 157. doi: 10.1007/BF02992111

Lal, R. and Pierce, M., 1991. *Soil Management for Sustainability*. Soil and Water Conservation, Ankeny.

Saha, S.K., 1996. Integrated use of Remote Sensing and GIS for soil erosion hazard modelling – a case study. *Proc. 17th Asian Conference on Remote Sensing*, Colombo.

Saha, S.K. and Pande, L.M., 1993. Integrated approach towards soil erosion inventory for environmental conservation using satellite and agrometeorological data. *Asian Pacific Remote Sensing Journal*, *5*(2), pp. 21–28.

Saha, S.K. and Pande, L.M., 1994. Modelling agro-ecosystem carbon dynamics using Remote Sensing and GIS technology – a case study. *Proc. International conference on Remote Sensing and GIS (ICORG-94)*, Hyderabad.

7 Flash Floods Cause and Remedial Measures for Their Control in Hilly Regions

Kusum Pandey
Punjab Agricultural University

Dinesh Kumar Vishwakarma
G. B. Pant University of Agriculture
and Technology (GBPUAT)

CONTENTS

7.1   INTRODUCTION

The mountain regions are more vulnerable to natural disaster due to its varying nature of relief where the developmental activities over the years has further accentuated the problem by upsetting the natural equilibrium of various physical processes operating in the mountain ecosystem. In India, most of the floods result from 75% of the annual average rainfall occurs during south–west monsoon season. During this season, the rainfall is highly non-uniform in time and space. The regions experiencing intense rainfall are affected due to flooding, whereas the regions having deficient rainfall face the problems of drought. There are incidents when one part of the country is experiencing flood while another is in the grip of severe drought. The flood problem varies from one river system to another. The rivers originating in the Himalayas carry a large amount of sediment, causing erosion of the banks in the upper reaches and over-topping in the lower segments. The Ganga–Brahmaputra–Meghana basin is one of the largest in the world. Flood is a common phenomenon in the Brahmaputra valley. All the districts of the Brahmaputra valley in Assam are inundated almost every year. An area of 30 out of 78 lakh ha, that is, about 45% of Assam's total area, is flood prone. In Ganga basin, in general, the flood problem increases from west to east and from south to north. There is a problem of drainage congestion in the extreme western and north-western parts. The rivers such as Gandak, Burhi Gandak, Bagmati and Kamla and other small rivers of the Adhwara Group, Kosi in lower reaches and Mahananda at the eastern end spill over their banks causing considerable damage. High floods also occur in the river Ganga in some of the years causing considerable inundation of the marginal areas in Bihar. The main rivers of North West region are the tributaries of river Indus, namely Sutlej, Beas, Ravi, Chenab, and Jhelum, all flowing from the Himalayas. These rivers carry quite substantial discharges during monsoon season and also large volumes of sediment. Compared to the Ganga and Brahmputra river regions, the flood problem is relatively less in this North West region (CWC, 1980; Allen et al., 2016).

Floods also occur in peninsular river basins which comprise the important rivers of Central India and Deccan river region including Narmada, Tapi, Mahanadi, Godavari, Krishna, and Cauvery. These rivers have adequate capacity within the natural banks to carry flood discharge except in the deltaic region. The Tapi and Narmada are flashy and occasionally in high floods. The systematic records of the flood occurrences, the damages associated and the measures taken to deal with them are available only from the year 1953 by which time the gravity of the problem passed by this natural phenomenon came to be realized by the government. As reported by the Central water Commission (CWC) under Ministry of Water Resources, the annual average area affected by floods is 7.382 million ha. This observation was based on data for the period 1953–2002, with variability ranging from 1.46 million ha in 1965 to 17.5 million ha in 1978 (Lohani and Jaiswal, 1996; Agrawal et al., 2004; Gupta and Uniyal, 2012; Jain et al., 2013). In the Himalayan river basins, flash floods are very common and caused by cloud burst, Landslide Lake Outburst Floods (LLOFs), Glacial lake outburst flood (GLOF), etc. In this paper, brief description of various flood management practices and techniques being used in India are presented. The status of the non-structural flood forecasting measures is also presented.

7.2 CAUSES OF FLOOD

Flooding occurs mostly from heavy rainfall when natural watercourses do not have the capacity to convey excess water. However, floods are not only caused by heavy rainfall but also resulted from other phenomena, particularly in coastal areas where inundation can be caused by a storm surge associated with a tropical cyclone, a tsunami or a high tide coinciding with higher than normal river levels (Mathur and Jain, 2004). Dam failure, triggered for example by an earthquake, will result in flooding of the downstream area, even in dry weather conditions. Causes of floods are presented in Table 7.1. Various factors which may contribute to flooding include:

- Volume, spatial distribution, intensity and duration of rainfall over a catchment;
- The capacity of the watercourse or stream network to convey runoff;
- Catchment and weather conditions prior to a rainfall event;
- Ground cover;
- Topography; and
- Tidal influences.

7.3 FLOOD PROBLEMS IN INDIA

Main problems in India with respect to floods are inundation, drainage congestion due to urbanization and bank erosion. The problems based on the river system, topography of the place and flow phenomenon. Being a vast country, the flood problems

TABLE 7.1
Causes of Flood

Climatic Factor	Geological Factor	Man-Made Factor
I. Change in solar irradiance	I. Loose/poor strata	I. Development activities
II. Change in global climate	II. Unfavorable slope angle	II. Effect of large growing population
III. Change in atmospheric pressure in local or regional		III. Unscienctific land use planning

in India may be visualized on regional basis. However, for the sake of simplicity, India may be broadly divided into four zones of flooding, viz. (i) Brahmaputra River Basin, (ii) Ganga River Basin, (iii) North-West Rivers Basin, and (iv) Central India and Deccan Rivers Basin. Flooding in these zones is presented in following subsections.

7.3.1 Brahmaputra River Basin

The first zone belongs to the basins of the rivers Brahmaputra and Barak with their tributaries. It covers the States of Assam, Arunachal Pradesh, Meghalaya, Mizoram, northern parts of West Bengal, Manipur, Sikkim, Tripura, and Nagaland. The catchments of these rivers receive large amount of rainfall. As a result, floods in this region take place very often and are severe by nature. The general tectonic up wrapping of North-East region has also significant effect on the river Brahmaputra. Almost all the Northern tributaries of Brahmaputra are affected by landslides in the upper catchment. Further, the rocks in the hills, where these rivers originated are friable and susceptible to erosion and thereby cause exceptionally a high silt charge in the rivers.

In addition, the region is subject to severe and frequent earthquakes causing numerous landslides in the hills, which upsets the regime of the rivers. Important problems in this region are flood inundation due to spilling of banks, drainage congestion due to natural as well as manmade structures and change of river flow. In recent years, the erosion along the banks of the Brahmaputra was enormous and has become a serious concern among the water resources engineers (Rangachari, 1986; Seth and Goel, 1985). Main problems of flooding in Assam are inundation caused by spilling of the rivers Brahmaputra and Barak as well as their tributaries. In addition, the erosion along the Brahmaputra is a serious problem. In Northern parts of West Bengal, the rivers Teesta, Torsa, and Jaldakha are in floods every year and inundate large areas. During flooding, these rivers carry large amount of silt and have a tendency to change their courses. The rivers in Manipur spill over their banks frequently. The lakes in the territory are filled up during the monsoon and spread to large marginal areas. In Tripura, flood problems occur due to the spilling and erosion by rivers.

7.3.2 Ganga River Basin

The Ganga and its many tributaries (the Yamuna, the Sone, the Ghaghra, the Gandak, the Kosi and the Mahananda) constitute the second zone. This zone covers Uttaranchal, Uttar Pradesh, Bihar, south and central parts of West Bengal, parts of Haryana, Himachal Pradesh, Rajasthan, Madhya Pradesh and Delhi. The normal annual rainfall of this region varies from about 60to –190 cm of which more than 80% occurs during the South-West monsoon. The rainfall increases from west to east and from south to north. The flood problem is mostly confined to the areas on the northern bank of the Ganga River. The damage is caused by the northern tributaries of the Ganga by spilling over their banks and changing their courses. Though the Ganga is carrying huge discharges (57,000–85,000 m³/s), the inundation and erosion problems are confined to some specific places only. In general, the flood problem increases from west to east and from south to north. There is the problem of drainage congestion in the north western parts of the region, and same problem exists in the southern parts of West Bengal. The problem becomes acute when the main river, in which the water is to be drained, already has high water level. The flooding and erosion problem is serious in Uttar Pradesh, Bihar and West Bengal, whereas in Rajasthan and Madhya Pradesh, the problem is not so serious. In Bihar, the floods are largely confined to the rivers of North Bihar and are an annual feature. Most of the rivers (e.g., the Burhi Gandak, the Bagmati, the Kamla Balan, other smaller rivers of the Adhwra Group, the Kosi in the lower reaches and the Mahananda at the eastern end) spill over their banks causing considerable damage to crops anddislocating traffic.

High floods occur in the Ganga occasionally causing considerable inundation of the marginal areas in Bihar. In the eastern districts of Uttar Pradesh, the flooding is frequent due to spilling of the Rapti, the Sarada, the Ghaghra and the Gandak. The problem of drainage congestion exists in the western and north-western areas of Uttar Pradesh, particularly in Agra, Mathura, and Meerut districts. Erosion is experienced in some places of the left bank of Ganga, on the right bank of the Ghaghra and on the right bank of the Gandak. In Haryana, flooding takes place in the marginal areas along the Yamuna and the problem of poor drainage exists in some of the south western districts. In Delhi, a small area along the banks of the Yamuna is subject to flooding as the river spills. In addition, local drainage congestion is experienced in some of the developing colonies during heavy rains. In the south and central parts of West Bengal, the Mahananda, the Bhagirathi, the Ajoy, and the Damodar cause flooding due to inadequacy in river channels and the tidal effect. The problem of erosion also exists in the banks of rivers and on the left and right banks of Ganga, upstream and downstream, respectively, of the Farakka barrage.

7.3.3 NORTH-WEST RIVER BASINS

The third zone comprises the basins of North-West rivers such as the Sutlej, Ravi, Beas, Jhelum, and Ghaggar. Compared to the second zone, the flood problem in this zone is less. The major problem is that of inadequate surface drainage which causes inundation and water logging. Another cause for flood has been the water logging in the irrigated area and changes in river regimes due to rise in ground water levels. At present, the problems in Haryana and Punjab are mostly of drainage congestion and water logging. The Ghaggar river used to disappear in the sand dunes of Rajasthan after flowing through Punjab and Haryana. Floods occur frequently in the Jhelum river in Kashmir causing a rise in the level of the Wullar Lake thereby submerging marginal areas of the lake.

7.3.4 CENTRAL INDIA AND DECCAN RIVERS BASIN

Important rivers in the fourth zone are the Narmada, the Tapi, the Mahanadi, the Godavari, the Krishna and the Cauvery. These rivers have mostly well-defined stable courses. They have adequate capacity within the natural banks to carry the flood discharge except in their lower reaches and in the delta area, where the average bed slope is very flat. The lower reaches of the important rivers on the East Coast have been embanked. This region covers all the southern states namely Andhra Pradesh, Chhattisgarh, Karnataka, Tamil Nadu, Kerala, Orissa, Maharashtra, Gujarat, and parts of Madhya Pradesh. The problems in this region are not serious except for some rivers in Orissa (the Brahmani, the Baitarni, and the Subarnarekha). The Delta areas of the Mahanadi, Godavari, and the Krishna rivers on the east coast periodically face flood and drainage problems, in the wake of cyclonic storms. The Tapi and the Narmada are occasionally affecting areas in the lower reaches of Gujarat due to high floods. Flood problems in Andhra Pradesh are confined to spilling by the smaller rivers and the submergence of marginal areas along the Kolleru Lake. Rivers like Budameru and the Thammileru not only overflow their banks along their courses to Kolleru Lake but also cause a rise in the level resulting in inundation of adjoining lands. In Orissa, damage due to floods is caused by the Mahanadi, the Brahmani and the Baitarani which have a common delta. Water from these rivers intermingles in the delta and results in a very high water level, which cause severe flooding in the region. The coastal districts are densely populated and receive heavy precipitation in the Eastern Ghat region. The silt deposited constantly by these rivers in the delta area raises the flood water level and, the rivers often overflow their banks or break through new channels causing heavy damage. The lower reaches of the Subarnarekha are affected by floods and drainage congestion. The small rivers of Kerala when in high floods often cause considerable damage. There is also the problem of mud-flow from the hills, which results in severe losses.

7.4   FLASH FLOOD

A flash flood is a rapid flooding of geomorphic low-lying areas: washes, rivers, dry lakes, and basins. Flash floods are not only associated with fast flowing water in steep terrain, but also with the flooding of very flat areas, where slope is too small to allow for the immediate runoff of storm water. Mostly flash floods are local events relatively independent of each other and scattered in time and space. They are produced by very intense rainfall over a small area in very short time. The ground is not usually saturated, but the infiltration rate is much lower than the rainfall rate. Although flash floods usually occur in a relatively small area and last only a few hours (sometimes minutes), they have an incredible potential for destruction. The frequent occurrence of flash floods in the Himalayan region poses a severe threat to lives, livelihoods, and infrastructure, both in the mountains and downstream. Flash floods carried high amount of debris than normal floods and as a result cause more damage to buildings, roads, bridges, hydropower stations, and other infrastructure (Figures 7.1 and 7.2). The reasons for flash floods can be cloud burst, that is, intense rainfall, the outburst of a landslide dam lake, the failure of artificial dam, or a glacial lake outburst. Flood events reported during the recent past in the hilly areas listed in Table 7.2.

FIGURE 7.1 Flash flood in Jammu-Kashmir.

FIGURE 7.2 Flash flood in Uttarakhand.

TABLE 7.2
Flood Events Reported During the Recent Past in the Hilly Areas

S. No.	Date/Year	Location	Cause of Flood	Damage
1.	1867 and 1880	Nainital, Uttarakhand	Flash flood	Two major landslides on the Sher-ka-Danda slope in Nainital. The 1,880 landslide took place due to rainfall and an earth tremor, destroying buildings, and permanently filled a portion of the Naini lake.

(Continued)

TABLE 7.2 (*Continued*)
Flood Events Reported During the Recent Past in the Hilly Areas

S. No.	Date/Year	Location	Cause of Flood	Damage
2.	1893	Alaknanda, Uttarakhand	Flash flood	Floods in the Birehi Ganga river near its confluence with the Alaknanda river triggered landslides, causing major blockage of the river with a 10–13 m afflux. A girder bridge was bypassed and another one was destroyed.
3.	1969, 1970, 1971, 1972 and 1985	Kaliasaur, Uttarakhand	Flash flood	Kaliasaur is one of the most persistent and regularly occurring landslides areas, located along the Rishikesh-Badrinath road. Landslides in this region results into frequent road blockage and land damage.
4.	1968	Rishi-Ganga, Uttarakhand	Landslide	The Rishi Ganga river in Garhwal was blocked due to landslide at Reni village.
5.	Jul 1970	Patal Ganga, Uttarakhand	Flash flood	The Patal Ganga (a tributary of the Alaknanda river) got choked and a reservoir was created. The bursting of this choked reservoir resulted in flash floods in the Alaknanda river, triggering many landslides; 200 people died.
6.	Aug 1978	Uttarkashi, Uttarakhand	Flash flood	The Kanauldia Gad, a tributary joining the Bhagirathi river upstream from Uttarkashi in the Uttarakhand formed a debris cone across the main river, impounding breaching caused flash floods, creating havoc. A river to a height of 30 m. Its 1.5 km long and 20 m deep lake was left behind as a result of the partial failure of the landslide dam.
7.	1971	Kanauldia Gad, Uttarakhand	Flash flood	A major landslide on the bank of the Kanauldia gad, a tributary of the Bhagirathi river upstream from Uttarkashi formed a debris cone which impounded water to a height of 30 m. Its breaching caused flash floods downstream.
8.	Sep 1989	Karanprayag, Chamoli, Uttarakhand	Flash flood	Three people died and two injured.

(*Continued*)

TABLE 7.2 (*Continued*)
Flood Events Reported During the Recent Past in the Hilly Areas

S. No.	Date/Year	Location	Cause of Flood	Damage
9.	Dec 1991	Uttarkashi, Uttarakhand	Flash flood	Three people died.
10.	July 10, 1994	Chaukhutia, Almora, Uttarakhand	Flash flood	Four people died.
11.	Aug 15, 1997	Chirgaon, Shimla district, Himachal Pradesh	Cloudburst	1,500 people died.
12.	Aug 17, 1998	Kali valley of the Kumaon division, Uttarakhand.	Cloudburst	Died 250 people including 60 Kailash Mansarovar pilgrims.
13.	July, 2001	Near Meykunda, Rudraprayag, Uttarakhand	Flash flood	Twenty seven people died.
14.	2002	Khetgaon, Pithoragarh, Uttarakhand	Cloudburst	Four people died.
15.	July15, 2003	Didihat, Pithoragarh, Uttarakhand	Flash flood	Four people died.
16.	Sep 2003	Varunavat Parvat, Uttarkashi, Uttarakhand	Landslide	Incessant rains triggered massive landslide in the area, causing the burial of numerous buildings, hotels, and government offices located at the foot of the hill slopes.
17.	July 16, 2003	Shilagarh in Gursa area of Kullu, Himachal Pradesh.	Cloudburst	Fourty people died.
18.	2004	Ranikhet, Uttarakhand	Cloudburst	One people died.
19.	21 May and 09 June 2004	Kapkot, Bageshwar, Uttarakhand	Flash flood	Six people died.
20.	July19, 2004	Badrinath, Chamoli, Uttarakhand	Landslide	Sixteen persons killed, 200 odd pilgrims stranded, 800 shopkeepers and 2,300 villagers trapped as cloudburst triggered massive landslides washed away nearly Badrinath road cutting off Badrinath area 200 m of road on the Joshimath.

(*Continued*)

TABLE 7.2 (*Continued*)

Flood Events Reported During the Recent Past in the Hilly Areas

S. No.	Date/Year	Location	Cause of Flood	Damage
21.	July 6, 2004	Alaknanda river basin	Cloudburst	At least 17 people killed and 28 injured when three vehicles were swept into the Alaknanda river by heavy landslides triggered by a cloudburst that left nearly 5,000 pilgrims stranded near Badrinath shrine area in Chamoli district, Uttarakhand.
22.	29–30 June 2005	Govindghat, Chamoli, Uttarakhand	Landslide	A cloudburst/landslide occurred in which was a huge quantity of debris and rock boulders were brought down along a seasonal nala. Eleven people were killed and property lost.
23.	21 Jul 2005	Vijaynagar, Rudraprayag, Uttarakhand	Flash flood	Four people died.
24.	Aug 13, 2007	Didihat, Pithoragarh, Uttarakhand	Flash flood	Four people died.
25.	Sep 06, 2007	Dharchula, Pithoragarh, Uttarakhand	Landslide	A landslide due to excessive rainfall resulted in 15 fatalities and loss of livestock.
26.	2007	Pithoragarh & Chamoli, Uttarakhand	Cloudburst	Twenty three people died.
27.	Aug 16, 2007	Bhavi village, Ghanvi, Himachal Pradesh	Cloudburst	Fifty two people died.
28.	2008	Pithoragarh, Uttarakhand	Cloudburst	One people died.
29.	2009	Munsiyari Tehsile, Pithoragarh, Uttarakhand	Cloudburst	Forty three people died.
30.	27 Jul–Sep, 2010	Kot, Pauri; Rudrapur, Udham Singh Nagar; Dehradun, Nainital, Chamoli, Champawat, Haridwar, Uttarakhand	Cloudburst Induced flash flood and Landslide	Fifty nine people died and 2 missing and 17 injured.

(Continued)

TABLE 7.2 (*Continued*)

Flood Events Reported During the Recent Past in the Hilly Areas

S. No.	Date/Year	Location	Cause of Flood	Damage
31.	Jul 21, 2010	Almora, Uttarakhand	Cloudburst	Thirty six people died in cloud burst induced flash flood.
32.	Aug 18, 2010	Kapkot, Bageshwar, Uttarakhand	Cloudburst	Eighteen school children were buried alive and eight injured due to massive cloudburst
33.	Aug 6, 2010	Leh town of Ladakh region in J & K	Cloudburst	Thousand persons dead and over 400 injured.
34.	May 6, 2011	Raipur, Dehradun, Uttarakhand	Flash flood	Three people died.
35.	Aug 15, 2011	Tuneda, Bageshwar, Uttarakhand	Flash flood	Twenty one people died and one injured.
36.	Jun 9, 2011	Jammu	Cloudburst	Two restaurants and many shops were washed away.
37.	Jul 20, 2011	Upper Manali in Himachal Pradesh	Cloudburst	Two people died and 22 missing.
38.	Aug 3, 2012	Asi Ganga Valley, Uttarkashi, Uttarakhand	Flash flood	The worst affected areas were Gangotri, Sangam, Chatti, and Bhatwari. About 7,389 people from 1,159 families in 85 villages were affected. Nearly 28 people were killed in flash floods and landslides.
39.	Sep 13–14, 2012	Okhimath, Rudraprayag	Cloudburst Induced flash flood &andLandslide	Sixty eight people killed in the landslides, which caused extensive damages to the buildings, agricultural lands and roads at several places. 39 people died.
40.	June16–17, 2013	Bageshwar, Chamol, Pithoragarh	Cloudburst induceed flash flood and Landslide	Rudraprayag & Uttarkashi Flash flood induced landslide. 68,026 people died, and 4,117 missing. Huge devastation to infrastructures and other properties mainly in five districts of Uttarakhand.
41.	Jun 15, 2013	Kedarnath and Rambada Region, Uttarakhand	Cloudburst	Over 1,000 killed.
42.	Jul 31, 2014	Tehri, Uttarakhand	Cloudburst	At least four people were reported dead.
43.	Sep 6, 2014	Kashmir valley	Cloudburst	Two hundred people killed.

7.4.1 CLIMATE CHANGE AND EXTREME RAIN EVENTS

Climate change is the biggest threat facing humankind; extreme weather events, droughts and rise in diseases have been forecast for many parts of the globe over the coming decades. The Intergovernmental Panel on Climate Change (IPCC), the scientific body that advises the UN and governments on global warming, has stated that the changes in extreme events (floods and droughts) could affect the frequency of natural hazards such as avalanches and mudslides.

As per IPCC, a major problem related to climate change in many mountain regions is the increased erosion and reduced slope stability. The combination of complex orography (physical geography of mountains and mountain ranges) with steep slopes, intense rainstorms, and, in some regions, frequent earthquakes, causes a high amount of mass movement, which eventually finds its way into rivers as heavy sediment load. In the mountains, climate change and environmental degradation have started showing some profound impacts. Mountain regions from the Andes to the Himalayas are warming faster than the global average under climate change. Climate change is a threat multiplier for instability in fragile Himalayan regions as well. This has led to most glaciers in the mountainous regions such as the Himalayas to recede during the last century and influence stream run-off of the Himalayan Rivers. Himalayan glaciers have been in a general state of recession since a long time. As per the report on "Snow and Glaciers of the Himalayas: Inventory and Monitoring" released by the Ministry of Environment and Forest (MOEF) in 2011, out of 2,700 glaciers which are monitored, 2,184 are retreating, 435 are advancing and 148 glaciers show no change. Existing studies of Himalayan glaciers indicate that many have exhibited an increased receding trend over the past few decades. Regular monitoring of a large number of Himalayan glaciers is important for improving our knowledge of glacier response to climate change. The widespread glacial retreat in the Himalayas has resulted in the formation of many glacial lakes. Glacier retreat and shrinking could form dangerous moraine lakes whose breaching may generate floods.

7.4.2 CLOUDBURSTS

A cloudburst is a weather event in which heavy rainfall (of the order of 100 mm/h or more) occurs over a localized area at very high intensity. In India, cloudbursts usually occur during the monsoon season over orographically dominant regions like the Himalayas and the Western Ghats. These can also occur over the plains, but such occurrences are rare (Figure 7.3)

It is believed that cloudbursts occur because of rapid lifting of clouds by the steep orography of the region. The clouds get vertically lifted and these convective clouds can extend up to the height of 15 km above the ground. This process is called the "cumulonimbus convection condition" which results in formation of towering vertical dense clouds. The lifting is usually dynamic and this causes thermodynamic instability resulting in rapid condensation. It is also believed that in the Himalayan region, the clouds which are being lifted rapidly are also accompanied by soil moistened by earlier precipitation. This soil perhaps acts as an additional source of

FIGURE 7.3 Cloudburst in Uttarakhand.

moisture and also have a role in the frequent cloudbursts in the region. Cloudbursts are more frequent during the monsoon season. A number of cloudburst incidents have been reported during the last few years.

7.4.3 GLACIAL LAKES

A glacial lake is defined as water mass existing in a sufficient amount and extending with a free surface in, under, beside, and/or in front of a glacier and originating from glacier activities and/or retreating processes of a glacier. The lakes at risk, however, are situated in remote and often inaccessible areas. To assess the possible hazards from glacial lakes it is, therefore, essential to have a systematic inventory of glacial lakes formed at the high altitudes. Remote sensing makes it possible to investigate simultaneously a large number of glaciers and glacial lakes in the inaccessible mountain region. However, high quality remote sensing images are difficult to obtain for mountain region because of cloud cover during the monsoon season and snow cover during winter. Besides making a temporal inventory, a repeat monitoring of these lakes is also required to assess the change in their nature and aerial extent. But sole reliance on remote sensing data is inadequate as it cannot furnish the necessary repeat bathymetric information, changes in the height of the damming moraine, or changes in lake level, which are also needed. Reliable determination of the degree of glacial lake instability, at least in most cases, will require detailed glaciological and geotechnical in situ field investigation. Whether glacial lakes become dangerous depends largely on their elevation relative to the spillway over the surrounding moraine. Triggering events for an outburst can be moraine failures induced by an earthquake, by the decrease of permafrost and increased water pressure, or a rock or snow avalanche slumping into the lake causing an overflow. Different triggering mechanisms of GLOF events depend on the nature of the damming materials, the position of the lake, volume of the water, the nature and position of the associated mother glacier, physical and topographical conditions, and other physical conditions of the surroundings. Interaction between the processes linked to above factors may strongly increase the vulnerability of the glacial lakes to GLOF hazard. The most significant chain reaction in this context is probably the danger from ice avalanches, debris flows, rock fall or landslides reaching a lake and thus provoking a lake outburst. Only Moraine Dammed Lakes, Ice Dammed Lakes and Ice cored Dammed Lakes are considered to be vulnerable from GLOF point of view. Since the beginning of last century, the number of GLOFs increased in the Himalaya. Previous studies showed already that the risk of lake development is highest where the glaciers have a low slope angle and a low flow velocity or are stagnant. The potentially dangerous lakes can be identified based on the condition of lakes, dams, associated mother glaciers, and topographic features around the lakes and glaciers.

The criteria used to identify these lakes are based on field observations, processes and records of past events, geomorphological and geo-technical characteristics of the lake and surroundings, and other physical conditions. Identification can also be done based on the condition of lakes, dams, associated glaciers, and topographic features around the lakes and glaciers. The findings, published in the reports by the Kathmandu-based International Centre for Integrated Mountain Development (ICIMOD) says that 20 glacial lakes in Nepal and 24 in Bhutan have become potentially dangerous as a result of climate change. As per the report, no glacial lake in Uttarakhnad Himalaya is vulnerable; however, there are 14 lakes in Tista river basin, 16 lakes in Himachal Pradesh which are potentially dangerous. A number of studies in National Institute of Hydrology, Roorkee have been carried out on GLOF for river basins of Tista, Dhauli Ganga (Gharwal Himalaya), Twang (Arunachal Pradesh) and Bhutan. As per these studies as such no lake is potentially dangerous in Dhauli Ganga, whereas some lakes are vulnerable in Bhutan, Tista and Twang basins. Breaching and the instantaneous discharge of water from glacial lakes can cause flash floods enough to create enormous damage in the downstream areas. Different type of lakes may have different level of hazard potential, for example, moraine dammed lakes located at snout of glacier have high probability of breaching thus may have high hazard potential; whereas erosion lakes have little chances of breaching hence

A glacial lake (left) and a glacial lake outburst flood (right)

FIGURE 7.4 Glacial lake outburst flood.

FIGURE 7.5 Landslide lake outburst flood.

have low hazard potential (Figure 7.4). These floods pose severe geomorphological hazards and can wreak havoc on all manmade structures located along their path. Much of the damage created during GLOF events is associated with large amounts of debris that accompany the floodwaters. GLOF events have resulted in many deaths, destruction of houses, bridges, entire fields, forests, and roads. Unrecoverable damage to settlements and farmland can take place at large distances from the outburst source. In most of the events livelihoods are disturbed for long periods. Flood events reported during the recent past in the hilly areas is shown in Table 7.2.

7.4.4 LANDSLIDE LAKE OUTBURST FLOODS

Landslide Lake Outburst Floods (LLOFs) are caused by breaching of lakes created by landslides. These floods are common in the Himalayan river basins. The active and palaeo landslide mapping along the Satluj and Spiti rivers indicate that these rivers were blocked and breached at many places during the Quaternary period. It has been observed that the loss of life and property due to these LLOFs is directly related to the disposition of the Quaternary materials and the different morphological zones observed in the area (Figure 7.5).

7.5 FLOOD MANAGEMENT AND CONTROL IN INDIA

In our country, flood management and control are necessary because the floods impose curse on the society. Structural measures involve the construction of flood control projects such as levees, dams and channel modifications. The general

TABLE 7.3

Structural and Non-Structural Measures for Flood Risk Management

Structural Measures			Catchment-wideinterventions (agriculture and forestry actions and water control work)
			River training
			Other flood control measures (passive control, water retention basins and river corridor enhancement, rehabilitation and restoration)
Non-structural Measures	Risk acceptance	Tolerence strategies	Tolerance
			Emergency response system
			Insurance
	Risk reduction	Prevention strategies	Delimitation of flood areas and securing flood plains
			Implementation of flood areas regulations
			Application of financial measures
		Mitigation Strategies	Reduction of discharge through natural retention
			Emergency action based on monitoring, warning, and response systems (MWRS)
			Public information and education

approach in the past has been one of adopting structural measures such as the construction of embankments and reservoirs. As more and more developments encroach the flood plains in India, the main thrust of current flood policy focuses on the non-structural flood management measure Non-structural measures include regulation of land use in the floodplain, acquisition and removal of flood-prone structures, restoration or protection of wetland areas, flood insurance, and flood warning. Some structural and non-structural measures for flood risk management are listed in Table 7.3.

7.5.1 NON-STRUCTURAL MEASURES

Various non-structural measures proposed for flood management include the real-time flood forecasting, flood plain zoning, Dam break flood simulation, flood hazard and flood risk mapping, etc. Such measures provide the information and input for planning the flood management program to regulate the developmental activities on the flood plain and to prepare evacuation plan during the emergency period of the flood. These measures are also used for formulating the legislation and acts to minimise the human intervention in the flood plain. Some of the popular non-structural measures are discussed briefly here under.

7.5.1.1 Flood Forecasting

Reliable forecasting and easily understandable warning information with sufficient lead time are of vital importance for evacuation. Of all the non-structural measures flood forecasting is gaining increased attention of the planners. In order to mitigate the damages from floods, a nationwide flood forecasting and warning system, as a non-structural measure, has been established by the Central Water Commission. This flood forecasting system issues forecasts at 171 stations in the country of which 144 stations are for stage forecast and 27 for inflow forecast. Warning of the approaching floods provides sufficient time for the authorities to:

1. Evacuate the affected people to the safer places.
2. Make an intense patrolling of the flood protection works such as embankments so as to save them from breaches, failures, etc.
3. Regulate the floods through the barrages and reservoirs, so that the safety of these structures can be taken care of against the higher return period floods.
4. Operate the multi-purpose reservoirs in such a way that an encroachment into the power and water conservation storage can be made to control the incoming flood.
5. Operate the city drains (out falling into the river) to prevent bank flow and flooding of the areas drained by them.

The various steps involved in the operation before issue of forecasts and warning are as follows:

- Observation and collection of hydrological and meteorological data.
- Transmission/Communication of data to the forecasting Centres.
- Analysis of data and formulation of forecasts.
- Dissemination of forecasts and warning to the Administrative and Engineering Authorities of the States.

7.5.1.2 Flood Plain Zoning

In areas with infrequent flooding, many people have either forgotten or have never known where flooding occurs. Flood hazard maps, therefore, have a useful informative role even in the absence of any land use regulations. These maps also form the basis for implementing land use regulations and zoning bylaws, etc. Such maps, if available to regional planners, would encourage an integrated approach to the development of river valleys. Emergency planners can use these maps to identify escape routes to be used when the inevitable flood event occurs. Also, these maps are a prerequisite for evaluating the economic viability of any proposed structural measures for flood protection. These maps provide information for land use planning, allow correct development for new urban areas, help evaluating the cost of flooding and risk reduction benefits, increase the public awareness about the risk due to flooding, prevent pollution due to flooding of sewage treatment plants and solid waste disposal sites locations on floodplains. Furthermore, toward the upper end of the spectrum, flood hazard maps could help in developing rational criteria for flood insurance as in the United States of America. Flood plains, the lands bordering rivers and streams, are normally dry round the year but get covered with water during floods. Floods can damage buildings or other structures like levees and embankments placed within the flood plains. On the other hand, the structures themselves can change the pattern of water flow and increase flooding by blocking the flow of water thereby increasing the spread, depth, or velocity of flood waters. Flood plain zoning restricts to use the land or any human activity in the flood plains of a river. Generally, the term "flood plain" includes water channel, flood channel and nearby low land areas susceptible to flooding by inundation.

Zoning involves the division of flood plain into specific zones and the regulation within these zones of:

  I. The use of structures and land,
  II. The height and bulk of structures, and
  III. The size of lots and density of use.

The flood plain regulations contained in a zoning ordinance consist of:

- A written text which sets forth the regulations which apply to each zone together with administrative provisions, and
- A map delineating the boundaries of the various use zones.

The important aspect of zoning is that it can be used to regulate what uses may be conducted and how uses are to be constructed or carried out. Zoning is also used to restrict riverine or coastal areas to particular uses, specify where the uses may be located and establish minimum elevation of flood proofing requirements for the uses. Flood hazard area boundary lines reflect the outer limits of hazard areas which may be defined with mathematical and engineering methods. As discussed above, the basic concept in flood plain zoning is to regulate the land use in flood plains to reduce/restrict the damage potential of floods. For regulating the land use in different flood zones, the following priorities are envisaged.

- **First Priority**
  The important installation like Defense installations, Industries, Public utilities like Hospitals, Electricity installations, Water supply, Telephone exchanges, Aerodromes, Railway stations, Commercial centers, etc. should be located in such a way that they are above the levels corresponding to 1 in 100-year flood or maximum observed flood levels. They should be above the levels corresponding to a 50-year rainfall and likely submergence due to drainage congestion.
- **Second Priority**
  Installation like Public institutions, Government offices, Universities, Public Libraries and Residential areas could be located above 25-year flood zones with the stipulation that they are built on stilts or far higher levels as indicated above.
- **Third priority**
  Parks, playgrounds, parking places could be located in areas vulnerable to frequent floods.

7.5.1.3 Decision Support System for Real-Time Flood Warning and Management

Decision support system for issuing the flood warning and managing the flood in real time is advanced software which is capable of providing the information to the decision makers for taking the necessary measures for managing the flood in real time (Dimri et al., 2017). Such system requires the spatial and temporal databases which include the basin characteristics, hydro metrological variables, social and economical data, etc. However, under the Hydrology Project II, which is likely to start very soon, the development of DSS for real-time flood forecasting for Bhakra reservoir system is proposed to be taken up by Bhakra Beas Management Board one of the important proposed activities.

7.6 WHO IS RESPONSIBLE?

Minimizing the risk to life and property is the foremost goal of risk management. To be successful, risk management must involve a wide range of individuals and institutions. Roles of different groups in flash flood risk management are given below.

I. Central administration:
- Develop national strategy,
- Create legal framework, and
- Create financial mechanisms.

II. River basin organisers:
- Long-term planning taking into account basin-wide conditions, development, and climate change scenarios,
- Create hazard/risk maps, and
- Forecasting and dissemination of warnings.

III. Provincial administration:
- Planning at provincial level,
- Implementing mitigation measures, and
- Linkage between national and local (basin and catchment) levels.

IV. Local administration:
- Formation of community-based flash flood management organisation Coordination with community-based organisations,
- Post-flash flood preparedness, and
- Local level early warning system.

V. Household:
- Securing household from flooding,
- Organizing life at home, and
- Preparing family for evacuation.

VI. Professionals/scientists:
- Support central administration in planning and strategy building,
- Prepare guidelines and practical solutions,

- Advice to government and academia, and
- Capacity building at policy level.

VII. Natl meteorological and hydrological services:
  - Create early warning and dissemination system, and
  - Research.

VIII. Regional organizations:
  - Knowledge transfer,
  - Capacity building,
  - Transboundary dialogue, and
  - Cooperation facilitation.

IX. Private sector:
  - Prepare action plan for damage minimisation,
  - Ensure safety of equipment and structures,
  - Insurance, and
  - Implementation of financial mechanisms.

X. Crisis management services:
  - Coordinate warning systems,
  - Identify vulnerable groups and their needs,
  - Planning response mechanisms, and
  - Post-flash flood activities.

XI. Spatial planners:
  - Create spatial planning,
  - Land zoning, and
  - Support regulation.

XII. Academia:
  - Flood education,
  - Research support, and
  - Advice to government.

XIII. Non Government Organizations (NGOs):
  - Awareness raising,
  - Capacity building,
  - Pressurize higher level, and
  - Post-event support.

XIV. Media:
  - Awareness raising,
  - Exert pressure,
  - Early warning,and
  - Post-event support and information dissemination.

7.7 FUTURE NEEDS IN FLOOD MANAGEMENT

7.7.1 FOCUSED APPROACH

Though there is a considerable progress in various flood management measures in India, the problems still persist. This may be attributed to inadequacy of planned investment. Flood management measures have to be more focused and targeted to achieve the goals within a stipulated time frame. A flood management measure needs the river basin as a whole to be considered as unit for integrated planning for optimum utilization of the resources.

7.7.2 LEGISLATION FOR FLOOD PLAIN ZONING

As discussed earlier, many flood-prone states have not adopted the recommendations regarding flood plain zoning. A detailed status for each state may be found in the report of the working group on flood management (CWC, 1980). Therefore, it is necessary to evolve methods in consultation with local bodies such that the legislation for flood plain zoning will be adopted. A task committee for the purpose is essential.

7.7.3 INADEQUACY OF FLOOD CUSHION IN RESERVOIRS

It is well recognized that long-term solution of flood problems lies in creating appropriate flood storage in reservoirs. The total live storage capacity of completed projects in India is $174 \, km^3$. However, provision of flood storage capacity is limited to a few reservoirs like Damodar Valley Corporation ($1.87 \, km^3$), Ukai reservoir ($1.33 \, km^3$) and some smaller projects. There is no flood storage provided in major reservoirs like Hirakud. Thus, the flood storage capacity as a percentage of live storage capacity is negligible. Therefore, in future, large flood space is required for a successful flood management program.

7.7.4 FLOOD INSURANCE

Though flood risk has been included in the list of items covered by General Insurance Companies in India, it is popular in urban areas only. The insurance companies have also not been able to arrive at different rates of insurance premiums for different areas. Therefore, there is a need to evaluate the risk associated with the flooding of different areas and modify the insurance schemes.

7.7.5 FLOOD DATA CENTER

Success of any research activities related to flood studies needs accurate and reliable data. Various flood management measures also depend on correct information about the flood. Therefore, an exclusive data center for flood management is required. Networking of nodal data centers is also required. Advanced techniques like remote sensing may be used to procure various data.

7.7.6 RESEARCH AND DEVELOPMENT

The Science and Technology Advisory Committee of the Ministry of Water Resources identifies and provides necessary guidelines and directions for Research and Development activities in water resources sector. Indian National Committee on Irrigation and

Drainage is presently entrusted with the work of coordinating R&D schemes in flood management sector. Advanced knowledge in science and technology like application of remote sensing data etc. has to be used in studies for flood management.

7.7.7 COMMUNITY PARTICIPATION

With floods being a common concern for all sectors of society, flood management needs active involvement and participation of all to fulfill its objective. Farmers, professional bodies, industries, women, voluntary organizations are to be aware of flood management. Voluntary organizations can act as an interface between the government and public. People's participation in preparedness, flood fighting, and disaster response is required. Media like TV and newspapers play an important role during flood management.

7.7.8 INTERNATIONAL COOPERATION

India is drained by a number of international rivers that originate beyond its borders and flow into India. India shares river systems with six neighboring countries, viz, Nepal, Bhutan, China, Myanmar, Bangladesh, and Pakistan. Bilateral cooperation for various flood management measures is essential for India and the concerned country. Government of India has already taken some initiatives in this regard. However, more active participation in the subject is required.

7.7.9 MODERNIZATION OF FLOOD FORECASTING SERVICES

In order to improve the warning time and the accuracy of the forecast, it is necessary to adopt latest technology for real-time collection and transmission of hydrological and hydro meteorological data and application of the high speed computer using hydrological and mathematical models.

7.8 CONTAINING THE DAMAGE IN THE HILLY BASINS

Damages by disasters triggered by cloudburst or GLOFs can be considerably reduced by forecasts and early warning systems. It appears that in the Kedarnath incident, a time lapse of at least a few hours was available between the initiation of intense rainfall and the time floods hit the Kedarnath area. If warning about intense rain or incoming flood were available, a large number of lives could have been saved. The decision making based on the scientific information should be given more attention and priority. What needs to be done? A number of initiatives are needed to reduce the damage due to natural disasters in the Himalayan region. First, the areas likely to be impacted by different types of disasters like intense rainfall, earthquake and GLOF need to be identified. The mountains may face landslides, lake overflows, movement of moraines, activation of dormant river channels, etc. As water, boulders, and debris move downstream, they have potential to cause damage and, therefore, it would be necessary to identify the areas which are likely to be impacted by extreme events of different magnitudes. The events of June 2013 have shown that

FIGURE 7.6   Kedarnath temple after flood.

extreme rainfall can indeed occur over a large geographical area and so the cumulative impact of flooding in various tributaries on a downstream location needs to be considered. Another notable aspect of the June 2013 event was that intense rainfall is not confined to the monsoon season and one has to be vigilant all the time. There are other areas in the Himalayas that are at risk of such events. Therefore, mapping of and monitoring of glacial lake, landslide prone area, etc. are urgently required. There is a need to prepare comprehensive reports of the entire Himalayan region for better planning and management of these kinds of disasters. Flood Forecasting and Early Warning System are also necessary in the Himalayan region so that timely actions can be taken to avoid such types of disasters. While nature's fury cannot be completely controlled, its impacts can certainly be moderated if principles of hydro-ecology and engineering are carefully employed in planning and design of infrastructure and disaster preparedness (Figure 7.6).

7.9   CONCLUSION

In India, both structural and non-structural measures are adopted for the flood management. In case of non-structural measures, flood management is being done by using real-time flood forecast based on statistical approach. For some pilot projects, network model and multiparameter hydrological models are used.

Conventional systems of communication are normally used for transmitting the data in real time. The automatic systems of data communication like Telemetry system are used in pilot projects on limited scale. In the hilly areas, flash floods are usually experienced. As such there is no system for formulating the flash flood forecast in the region. It results in heavy losses of lives and properties. There is a need for significant improvement of the real-time flood forecasting systems in India.

The information about the flood is to be disseminated well in advance to the people likely to be affected so that an emergency evacuation plan may be prepared and properly implemented. Flash flood forecasting is an important area which requires immediate attention.

REFERENCES

Agrawal, S.K., Singh, N.J. and Roy, V.D., 2004. Flood management in India. In *Workshop Flood and Drought Management*, pp. 16–17.

Allen, S.K., Rastner, P., Arora, M., Huggel, C. and Stoffel, M., 2016. Lake out burst and debris flow disaster at Kedarnath, June 2013: hydrometeorological triggering and topographic predisposition. *Landslides, 13*(6), pp. 1479–1491.

CWC (Central Water Commission). 1980. *Manual on Flood Forecasting*, Central Flood Forecasting Organisation, Atna.

Dimri, A.P., Chevuturi, A., Niyogi, D., Thayyen, R.J., Ray, K., Tripathi, S.N., Pandey, A.K. and Mohanty, U.C., 2017. Cloudbursts in Indian Himalayas: a review. *Earth-Science Reviews, 168*, pp. 1–23.

Gupta, P. and Uniyal, S., 2012. Landslides and flash floods caused by extreme rainfall events/cloudbursts in Uttarkashi district of Uttarakhand. *Journal of South Asian Disaster Studies*, p. 77.

Jain, S.K., Lohani, A.K. and Jain, S.K., 2013. *Flash Floods! Threatening the Himalayan Region*, NISCAIR-CSIR, pp. 12–18.

Lohani, A.K. and Jaiswal, R.K., 1996. Water logged and drainage congestion problem in Mokama Tal area, Bihar. *Report No. CS (AR), 194*. National Institute of Hydrology, Roorkee.

Mathur, G.N. and Jain, D.K., 2004. Workshop flood and drought management, 16–17 September 2004, New Delhi. In *Workshop on Flood and Drought Management, Central Board of Irrigation and Power*, New Delhi, pp. 13–20.

Rangachari, R., 1986. *Flood Forecasting and Warning Network in Interstate Rivers of India*, Central Water Commission, New Delhi.

Seth, S.M. and Goel, N.K., 1985. Flood Forecasting Models. *RN-35*. National Institute of Hydrology, Roorkee.

8 Role of Crop Modeling in Mitigating Effects of Climate Change on Crop Production

Lal Singh
Sher-e-Kashmir University of Agricultural Sciences
and Technology of Kashmir (SKUAST-K)

CONTENTS

8.1 INTRODUCTION

Agriculture is one of the sectors most affected by ongoing climate change. In fact, climate change can affect different agricultural dimensions, causing losses in productivity, profitability, and employment. Food security is clearly threatened by climate change (Siwar et al., 2013) due to the instability of crop production, as well as induced changes in markets, food prices, and supply chain infrastructure. The ongoing effects of climate change require the individuation of mitigation policies to reduce greenhouse gas emissions and identify appropriated adaptation strategies that aim to contain agricultural losses both in market goods and

101

environmental services (such as protection of biodiversity, water management, landscape preservation, and so on). Crop models are considered "agriculture-oriented" because the analysis of these models focuses on the biological and ecological consequences of climate change on crops and soil. In these models, farmers' behavior is not captured and the management practice is considered fixed. Moreover, they are crop and site-specific and were calibrated only for the major grains and a limited number of places (Shawcroft et al., 1975; Mendelsohn and Dinar, 2009).

Crop production is an aggregation of individual plant species grown in a unit area with the aim of an irreversible increase in the growth, size, and volume of seeds or consumables from these plants, which are harvested for economic purposes. Ultimately, the breeders can anticipate future requirements based on climate change by simulating the characteristics of the natural environmental system studied in an abbreviated time scale through an appropriate model. A model is a schematic representation of the conception of a system or a mimicry or a set of equations representing the behavior of a system, with the purpose of aiding, understanding, and improving the performance of the system. Crop models can be used to understand the effects of climate change such as elevated carbon dioxide, changes in temperature, and rainfall on crop development, growth, and yield. For example, a change in weather to warm and humid may lead to more rapid development of plant diseases, a loss in crop yield, and consequent financial adversity for individual farmers and for the people of the region. Most natural systems are complex and many do not have boundaries. It is a difficult task to produce a comprehensible, operational representation of a part of reality, which grasps the essential elements and mechanisms of that real-world system. This is even more demanding when complex systems are encountered in environmental management (Murthy, 2002).

8.2   TYPES OF MODELS

Depending upon the purpose for which it is designed, models are classified into different groups or types. Of them a few are:

i. **Statistical models**: These express the relationship between yield or yield components and weather parameters. In these models, relationships are measured in a system using statistical techniques. Examples include step-down regression, correlation, etc.
ii. **Mechanistic models**: These explain not only the relationship between weather parameters and yield but also the mechanism of these models (explains the relationship of influencing dependent variables). These models are based on physical selection.
iii. **Deterministic models**: These estimate the exact value of the yield or dependent variable. These models also have defined coefficients.
iv. **Stochastic models**: In these, a probability element is attached to each output. For each set of inputs different outputs are given along with probabilities. These models define yield or the state of dependent variable at a given rate.

v. **Dynamic models**: In these, time is included as a variable. Both dependent and independent variables have values that remain constant over a given period of time.

vi. **Static models**: In these, time is not included as a variable. Dependent and independent variables remain constant over a given period of time.

vii. **Simulation models**: Computer models, in general, are a mathematical representation of a real world system. One of the main goals of crop simulation models is to estimate agricultural production as a function of weather and soil conditions as well as crop management. These models use one or more sets of differential equations and calculate both rate and state variables over time, normally from planting until harvest maturity or final harvest.

viii. **Descriptive models**: These define the behavior of a system in a simple manner. The model reflects little or none of the mechanisms that are the causes of phenomena. However, it consists of one or more mathematical equations. An example of such an equation is the one derived from successively measured weights of a crop. The equation helps determine the weight of the crop where no observation was made.

ix. **Explanatory models**: These consist of quantitative description of the mechanisms and processes that cause the behavior of the system. To create this model, a system is analyzed and its processes and mechanisms are quantified separately. The model is built by integrating these descriptions for the entire system. It contains descriptions of distinct processes such as leaf area expansion, tiller production, etc. Crop growth is a consequence of these processes.

Different crop growth simulation models for different crop are summarized in Table 8.1.

8.3 CLIMATE CHANGE AND CROP MODELING

8.3.1 CLIMATE CHANGE

Climate change is defined as "Any long-term substantial deviation from present climate because of variations in weather and climatic elements."

8.3.1.1 The Causes of Climate Change

1. Natural causes such as changes in earth revolution, changes in area of continents, variations in solar system, etc.

2. Due to human activities the concentrations of carbon dioxide and certain other harmful atmospheric gases are increasing. The present level of carbon dioxide is 325 ppm and is expected to reach 700 ppm by the end of this century because of the present trend of burning forests, grasslands, and fossil fuels. Few models predicted an increase in the average temperature of 2.3°C–4.6°C and precipitation per day from 10% to 32% in India.

TABLE 8.1

Crop Growth Simulation Models for Different Crop/Processes

Model Name	Crop	Processes Involved	Reference
CERES-Sorghum	Sorghum	Growth and development, grain yield	Ritchie and Alagarswamy (1989)
RESCAP	Sorghum	Dry matter, water use, grain yield, radiation interception	Monteith et al. (1989)
RESCAP	Pearl Millet	Dry matter accumulation, water use, grain yield	Monteith et al. (1989)
SIMAIZ	Maize	Growth, grain yield	Duncan (1975)
CERES-Maize	Maize	Growth and development, grain yield	Stapper and Arkin (1980)
COTTAM	Cotton	Growth, development, soil water budget, morphology	Jackson et al. (1988)
GOSSYM	Cotton	Growth, yield	Baker et al. (1983)
SUBSTOR	Potato	Growth, development, yield	Hodges et al. (1989)
IRRIMOD	Rice	Growth, development, yield	Angus and Zandstra (1980)
CERES-Rice	Rice	Growth, yield, phenology	Ritchie et al. (1986)
AUSCANE	Sugarcane	Growth, development, cane yield	Jones et al. (1989)
SOYGRO	Soybean	Vegetative, reproductive, grain yield, phenology	Wilkerson et al. (1985)
SOYMOD	Soybean	Growth, development, yield	Curry et al. (1975)
BEANGRO	Dry bean	Growth, development, phenology, grain yield	Hoogenboom et al. (1990)
PNUTGRO	Peanut	Growth, development, phenology, yield	Boote et al. (1989)
POTATO	Potato	Growth, yield	Hoogenboom et al. (1990)
SIMTAG	Wheat	Genotypes growth and development, grain yield	Stapper (1984)
CERES-Wheat	Wheat	Growth, development, grain yield	Ritchie et al. (1985)
	Wheat	Water use, nitrogen nutrition, growth, grain yield	Van Keulen and Seligman (1987)
	Any living systems	Plant growth and crop production	de Wit (1982)
EPIC	Any crop and cropping	Soil productivity, erosion, plant growth processes, yield	Williams et al. (1984)
SPAW	Any crop	Plant environment interaction, microclimate, dry matter, grain	Shawcroft et al. (1974)
ALMANAC	Crop-weed competitions	Crop-weed competition	Kiniry et al. (1992)
PLANTGRO	Any crop	Evapotranspiration and grain yield	Retta and Hanks (1980)

8.3.1.2 Greenhouse Effect

Greenhouse effect causes the earth to heat up more than expected due to the presence of atmospheric gases like carbon dioxide, methane, as well as other tropospheric gases. Shortwave radiation can pass through the atmosphere easily, however, the resultant outgoing terrestrial radiation cannot escape because atmosphere is opaque to this radiation. This phenomenon conserves heat which in turn increases temperature.

8.3.2 Effects of Climate Change

1. The increased concentration of carbon dioxide and other greenhouse gases are expected to increase earth's temperature.
2. Crop production is highly dependent on variation in weather, and therefore, any change in global climate will have major effects on crop yields and productivity.
3. Elevated temperature and carbon dioxide affects the biological processes like respiration, photosynthesis, plant growth, reproduction, water use, etc. In case of rice, increased carbon dioxide levels results in larger number of tillers, greater biomass, and grain yield. Similarly, in groundnut, increased carbon dioxide levels result in greater biomass and pod yields.
4. However, in tropics and sub-tropics, the possible increase in temperatures may offset the beneficial effects of carbon dioxide and results in significant yield losses and water requirements.
5. Proper understanding of the effects of climate change helps scientists guide farmers to make crop management decisions such as selection of crops, cultivars, sowing dates, and irrigation scheduling to minimize the risks.

8.4 ROLE OF CROP MODELING IN AGRICULTURE ON CLIMATE CHANGE

The increased concentration of carbon dioxide and other greenhouse gases is expected to increase earth's temperature. Crop production is highly dependent on variation in weather, and therefore, any change in global climate will have major effects on crop yields and productivity. Elevated temperature and carbon dioxide affects the biological processes like respiration, photosynthesis, plant growth, reproduction, water use, etc (Murthy, 2002). However, in tropics and sub-tropics, the possible increase in temperatures may offset the beneficial effects of carbon dioxide, resulting in significant yield losses and water requirements.

Proper understanding of the effects of climate change helps scientists guide farmers to make crop management decisions such as selection of crops, cultivars, sowing dates, and irrigation scheduling to minimize the risks. In recent years, there has been a growing concern that changes in climate will lead to significant damage to both market and non-market sectors. Climate change will have a negative effect in many countries. The farmer's adaptation to climate change, through changes in farming practices, cropping patterns, and use of new technologies will help ease the impact. The variability of our climate and especially the associated weather extremes is currently one of the concerns of the scientific as well as general community.

The application of crop models to study the potential impact of climate change and climate variability provides a direct link between models, agro-meteorology, and society's concerns. As climate change deals with future issues, the use of general circulation models (GCMs) and crop simulation models provide a more scientific approach to study the impact of climate change on agricultural production and world food security compared to other surveys. Cropgro (DSSAT) is one of the first packages that modified weather simulation generators and introduced a package to evaluate the performance of models for climate change situations. Irrespective of the limitations of GCMs, it would be in the larger interest of farming community of the world that these DSSAT modelers follow GCMs for more accurate and acceptable weather generators for use in models. This will help in finding solutions to crop production under climate changes conditions, especially in underdeveloped and developing countries.

8.4.1 Applications and Uses of Crop Growth Models in Agriculture

Crop growth models are being developed to meet the demands under the following situations in agricultural meteorology:

a. When farmers have the difficult task of managing their crops on poor soils in harsh and risky climates.
b. When scientists and research managers need tools that can assist them in adopting an integrated approach to finding solutions in the complex problem of weather, soil, and crop management.
c. When policy makers and administrators need simple tools that can assist them in policy management in agricultural meteorology.

The potential uses of crop growth models for practical applications are discussed below (Sivakumar and Glinni, 2000).

8.4.1.1 On-Farm Decision-Making and Agronomic Management

Models allow evaluation of one or more options that are available with respect to one or more agronomic management decisions like

- Determine optimum planting date
- Determine best choice of cultivars
- Evaluate weather risk
- Investment decisions.

Crop models help in testing scientific hypothesis, highlight where information is missing, organize data, and integrate across disciplines. Crop growth models can be used to predict crop performance in regions where the crop has not been grown before or not grown under optimal conditions. Such applications are of value for regional development and agricultural planning in developing countries (Kiniry and Bockholt, 1998).

A model can calculate probabilities of grain yield levels for a given soil type based on rainfall (Kiniry and Bockhot, 1998). Investment decisions like purchase of irrigation systems (Boggess and Amerling, 1983) can be taken with an eye on long-term usage of the equipment, through the predictions from growth models. Crop models also can be used to understand the effects of climate change such as

a. consequences of elevated carbon dioxide, and
b. changes in temperature and rainfall on crop development, growth, and yield. Ultimately, the breeders can anticipate future requirements based on the climate change.
c. Crop simulation models assist in genetic improvement;
   - Evaluate optimum genetic traits for specific environments.
   - Evaluate cultivar stability under long term weather.

The environmental policy integrated climate (EPIC), agricultural land management alternatives with numerical assessment criteria (ALAMANC), cropping systems simulation model (CROPSYST), World Food Studies (WOFOST), and architectural model of development (ADEL) models are being successfully used to simulate maize crop growth and yield. The Sorghum (SORKAM), SorModel, SORGF field (SORGF), and ALMANAC models are being used to address the specific tasks of sorghum crop management. Clouds and earths radiant energy system (CERES), pearl millet model, CROPSYST, and Penman–Montieth (PM) models are being used to study the suitability and yield simulation of pearl millet genotypes across the globe. Similarly, the two most common growth models used in application for cotton are the dynamic cotton (*Gossypium hirsutum* L.) crop simulation model (GOSSYM) and Cotton (COTONS) models. On the same analogy, the PNUTGRO for groundnut, CHIKPGRO for chick pea, WTGROWS for wheat, SOYGRO for soybean, and QSUN for sunflower are in use to meet the requirements of farmers, scientists, decision makers, etc., at present.

8.4.2 Strategies to Mitigate the Impacts of Climate Change

To cope with the changes in regional climate due to climatic changes, agriculture may have to adopt some changes. These adjustments will depend on future developments in technology, demand, prices, and national policy, and such changes cannot be predicted in present conditions. However, several adaptations can be instituted.

- The key step for an agriculture adaptation strategy could be the choice of suitable crops and cultivars.
- Shifts on sowing date of spring crops will allow more effective use of the soil moisture content formed by snow melting.
- Optimum use of fertilizers and ecologically clean agrotechnologies would be beneficial for agriculture.
- Conservation of water used for irrigated agriculture, therefore, should be given priority attention in rainfed condition.
- With increased evapotranspiration, any adaptation strategy in agriculture should be oriented toward a shift from conventional crops to types of agriculture that are not vulnerable to evapo-transpiration.

- Protection of soils from degradation should be given serious consideration.
- Other adaptive options included developing cultivars resistant to climate change; adopting new farm techniques that will respond to the management of crop under stressful conditions, plant pests and disease; design and development of efficient farm implements.
- An adaptive response in the agricultural sector should be an effort to breed heat-resistant crop varieties by utilizing genetic resources that may be better adapted to warmer and drier conditions.
- Improvement in farming systems, fertilizer management, and soil conservation from major adaptation strategies.
- Crop architecture and physiology may be genetically altered to adapt to warmer environmental conditions.

8.4.3 Future Issues Related to Weather on Crop Modeling

For any application of a crop model weather data are an essential input and continue to play a key role.

1. There is an urgent need to develop standards for weather station equipment and sensor installation and maintenance.
2. It is also important that a uniform file format is defined for the storage and distribution of weather data, so that they can easily be exchanged among agrometeorologists, crop modelers, and others working in climate, and weather aspects across the globe.
3. Easy access to weather data, preferably through the internet and the World Wide Web, will be critical for the application of crop models for yield forecasting and tactical decision making.
4. Previously one of the limitations of the current crop simulation models was that they can only simulate crop yield for a particular site. At this site weather (soil and management) data also must be available. It is a known fact that the weather data (and all these other details) are not available at all locations where crops are grown. To solve these problems the geographical information system (GIS) approach has opened up a whole field of crop modeling applications at spatial scale. From the field level for site-specific management to the regional level for productivity analysis and food security the role of geographical information system is going to be tremendous (Hoogenboom, 2000).

8.4.4 Minimum Dataset Required for Crop Weather Relations and Crop Simulation Models—Rice

Station details (Latitude, Longitude, and Altitude) and daily weather data viz., T_{max} & T_{min} (°C), rainfall (mm), sunshine hours (h), solar radiation (MJ/m^2), wind speed (km/h), RH (%—I & II), and pan evaporation (mm) are needed throughout the year.

I. General information
   1. Country:
   2. Site name:
   3. Latitude:
   4. Longitude:
   5. Soil data source
   6. Soil series name
   7. Soil classification
II. Surface information
   1. Color (a) brown (b) red (c) black (d) gray (e) yellow
   2. Drainage (a) very excessive (b) excessive (c) somewhat excessive (d) well (e) moderate well (f) somewhat poorly (g) poorly (h) very poorly
   3. % slope
   4. Runoff potential (a) lowest (b) moderately low (c) moderately high (d) highest
   5. Fertility factor (0 to 1)
   6. Runoff curve number
   7. Albedo
   8. Drainage rate
III. Layer-wise soil information: Number of layer depends on the location. Here layers up to 120 cm depth are shown as a sample.

Periodic measurements of **leaf area index, dry matter production** (leaf, stem, root, ear, and total), **plant height, specific leaf area, relative growth rate, and PAR inside and outside canopy** are to be taken throughout crop season. Measurements should be started at 30 days after sowing and continued at 15 days interval up to physiological maturity. In addition to these observations, any biotic stress (heavy weed infestation, insect, and disease) and abiotic stress should be noted down during crop season.

8.5 CONCLUSION

Crop production practices enhance the capacity of the soil to conserve and accumulate soil organic carbon. Physiology-based crop simulation models have become a key tool in extrapolating the impact of climate change from limited experimental evidence to broader climatic zones. Different models are a simplification of the reality; they allow a first assessment of the complexity of climate change impact in agriculture. They are playing an increasingly important role in assisting agriculture to adapt to climate change. In order to meet the increasing demand for assessment of climate change impact, crop models need to be further improved and tested with climate change scenarios involving various changes in ambient temperature and CO_2 concentration. These models have facilitated establishment of new hypotheses for climate change studies, stimulated investigations into climate change adaptation, and assisted in communicating to the public and policy makers that continued climate change could have devastating impacts on food supply.

REFERENCES

Angus, J.F. and Zandstra, H.G., 1980. Climatic factors and the modeling of rice growth and yield. *Proceedings of a Symposium on the Agrometeorology of the Rice Crop: World Meteorological Organization.* International Rice Research Institute, IRRI, Los Banos, pp. 189–199.

Baker, D.N., Lambert, J.R. and McKinion, J.M., 1983. GOSSYM: A simulator of cotton crop growth and yield. *South Carolina. Agricultural Experiment Station. Technical bulletin (USA). Clemson Univ., Clemson, SC,* p. 1089.

Boggess, W.G. and Amerling, C.B., 1983. A bioeconomic simulation analysis of irrigation investments. *Journal of Agricultural and Applied Economics, 15*(2), pp. 85–91.

Boote, K.J., Jones, J.W., Hoogenboom, G., Wilkerson, G.G. and Jagtap, S.S., 1989. *PNUTGRO, Peanut Crop Growth Simulation Model. User's Guide.* Departments of Agronomy and Agricultural Engineering. Journal No. 8420. University of Florida, Gainesville, FL, pp. 1–76.

Curry, R.B., Baker, C.H. and Streeter, J.G. 1975. SOYMODI: A dynamic simulation of soybean growth and development. *Texas Society of Association Executives, 8*, pp. 963–968.

de Wit, C.T., 1982. Simulation of Living Systems. In: F.W.T.P. de Vries and H.H. Van Laar (Eds.). *Simulation of Plant Growth and Crop Production. Simulation Monograph Series.* PUDOC, Wageningen, pp. 3–8.

Duncan, W.G., 1975. SIMAIZ: A model simulating growth and yield in corn. In: D.N. Baker, P.B. Creech and F.G. Maxwell (Eds.). *The Application of Systems Methods to Srop Production.* Mississippi Agricultural and Forestry Experimental Station, Mississippi State University, Starkville, MS, pp. 32–48.

Hodges, T., Johnson, B.S. and Manrique, L.A., 1989, October. SUBSTOR: A model of potato growth and development. Abs. 1989 ASA Meeting, Las Vegas, NV, October 15–19, 1989. p. 16.

Hoogenboom, G., 2000. Contribution of agrometeorology to the simulation of crop production and its applications. *Agricultural and Forest Meteorology, 103*(1–2), pp. 137–157.

Hoogenboom, G., White, J.W., Jones, J.W. and Boote, K.J., 1991. *BEANGRO V1.01. Dry Bean Crop Growth Simulation Model.* Department of Agricultural Engineering, University of Florida; International Bechmark Sites Network for Agrotechnology Transfer; Cali, CO; Centro Internacional de Agricultura Tropical (CIAT), Gainesville, FL, p. 122.

Jackson, B.S., Arkin, G.F. and Hearn, A.B., 1988. The cotton simulation model. *Transactions of the American Society of Agricultural Engineers, 31*(3), pp. 846–854.

Jones, C.A., Wegener, M.K., Russell, J.S., McLeod, I.M. and Williams, J.R., 1989. AUSCANE-Simulation of Australian sugarcane with EPIC. CSIRO, Division of Tropical crops and pastures, Tech. Paper no. 29.

Kiniry, J.R. and Bockholt, A.J., 1998. Maize and sorghum simulation in diverse Texas environments. *Agronomy Journal, 90*(5), pp. 682–687.

Kiniry, J.R., Williams, J.R., Gassman, P.W. and Debaeke, P., 1992. A general, process-oriented model for two competing plant species. *Transactions of the American Society of Agricultural Engineers, 35*(3), pp. 801–810.

Mendelsohn, R. and Dinar, A., 2009. *Climate Change and Agriculture-An Economic Analysis of Global Impacts, Adaptation and Distributional Effect.* New Horizons in Environmental Economics. Edward Elgar, Cheltenham.

Monteith, J.L., Huda, A.K.S. and Midya, D. 1989. Modelling sorghum and pearl millet. *International Crops Research Institute for the Semi-Arid Tropics, India, 12*, pp. 30–39.

Murthy, V.R.K., 2002. *Basic Principles of Agricultural Meteorology.* Book Syndicate Publishers, Koti, Hyderabad.

Retta, A. and Hanks, R.J., 1980. *Manual for Using Model PLANTGRO.* Utah State University, Logan, pp. 1–14.

Ritchie, J.T. and Alagarswamy, G., 1989. Physiology of sorghum and pearl millet. *Modeling the Growth and Development of Sorghum and Pearl Millet, International Crops Research Institute for the Semi-Arid Tropics, India, 12*, pp. 24–29.

Ritchie, J.T., Alocilja, E.C., Singh, U., and Uchara, G., 1986. CERES-Rice Model. *Proc. in workshop on Impact of Weather Parameters on Growth and Yield of Rice IRRI, Los Banos.*

Ritchie, J.T., Godwin, D.C., and Otter-Nacke, S. 1985. *CERES – Wheat. A Simulation Model of Wheat Growth and Development.* Texas A&M University Press, College Station, Texas.

Shawcroft, R.W., Lemon, E.R., Allen, L.H., Stewart, D.W. and Jensen, S.E., 1974. The soil-plantatmosphere model and some of its predictions. *Agricultural and Forest Meteorology, 14*, pp. 287–307.

Shawcroft, R.W., Lemon, E.R., Allen Jr, L.H., Stewart, D.W. and Jensen, S.E., 1975. The soil–plant–Atmosphere model and some of its predictions. Developments in Agricultural and Managed Forest Ecology, Vol-1, 287–307. Elsevier, Amsterdam.

Sivakumar, M.V.K. and Glinni, A.F., 2002. Applications of Crop Growth Models in the Semiarid Regions. In: L.R. Ahuja, L. Ma, and T.A. Howell (Eds.). *Agricultural System Models in Field Research and Technology Transfer.* Lewis Publishers, CRC Press Company, Boca Raton, FL, pp. 177–205.

Siwar, C., Ahmed, F. and Begum, R.A., 2013. Climate change, agriculture and food security issues: Malaysian perspective. *Journal of Food, Agriculture and Environment, 11*(2), pp. 1118–1123.

Stapper, M. 1984. *SIMTAG. A Simulation Model of Wheat Genotypes.* Department of Agron & Soil Science, University of New England, Armidale.

Stapper, M. and Arkin, G.F., 1980. CORNF: A dynamic growth and development model for maize (Zea mays L.) Texas A&M, BRC, documentation no. 80–82, pp. 1–83.

Van Keulen, H. and Seligman, N.G., 1987. *Simulation of Water Use, Nitrogen Nutrition and Growth of a Spring Wheat Crop.* PUDOC, Wageningen.

Wilkerson, G.G., Jones, J.W., Boote, K.J., and Mishoe, J.W., 1985. *SOYGRO. Soybean Crop Growth and Yield model. Tech. Documentation.* University of Florida, Gainesville, pp. 1–253.

Williams, J.R., Jones, C.A. and Dyke, P.T. 1984. A modeling approach to determining the relationship between erosion and soil productivity. *Transactions of the American Society of Agricultural Engineers, 27*(1), pp. 129–144.

9 Forestry and Climate Change

K. N. Qaisar

Sher-e-Kashmir University of Agricultural Sciences
and Technology of Kashmir (SKUAST-K)

CONTENTS

9.1 INTRODUCTION

The word "Forest" is derived from the Latin word "foris" meaning outside the village boundary or away from inhabited land. Generally, forest is referred to an area occupied by different kinds of trees, shrubs, herbs, and grasses and maintained as such. Technically, forest is an area set aside for the production of timber and other forest produce, or maintained under woody vegetation for certain direct and indirect benefits which it provide, e.g., climatic and protective. Five main features of forest are as follows:

- It is an uncultivated land area,
- The land area should be occupied by different kinds of natural vegetation essentially by trees or it is proposed to establish trees and other forms of vegetation,
- Trees should form a closed or partially closed canopy,
- Trees and other forms of vegetation should be managed for obtaining forest produce and/ or benefits, and
- It should provide shelter to wildlife, birds, and other fauna.

Trees are the dominating feature of a forest; trees and their communities are unique in many respect.

- They are the tallest and longest (oldest) living being on earth,
- They exceed others in productivity per unit area and intense competition
- They influence the environment.

Forests in India cover about 24.16% of the total geographical area, covering 79.42 million ha (70.14 million ha. forest cover −21.37% + 9.26% million ha. tree cover −2.82%) as against 33% enunciated in the National Forest Policy, 1988 (FSI, 2015). An increase of 3.77 million ha as per the SFR, 2015 (Table 9.1). India ranks 10th in the list of most forested nations in the world (Global Forest Resources Assessment, 2005). Nearly 4 crore people are living in 1.73 lakh villages in or around the forests and as per MoEF reports. The livelihood of around 7 crores tribal and more than 20 crores non-tribal population is linked with forests (23%). They derive 40%–60% of their livelihood from forests. In India, per capita forest area is only 0.064 ha against the world average of 0.64 ha. The productivity of our forests is only 1.34 m³/ha/yr against the world average of 2.1 m³/ha/yr. While 78% of the forest area is subjected to heavy grazing and other unregulated uses, adversely affecting productivity and regeneration. India annually imports 2.74 billion $ of forest produce to meet the demand (GAIN, 2014).

TABLE 9.1

Demand and Supply of Major Forest Products (Aggarwal et al., 2009)

Forest Product	Demand MT	Supply MT	Gap/Unsustainable Use MT
Fuelwood	228	128	100 (44%)
Fodder (green & dry)	1,594	741	853 (54%)
Timber	55	41	14 (26%)

Bold represents percentage value of total supply of demand.

9.2   ROLE OF FORESTS

A. Productive
B. Protective
C. Ameliorative
D. Recreational
E. Developmental

All these functions are discussed below:

9.2.1   PRODUCTIVE FUNCTIONS OF THE FORESTS

1. Forests are valuable natural resources. The goods provided by the forests are of immense importance to animals and mankind. Wood is a major forest produce and it is extensively used for various purposes. In India, most of wood produced is used for construction of houses, agriculture implements, bridges, sleepers, etc. Many species, e.g., teak, sal, deodar, sissoo, babul, chir, haldu, axelwood, rosewood, dipterocarps, etc. yield valuable timber.

2. Wood is a universal fuel. For about thousands of years until the advent of coal, oil, gas, electricity, etc. wood constituted man's chief source of fuel. Even today, more than half of the total world consumption of wood is for fuel. Wood remains the major source of domestic fuel in India.

3. Forest provides raw material to a large number of industries, e.g., paper and pulp, plywood and other boards, saw mills, furniture making, packing cases, match, and toys, etc.

4. A large number of nonwood forest produces are also available from forests:
   - **Fibers and flosses:** Fibers are obtained from bast tissues of certain woody plants which are used for making ropes (*Bombax cieba)* and kapok (*Cieba pentandra)*
   - **Grasses and bamboos:** A large variety of grasses are found in the forests. Among valuable grasses, sabazi (*Eulaliopsis binata*) is harvested annually about 80,000 tonnes. About 5.5 million tonnes of bamboo is harvested from our forests every year. Approximately 10% of the bamboo is used for housing, 25% for rural/agricultural works, 25% for paper pulp, and remaining quantity for packaging and other uses.
   - **Essential oils:** India produced about 32,000 tonnes of essential oils during 2011. It is utilized in making soaps, perfumes, detergents, and chemicals. Many spices, e.g. *Eucalyptus* spp., *Bursera* spp., *Cymbopogon* spp., *Santalum album*, etc. produce these oils.
   - **Oil seeds:** Many tree species, e.g., Jatropha, Apricot, *Madhuca indica*, *Pongamia pinnata*, *Shorea robusta*, *Azadirachta indica*, *Schleichera oleosa*, *Vateria indica*, etc. produce oil bearing seeds which are commercially important. Some of these oils can be made fit for human consumption. Presently these seeds are used in soap industry and for biodiesel. Tribals use these oils for various purposes. The production

potential for bio-diesel is nearly 20 million tonnes per annum. Only a few million tonnes have been utilized (due to lack of demand). Estimated potential varies from 0.1 to 20 million tonnes, out of which 20%–25% has been utilized. About 150 nonedible tree borne oilseeds exist in India. The oil content varies between 21% and 73% in these species. (Menzel, 2000; Datta and Nischal, 2010; Dhyani et al., 2015).

• **Tans and dyes:** A variety of vegetable tanning materials are produced in the forests. Important vegetable tanning materials are the myrobalan nuts and bark of wattles *(Acacia mearnsii*, A. *decurrens*, A. *nilotica*, *Cassia auriculata*, etc.). Other tanning materials include, leaves of *Emblica offcinalis*, bark of *leistanthus collinus*, fruits of *Zizyphus xylocarpa*, bark of *Cassia fistula*, *Terminalia alata*, T. *arjuna*, etc. Katha and cutch are obtained from *acacia catechu* trees.

• **Gums and resins:** Gums and resins are exuded by trees as a result of wound or injury to the bark of wood. Gums are collected from several tree species, viz. *Sterculia urens*, *Anogeissus latifolia*, *Lannea coromandelica Acacia nilotica*, *Cochlospermum religiosum*, *Pterocarpus marsupium*, *Butea monosperma*, etc. Resin is obtained from *Pinus roxburghii.*

• **Drugs, spices:** Out of the total of about 2,000 items of drugs mentioned in Indian Materia Medica, over 1,800 are of vegetable origin. A large number of these are obtained from forests. All parts of the plants, e.g. roots, shoots, leaves, fruits, seeds, barks, etc. are used for drugs including, *Rauvolfia serpentine*, *Ephedra* spp., *Hemidesmus cus*, *Swertia chirata*, *Dioscorea* spp., *Podophyllus* spp., *Atropa* spp., *Datura innoxia*, etc. A variety of spices like kalajira (seeds of *Carum carvi)*, dalchini (bark of *Cinnamomum zeylanicun)*, cardamom (dried capsules of *Ellatteria cardamomum*), tejpat (leaves of *Cinnamomum tamala*), etc. are obtained from forest. Pyrethrum and neem are used as insecticides.

• **Tendu leaves:** Tendu leaves are used to produce bidi and therefore called bidi leaves. Annual collection of tendu (*Diospyros melanoxylon)* leaves is about 90,000 tonnes in the country. In Madhya Pradesh alone, about 45% of this quantity is produced. Leaves of trees such as, *Bauhinia* spp. and *Butea* spp., are used for making plates and dona.

• **Edible products:** Fruits, flowers, seeds, tubers, etc. of several forest species are eaten. Fruits and seeds of *Anacardium occidentale*, *Tamarindus indica*, *Syzygium cumini*, *Emblica officinalis Buchanania lanzan*, etc. and flowers of *Madhuca indica*, green pods of *moringa oleifera*, new shoots of bamboo, etc. are in great demand. There are more than 213 species of trees, 14 species of palms, 128 species of shrubs, 116 species of herbs, 4 species of ferns, and 15 species of fungi in our forests which yield edible products.

• **Lac and other products:** Lac is a resinous secretion of the lac insects which feed on forests trees, particularly on *Butea monosperma*. Similarly, silk is another important product from forests. It is obtained from the cocoons of silk worm. Silk worm is raised on *Terminalia alata* and *Morus alba* plantations for obtaining silk. Honey is another product which is obtained from forests.

- **Fodder and grazing:** Forests provide fodder leaves and grazing facility to the rural animals. It is estimated that more than 270 million cattle graze in the forests against the carrying capacity of 30 million (ICFRE, 2011). Leaf fodder of several tree species is almost as nutritious as that of agricultural fodder crops. Good fodder yielding tree species include; *Ailanthus excelsa, Moringa oleifera, Sesbania* spp. *Morus alba, Albizia lebbeck, Leucaena leucocephala, Pongamia pinnata, Hardwickia binata,* etc.

9.2.2 PROTECTIVE AND AMELIORATIVE FUNCTIONS OF THE FORESTS

Forests play a significant role in maintaining the CO_2 balance in the atmosphere. Without sufficient forest cover, the CO_2 which is released in the atmosphere will not be utilized completely resulting in higher % of the CO_2 in the atmosphere. Increased CO_2 (50% of GHG) results in warming of world temperature, disturbance in the climate, melting of polar ice caps, increasing sea levels, etc. Forests increase local precipitation by about 5%–10 % due to their orographic and micro-climatic effects. These create conditions favorable for the condensation of the clouds. Forests reduce temperature and increase humidity. Temperature in the forests is 3°C–8°C less than the adjoining open areas. A tree release 10 times more moisture into atmosphere than the equivalent area of the ocean. It also reduces evaporation losses. The effects of forests on temperature are not limited to forested areas but are also experienced in nearby areas. Forests maintain the productivity of the soil by adding a large quantity of organic matter and re-cycling of nutrients (return 40%–60%). The leaves of trees are used as manure. Supply of fire wood from forests releases dung for the use as manure. Tree crowns reduce the violence of rain and checks splash erosion. Forests increase infiltration and water holding capacity of the soil resulting in much lower surface run-off. This in turn results in checking of soil erosion. Forests check floods, they intercept 15%–30% of the total rainfall. They increase the infiltration rate and water holding capacity of the soil. This results in reduced surface run-off and checks erosion. Mostly floods are caused due to siltation of river channels, caused due to erosion and higher peak discharges, and due to greater surface run-off. Forests conserve soil and water both. They prolong the water cycle from its inception to the final disposal into streams and ocean. The longer the water is retained in the land, the greater is the usefulness in nurturing crops and trees and in maintaining regular supply of water in stream slower than surface run-off, and also it does not cause erosion. Forests and trees reduce wind velocity considerably. Reduction of wind velocity causes considerable reduction in wind erosion, checks shifting of sand dunes, and halts the process of desertification. Forests are the store house of genetic diversity. Several unknown plants may have potential for medicines and food. Deforestation eliminates several species of plants and animals. About 5,000 species of plants, 180 species of mammals, 180 species of birds, and several thousand species of insects are already on the verge of extinction due to loss of forests. Forests protect from physical, chemical, and noise pollution. Dust and the other particulate and gaseous pollutants cause serious problems. Forests protect us from such pollutants (Perala, 1985; Shah et al., 2015).

9.2.3 RECREATION AND EDUCATIONAL FUNCTIONS OF THE FORESTS

Forests provide recreational facilities to the people. A large variety of trees and shrubs, animals, and birds attract a large number of people toward them. National parks and sanctuaries, which are rich in flora and fauna, are visited by a large number of people. Forests provide experimental field and laboratory for learning to college and university students. They provide sites for ecological studies. Forests provide a natural healing effect for a number of diseases. We have a number of sanatoria established in well wooded areas.

9.2.4 FORESTS' DEVELOPMENTAL FUNCTIONS

Forests provide employment to a large number of people. Forests have provided employment to about 10 million people in the primary sector. Much larger employment was generated in secondary and tertiary sectors. Livelihood of nearly 23% population is linked with the forests of the country. Forests and various forest activities help tribals in their socio-economic condition through collection, processing, and marketing of various forest products and by providing gainful employment. Forests help to earn a good sum of revenue to the government which is used for various developmental works. Forestry and logging contributed Rs. 23,798 crores in 2001–2002 which was roughly 1.5% of the total GDP of the country (2.37% as per Indian Institute of Economic Growth, 2002; Chopra et al., 2001).

9.3 WEATHER AND CLIMATE?

Weather is the short-term changes we see in temperature, clouds, precipitation, humidity, and wind in a region or a city. Weather can vary greatly from 1 day to the next, or even within the same day. In the morning the weather may be cloudy and cool. But by afternoon it may be sunny and warm. The climate of a city, region, or the entire planet changes very slowly. These changes take place on the scale of tens, hundreds, and thousands of years.

9.3.1 CLIMATE CHANGE

Climate change refers to any significant change in the measures of climate lasting for an extended period of time. In other words, climate change includes major changes in temperature, precipitation, or wind patterns, among other effects, that occur over several decades or longer. *Global warming* refers to the recent and ongoing rise in global average temperature near Earth's surface. It is caused mostly by increasing concentrations of greenhouse gases (GHGs) in the atmosphere. Global warming is causing climate patterns to change. However, global warming itself represents only one aspect of climate change.

9.3.2   Evidences of Climate Change

i. **Sea Level Rise**: Global sea level rose about 17 cm (6.7 inches) in the last century. The rate in the last decade, however, is nearly double that of the last century. Oceans have warmed, the amounts of snow and ice have diminished and sea level has risen. From 1901 to 2010, the global average sea level rose by 19 cm as oceans expanded due to warming and ice melted. The Arctic's sea ice extent has shrunk in every successive decade since 1979, with 1.07 million km² of ice loss every decade.

ii. **Global Temperature Rise:** All three major global surface temperature reconstructions show that Earth has warmed since 1880. Most of the warming has occurred since the 1970s with the 20 warmest years having occurred since 1981 and with all the 10 warmest years occurring in the past 12 years. From 1880 to 2012, average global temperature increased by 0.85°C.

iii. **Erratic Precipitation:** There is an increased variance of precipitation everywhere. The wet areas become wetter, and dry and arid areas become more so and, increased precipitation in high latitudes (Northern hemisphere).

iv. **Extreme Events:** The frequency and number of extreme events globally and locally have been increasing. High temperatures, very cold temperatures, intense rainfall, snowstorms, hailstorms, windstorms and resulting floods, droughts, and landslides are frequent phenomenon now. India has also witnessed increased numbers of extreme events in recent times.

v. **Ocean Acidification:** The acidity of surface ocean waters has increased by about 30%. This increase is the result of humans emitting more carbon dioxide (CO_2) into the atmosphere and hence being more absorbed into the oceans. The amount of carbon dioxide absorbed by the upper layer of the oceans is increasing by about 2 billion tons per year (Figure 9.1).

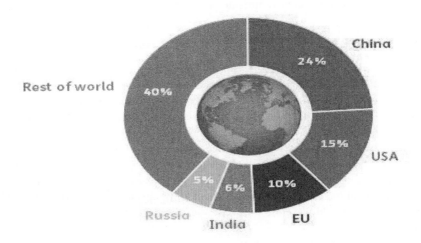

FIGURE 9.1   Major contributors of green house gasses.

The average annual CO_2 emission per person in India is 1.8 tons, USA is 16.5 tons, China is 7.6 tons, and 0.9 tons in J&K.

On Earth, human activities are changing the natural greenhouse. Over the last century the burning of fossil fuels like coal and oil has increased the concentration of atmospheric CO_2. The clearing of land for agriculture, industry, and other human activities has increased concentrations of greenhouse gases. Now it is 95% probability that human activities over the past 250 years have warmed our planet. The industrial activities that our modern civilization depend upon have raised atmospheric CO_2 levels from 280 parts per million to 400 parts per million in the last 150 years.

9.4 EFFECTS OF CLIMATE CHANGE

  i. The potential future effects of global climate change include more increase in temperature, intense rainfall, floods, wildfires, longer periods of drought, windstorms, snowstorms, hailstorms, landslides, floods, droughts, disease, etc.
  ii. Global climate change has already had observable effects on the environment. Glaciers have shrunk, ice on rivers and lakes is breaking up earlier.
 iii. Plant and animal ranges have shifted and trees are flowering sooner.
  iv. Change in the behavior of migratory birds.
   v. Invasion of alien species.
  vi. Water-scarcity.
 vii. Food-insecurity.
viii. Effects that scientists had predicted in the past would result from global climate change are now occurring: loss of sea ice, accelerated sea level rise, and longer, more intense heat waves.
  ix. According to the IPCC, the extent of climate change effects on individual regions will vary over time and with the ability of different societal and environmental systems to mitigate or adapt to change.

9.5 CLIMATE CHANGE IN JAMMU AND KASHMIR

In the context of Jammu and Kashmir, which nestles in fragile Himalayan Ecosystem, there indicators of climate change are evident now. As per United Nations Environment Programme (UNEP) report some parts of the state are moderate to highly vulnerable. As per Indian Network for Climate Change Assessment (INCCA) assessment the number of rainy days in the Himalayan region in 2030s may increase by 5–10 days on an average. The intensity of rain fall is likely to increase by 1–2 mm/day. This is likely to impact some of the horticultural crops. The rate of recession of glaciers is reportedly varying which is being attributed to winter precipitation, climate warming, and anthropogenic elements. Temperature, precipitation, and cold wave are most likely to significantly impact the agriculture sector.

Long term weather data of Kashmir valley revealed that there is increasing trend in temperature. Average maximum temperature has increased by 1°C in last 30 years.

Consequently average minimum temperature has increased by 0.5°C. Precipitation trend is decreasing erratically (Singh, 2014).

Deficit in food production is growing in Jammu & Kashmir. With the reduction in rainfall, the rain-fed agriculture will suffer the most. Horticultural crops like apples are also showing decline in production particularly due to decline in snowfall. About 34% and 39% of the forested grids are likely to undergo shifts in vegetation type with a trend toward increased occurrence of the wetter forest types. Climate change has impact on human health, e.g., impacts of thermal stress, vectors, water-borne pathogens, water quality, air quality, food availability, etc. It is projected that the spread of malaria, tick-borne diseases, etc. will be enhanced.

9.6   CLIMATE CHANGE AND FORESTS

Vegetation patterns (distribution, structure, and ecology of forests) across globe are controlled mainly by the climate. Even with global warming of 1°C–2°C, much less than the most recent projections of warming during this century, most ecosystems, and landscapes will be impacted through changes in species composition, productivity, and biodiversity. It has been demonstrated that upward movement of plants will take place in the warming world. Due to increase in temperatures, change in vegetation, rapid deforestation and scarcity of drinking water, habitat destruction, and corridor fragmentation may lead to a great threat to extinction of wild flora and fauna. It is expected that with the increase in drought cycles and concomitant increase in forest fires, some pine species will encroach upon the other species and will play a pivotal role in reducing the yield of non-timber forest products from these forests. The ecosystem services generated would also alter. Inevitably, any change in the forest (distribution, density, and species composition) under CC would immensely influence economies like forestry, agriculture, livestock husbandry, non-timber forest products, and medicinal plant-based livelihoods.

Forest sector is important in the context of climate change due to three reasons namely,: (i) deforestation, forest degradation, and land-use change contributes to about 20% of global CO_2 emissions, (ii) forest sector provides a large and low-cost opportunity to mitigate climate change, and (iii) forest ecosystems are projected to be adversely impacted by climate change, affecting biodiversity, biomass production, and forest regeneration. Climate is one of the most important determinants of forest vegetation patterns, and it is likely that changes in climate would alter the configuration of forest ecosystems in India (Planning commission, 2011).

Ecosystems are increasingly being subjected to other human-induced pressures, such as extractive use of goods and increasing fragmentation and degradation of natural habitats (e.g., Bush et al., 2004). In the medium-term (i.e., decades) especially, climate change will increasingly exacerbate these human-induced pressures, causing a progressive decline in biodiversity (Lovejoy and Hannah, 2005). However, this is likely to be a complex relationship that may also include some region-specific reductions in land use pressures on ecosystems (e.g., Goklany, 2005; Rounsevell et al., 2006). Projected future climate change and other human-induced pressures are virtually certain to be unprecedented (Forster et al., 2007) compared with the past several hundred millennia (e.g., Petit et al., 1999; Siegenthaler et al., 2005). An understanding of time lags in ecosystem responses is still developing, including, for example, broad-scale biospheric responses or shifting species geographical ranges. Many ecosystems may take several centuries (vegetation) or even possibly millennia (where soil formation is involved) before responses to a changed climate are played out (Lischke et al., 2002). A better understanding of transient responses and the functioning of ecosystems under continuously changing conditions are needed to narrow uncertainties about critical effects and to develop effective adaptation responses at the timescale of interest to human society. Species extinctions and especially global extinction as distinct from local extinctions are key issues that need to be addressed, as the former represents irreversible change. This is crucial, especially because of a very likely link between biodiversity and ecosystem functioning in the maintenance of ecosystem services (Duraiappah et al., 2005; Hooper et al., 2005; Díaz et al., 2006; Worm et al., 2006), and thus extinctions critical for ecosystem functioning, be they global or local, are virtually certain to reduce societal options for adaptation responses.

Research works carried out elsewhere have shown that the community composition has changed at high alpine sites, and tree line species have responded to climate warming by invasion of the alpine zone or increased growth rates during the last decades (Paulsen et al., 2000). Several field studies in different parts of the world indicate that climate warming earlier in the 20th century (up to the 1950s and 1960s) has caused advances in the tree limit (Kullman, 1998). Yet another study has reflected climate sensitivity of high-altitude forests of the western Himalayas.

The study by Borgaonkar et al. (2011) has reported an unprecedented enhancement in growth during the last few decades in the five tree-ring width chronologies of Himalayan conifers (*Cedrus deodara D. Don.*; *Picea smithiana Boiss*) from the high-altitude areas of Kinnaur (Himachal Pradesh) and Gangotri (Uttarakhand) regions. Analysis of surface air temperature data over the region indicates significant and increasing trend during the last century, with recent noticeable enhanced warming in the past four decades. The time series of annual highest values of daily maximum and minimum temperatures also show an increasing trend. As such, the study partially attributes the accelerated tree-ring growth during the last few decades to the overall warming trend seen over the region.

9.6.1 The Effects of Climate Change on Plant Phenology

Change in plant phenology may be one of the earliest observed responses to rapid global climate change and could potentially have serious consequences for the plants and animals that depend on periodically available resources. The phenology and development of most organisms generally follow a time scale, which is temperature dependent (Allen, 1976). Phenological changes have been studied by many scientists at global level. It has been observed that in certain plant species leaf unfolding in spring has advanced by up to 6 days. Whereas the autumn leaf coloring is delayed by 4–8 days (Menzel and Fabian, 1999). Abu-Asab et al. (2001) have studied changes in first flowering times of over 100 plant species representing 44 families of angiosperms for 29 years (1970–1999) in Washington, D.C. They observed that most of the trees now flower 3–5 days earlier than they did some years ago. Another study was conducted for the phenodynamics of ten MPTs of Kashmir under Shalimar conditions during 2009–2010. Phenophases of ten MPTs namely *Populus deltoides*,

P. alba, P. Nigra, P. Balsemifera, Salix alba, S. caprea, Albizia julibrissin, Quercus rober, Fraxinus floribunda, and *Morus alba* were recorded. The longest growing period was observed in *Salix caprea*, i.e. from January 20th to November 20th in 2009 and increased 10 days in 2010 and shortest growing period was recorded in *Fraxinus floribunda* from March 10th to October 30th in 2009 and from 20th March to 20th October in 2010. Earliest bud break was observed in *Salix caprea* and late in *Albizia julibrissin* in 2009 and in *Quercus robur* in 2010 (Qaisar et al., 2016)

9.6.2 IMPACTS ON INDIAN FOREST VEGETATION TYPES

Studies by Ravindranath and Sudha (2004), Ravindranath et al. (2006) made an assessment of the impact of projected climate change on forest ecosystems in India. This assessment was based on climate projections of the Regional Climate Model of the Hadley Centre (HadRM3) using the A2 (740 ppm CO_2) and B2 (575 ppm CO_2) scenarios and the BIOME4 vegetation response model. The main conclusion was that under the climate projection for the year 2085, 77% and 68% of the forested grids in India are likely to experience shift in forest types under A2 and B2 scenario, respectively. A recent assessment (Chaturvedi et al., 2011) of the impact of projected climate change on forest ecosystems India was made using a dynamic global vegetation model. Using climate projections of the regional climate model of the Hadley centre and the dynamic global vegetation model IBIS for A2 and B2 scenarios, it has been projected that 39% of forest grids are likely to undergo vegetation type change under the A2 scenario and 34% under the B2 scenario by the end of this century. The study concluded that the impacts varied with the region and forest types. Thus there is need for regional studies particularly for the ecologically sensitive regions and for short-term periods such as the 2030s. In another study by Gopalakrishnan et al. (2011) an assessment of the impact of projected climate change on forest ecosystems in India based on climate projections of the regional climate model of the Hadley centre and the global dynamic vegetation model IBIS for A1B scenario is conducted for short-term (2021–2050) and long-term (2071–2100) periods. Based on the dynamic global vegetation modelling, vulnerable forested regions of India have been identified to assist in planning adaptation interventions. The assessment of climate impacts showed that at the national level, about 45% of the forested grids are projected to undergo change. Vulnerability assessment showed that such vulnerable forested grids are spread across India. However, their concentration is higher in the upper Himalayan stretches, parts of Central India, northern Western Ghats, and the Eastern Ghats. In contrast, the northeastern forests, southern Western Ghats, and the forested regions of eastern India are estimated to be the least vulnerable. Low tree density, low biodiversity status as well as higher levels of fragmentation, in addition to climate change, contribute to the vulnerability of these forests. The mountainous forests (sub-alpine and alpine forest, the Himalayan dry temperate forest, and the Himalayan moist temperate forest) are susceptible to the adverse effects of climate change. This is because climate change is predicted to be larger for regions that have greater elevations. Marked expansion (11%) in temperate deciduous, cool mixed, and conifer forests at the cost of alpine pastures are likely to shrink. The unusual trends of winter migration of birds in the wetlands of Jammu, Kashmir, and Ladakh decline in socio-economic important species like deodar, fir, and spruce and increase in Blue Pine in Kashmir Valley and Chir Pine in Jammu. The spread of invasive alien species like *Parthenium, Lantana, Ageratum, Xanthium, Anthemis, etc.*, decreases tree density and forest fragmentation. Increased net primary productivity increased incidences of forest fires; these are some of the observations reported from J&K. (ENVIS, 2015).

9.7 MANAGEMENT OPTIONS

A. **Adoption:** Two main types of adaptation are autonomous and planned. Autonomous adaptation is the reaction, for example, a farmer to change precipitation patterns, in that they change crops or uses different harvest and planting/sowing dates. Planned adaptation measures are conscious policy options or response strategies, often multi sectoral in nature, aimed at altering the adaptive capacity of the agricultural system, or facilitating specific adaptations. For example, deliberate crop selection and distribution strategies across different agroclimatic zones, substitution of new crops for old ones, and resource substitution induced by scarcity. Over the period of time we will need to
   * increase the efficiency in the use of natural resources,
   * make best use of climate and weather predictions,
   * enhance forest regeneration,
   * rehabilitation of degraded forests on war footing,
   * massive afforestation in river catchments,
   * in situ conservation of RET species, wild edible, and medicinal plants,
   * ecotourism promotion and generation of sustainable,
   * forest fire management, promotion of agro forestry, adaptive management with suitable species, and silvicultural practices.

Some examples of the "win-win" adaptation practices are as follows (Murthy et al., 2011):

* Expand protected areas and link them wherever possible to promote migration of species.
* Promote forest conservation since biodiversity rich forests are less vulnerable due to varying temperature tolerance of plant species.
* Anticipatory planting of species along latitude and altitude.
* Promote assisted natural regeneration.
* Promote mixed species forestry.
* Promote species mix adapted to different temperature tolerance regimes.
* Develop and implement fire protection and management practices.
* Adopt thinning, sanitation, and other silvicultural practices.
* Promote in situ and ex situ conservation of genetic diversity.
* Develop temperature, drought, and pest resistance in commercial tree species.
* Develop and adopt sustainable forest management practices.
* Conserve forests and reduce forest fragmentation to enable species migration.
* Adoption of energy efficient fuelwood cooking devices to reduce pressure of forests.

B. **Mitigation**
   a. *Reducing emissions*: The fluxes of GHGs can be reduced by managing more efficiently the flows of carbon and nitrogen in agricultural systems. The exact approaches, that best reduce emissions, depend on local conditions and therefore vary from region to region.
   b. *Enhancing removals*: Forest ecosystems hold large reserves of C, mostly in soil organic matter. Any practice, that increases the photosynthetic input of C or slows the return of stored C via respiration, will increase stored C, thereby "sequestering" C or building C "sinks."
   c. *Avoiding emissions*: Using bioenergy feed-stocks would release CO_2-C of recent origin and would, thus, avoid release of ancient C through combustion of fossil fuels. Emissions of GHGs can also be avoided by agricultural management practices that forestall the cultivation of new lands.

9.8   AGROFORESTRY AS AN ALTERNATE LAND USE SYSTEM

Agroforestry is an intensive land management system that optimizes the benefits from the biological interactions created when trees and/or shrubs are deliberately combined with crops and/or livestock. Agroforestry is a collective name for land use systems in which woody perennials are grown with arable crops and animals on the same unit of land. The components of agroforestry viz., trees, crops, and animals are compatible with each other; they have environmental and economic interactions. Trees conserve soil and moisture by resisting erosion. Agroforestry systems are specific to sites. They have received due emphasis in the recent years because of their potential to yield fodder, fuel wood, and small timber in addition to food besides enhancing the microclimate and environmental quality (Mugloo et al., 2017).

Agro-forestry technologies, when used appropriately, help attain sustainable agricultural land-use systems in many ways. Specifically, agro-forestry technologies

   i. provide protection for valuable topsoil, livestock, crops, and wildlife,
   ii. increase productivity of agricultural and horticultural crops,
   iii. reduce inputs of energy (physical, chemical, or biological) and chemicals,
   iv. increase water-use efficiency of plants and animals,
   v. provide carbon sequestration and improve water quality,
   vi. diversify local economies, and
   vii. enhance biodiversity and landscape diversity and, ultimately, the quality of life for people.

Adaptation capabilities of agroforestry to climate change

   • **Drought:** Tree components through their deep roots explore a large soil volume of water and nutrients which help to maintain production during drought seasons.
   • **High rainfall:** Pumping excess water out of the soil profile more rapidly by higher evapo-transpiration and maintain aerated soil conditions.

- **Temperature:** Increased soil cover and multi strata cropping pattern system utilize the light resource efficiently and guard the soil from direct sunlight which leads to a reduction in soil temperature.

9.9 CARBON SEQUESTRATION POTENTIAL OF AGROFORESTRY

Agroforestry for carbon sequestration is attractive because (i) it sequesters carbon in vegetation and in soils depending on the pre-conversion soil C, (ii) the more intensive use of the land for agricultural production reduces the need for slash-and-burn or shifting cultivation, (iii) the wood products produced under agroforestry serve as substitute for similar products unsustainably harvested from the natural forest, (iv) to the extent that agroforestry increases the income of farmers, it reduces the incentive for further extraction from the natural forest for income augmentation, and finally, (v) agroforestry practices may have dual mitigation benefits as fodder species with high nutritive value can help to intensify diets of methane-producing ruminants while they can also sequester carbon.

Agroforestry has importance as a carbon sequestration strategy because of carbon storage potential in its multiple plant species and soil as well as its applicability in agricultural lands and in reforestation. The potential seems to be substantial; average carbon storage by agroforestry practices has been estimated as 9, 21, 50, and 63 Mg C/ha in semiarid, subhumid, humid, and temperate regions. For smallholder agroforestry systems in the tropics, potential carbon sequestration rate ranges from 1.5 to 3.5 Mg C/ha/yr (Montagnini and Nair, 2004). Agroforestry can also have an indirect benefit on carbon sequestration when it helps to decrease pressure on natural forests, which are the largest sinks of terrestrial carbon. Another indirect avenue of carbon sequestration is through the use of agroforestry technologies for soil conservation, which could enhance carbon storage in trees and soils. The available estimates of C stored in agroforestry range from 0.29 to 15.21 Mg C/ha/yr above ground, and 30–300 Mg C/ha up to 1 m depth in the soil (Nair et al., 2010). Carbon compounds are sequestered or accumulated by plants to build their structure and maintain their physiological process.

The energy captured in the molecular bonds of carbon compounds generally present between 2% and 4% of the radiation absorbed by the tree canopy. Stem wood growth often accounts for less than 20% of the dry matter produced in a year, the rest being used by foliage most of which is shed during leaf fall which is an important pathway for the flow of organic matter and energy from the canopy to the soil. Only green plant can assimilate carbon on the earth. Perennial systems like home gardens and agroforestry can store and conserve considerable amounts of carbon in living biomass and also in wood products.

For increasing the carbon sequestration potential of agroforestry systems practices such as conservation of biomass and soil carbon in existing sinks; improved lopping and harvesting practices; improved efficiency of wood processing; fire protection and more effective use of burning in both forest and agricultural systems; increased use of biofuels; and increased conversion of wood biomass into durable wood products need to exploited to their maximum potential. Agroforestry practices such as agrisilviculture or agrihorticulture systems for food and wood/fruit production; boundary and contour planting for wind and soil protection; silvipasture system for fodder production as well as soil and water conservation; complex agroforestry systems, viz. multi strata tree gardens, home gardens, agri-silvi-horticulture, and horti-silvipasture systems for food, fruits, and fodder especially in hill and mountain regions and coastal areas and bio-fuel plantations are suitable for sequestering atmospheric carbon and act as the potential sinks for sequestering surplus carbon from the atmosphere.

Agroforestry has such a high potential, not because it is the land use practice with the highest carbon density, but because there is such a large area that is susceptible for the land use change. The potential carbon sequestration in different land use is given in Figure 9.2.

9.9.1 CASE STUDIES

Carbon mitigation potential of an agri-horti-pasture (AHP) system under temperate conditions of Kashmir valley was evaluated with five treatments viz: Apple +Lucerne (T_1), Apple + Beans-Pea (T_2), Apple +White clover (T_3), Apple + Orchard grass (T_4), and Apple + natural sward (T_5). The experiment was laid in an established apple orchard of about 20 years of age at Sher-e-Kashmir University of Agricultural Sciences and Technology at its Shalimar campus. A single row of six trees per plot (net-cropped area 81 m^2) was replicated six times for each treatment with tree to tree spacings at 4.5 m. The average density of trees was 610/ha. The total biomass of the agri-horti pasture system has been assessed and evaluated by the non-destructive method. Total biomass and carbon stock of the system (tree + crop + fruit) were estimated and Apple + Lucerne (T_1) gained maximum total biomass 33.48 t/ha and carbon stock 15.06 t C/ha and e CO_2 55.23 t/ha, whereas apple + natural sward (T_5) registered lowest carbon stock 10.33 t C/ha and 37.99 t/ha e CO_2. Potential of carbon mitigation through agri-horti-pasture system has been foreseen under temperate

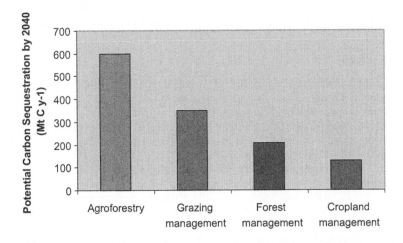

FIGURE 9.2   Potential carbon sequestrations in different land use. (Adapted from IPCC, 2000)

Carbon Stock of Different Land Use of Kashmir (Zaffar, 2013) in Tons/ha

Natural forest (blue pine-45 years)	556.73
Plantation (poplar-21 years)	407.70
AFS agri-silvi.(16 years)	80.16
AFSH orti-agri. (-do-)	64.17
Apple cultivation alone (19 years)	61.63
Agriculture rice-oat	5.04
Grassland	1.19
Wasteland	0.84

Tree Species	Carbon Biomass (t/ha)	Potetial of CO_2 Mitigation (t/ha)
Deodar (19 years) 30–40 cm	89.06	326.63
Elm (19 years) 30–40 cm	212.49	779.83
Fraxinus (19 years) 20–30 cm	33.90	124.41
Salix alba (>50 cm)	292.98	1,075
Natural forest (fir-spruce-pine-deodar)	95.35	349.93

Sources: 1. Wani et al., 2014a., 2. Wani et al., 2014b., 3. Wani et al., 2014c., 4. and 5. Wani et al., 2015.

Kashmir valley conditions, where substantial area 140,156 ha is under apple orchards. If the best treatment of the study (Apple + Lucerne) is adopted under apple orchards of Kashmir, it would conserve 7.74 million tons of e CO_2. (Qaisar, 2010). Some other studies carried out by our faculty for carbon sequestration through forest land use/trees are as follows:

9.10 CONCLUSION

Forestry is an important land use supporting large population of the country having multiple benefits directly or indirectly. Forest trees are also vulnerable to climate change and have an added advantage of playing an important role in climate change mitigation through locking the carbon through trees and ameliorating the environment.

REFERENCES

Abu-Asab, M.S., Peterson, P.M., Shetler, S.G. and Orli, S.S., 2001. Earlier plant flowering in spring as a response to global warming in the Washington, DC, area. *Biodiversity & Conservation, 10*(4), pp. 597–612.

Aggarwal, A., Paul, V. and Das, S., 2009. Forest resources: degradation, livelihoods, and climate change. In: Datt, D. and Nischal, S. (Eds.), *Looking Back to Change Track* (pp. 91–108). TERI, New Delhi.

Allen, J.C., 1976. A modified sine wave method for calculating degree days. *Environmental Entomology, 5*(3), pp. 388–396.

Borgaonkar, H.P., Sikder, A.B. and Ram, S., 2011. High altitude forest sensitivity to the recent warming: a tree-ring analysis of conifers from Western Himalaya, India. *Quaternary International, 236*(1–2), pp. 158–166.

Bush, M.B., Silman, M.R. and Urrego, D.H., 2004. 48,000 years of climate and forest change in a biodiversity hot spot. *Science, 303*(5659), pp. 827–829.

Chaturvedi, R.K., Gopalakrishnan, R., Jayaraman, M., Bala, G., Joshi, N.V., Sukumar, R. and Ravindranath, N.H., 2011. Impact of climate change on Indian forests: a dynamic vegetation modeling approach. *Mitigation and Adaptation Strategies for Global Change, 16*(2), pp. 119–142.

Chopra, K., Bhattacharya, B.B. and Kumar, P., 2001. *Contribution of Forestry Sector to Gross Domestic Product in India.* Institute of Economic Growth, Delhi University Enclave, New Delhi.

Datta, D. and Nischal, S., 2010. *Looking Back to Change Track.* The Energy and Resources Institute (TERI), New Delhi.

Dhyani, S.K., Vimala Devi, S. and Handa, A.K., 2015. Tree borne oilseeds for oil and biofuel. *Technical Bulletin, 2*(2015), p. 50.

Díaz, S., Fargione, J., Chapin III, F.S. and Tilman, D., 2006. Biodiversity loss threatens human well-being. *PLoS Biology, 4*(8), p. e277.

Duraiappah, A., Naeem, S., Agardi, T., Ash, N., Cooper, D., Dnz, S., Faith, D.P., Mace, G., McNeilly, J.A., Mooney, H.A., Oteng-Yeboah, A.A., Pereira, H.M., Polasky, S., Prip, C., Reid, W.V., Samper, C., Schei, P.J., Scholes, R., Schutyser, F. and van Jaarsveld, A., 2005. *Ecosystems and Human Well-being: Biodiversity Synthesis.* Island Press, Washington, DC, p. 100.

ENVIS. Oct-Dec, 2015. Climate Change and concerns of J&K.ENVIS news letter. Department of Ecology Environment and Remote Sensing, J&K.

Forest Survey of India. 2015. India State Forest Report -2015, Forest Survey of India Race Course, Dehradun, Uttarakhand 248001.

Forster, P., Ramaswamy, V., Artaxo, P., Berntsen, T., Betts, R., Fahey, D.W., Haywood, J., Lean, J., Lowe, D.C., Myhre, G. and Nganga, J., 2007. Changes in atmospheric constituents and in radiative forcing. In Solomon, S., Qin, D., Manning, M., Chen, Z., Marquis, M., Averyt, K.B., Tignor, M. and Miller, H.L. (Eds.), *Climate Change 2007. The Physical Science Basis.* The Physical Science Basis. Contribution of Working Group I to the Fourth Assessment Report of the Intergovernmental Panel on Climate Change (pp. 130–234). Cambridge University Press, Cambridge.

Global Agriculture Information Network. 2014. Wood and Wood products in 2014. GAIN Report No: IN4049. USDA, Foreign Agriculture Service.

Global Forest Resources Assessment. 2005. Progress towards Sustainable Forest Management. FAO Forestry Paper 147, Food and Agriculture Organization of the United Nations, Rome, Italy.

Goklany, I.M., 2005. A climate policy for the short and medium term: stabilization or adaptation? *Energy & Environment, 16*(3), pp. 667–680.

Gopalakrishnan, R., Jayaraman, M., Bala, G. and Ravindranath, N.H., 2011. Climate change and Indian forests. *Current Science, 101*(3), pp. 348–355.

Hooper, D.U., Chapin, F.S., Ewel, J.J., Hector, A., Inchausti, P., Lavorel, S., Lawton, J.H., Lodge, D.M., Loreau, M., Naeem, S., Schmid, B., Setala, H., Symstad, A.J., Vandermeer, J. and Wardle, D.A., 2005. Effects of biodiversity on ecosystem functioning: a consensus of current knowledge. *Ecological Monographs, 75*(1), pp. 3–35.

ICFRE, 2011. Forestry Statistics India 2011. Division of Statstics, Directorate of Extension, Indian Council of Forestry Research and Education, New Forest, Dehradun, p. 113.

IPCC. 2000. Emmission Scenario. A special report of IPCC special working group. Nebojsa Nakicenovic and Rob Swart (Eds.) Cambridge University Press, UK, p. 570.

Kullman, L., 1998. Tree-limits and montane forests in the Swedish Scandes: sensitive biomonitors of climate change and variability. *Ambio, 27*, pp. 312–321.

Lischke, H., Lotter, A.F. and Fischlin, A., 2002. Untangling a Holocene pollen record with forest model simulations and independent climate data. *Ecological Modelling, 150*(1–2), pp. 1–21.

Lovejoy, T.E. and Hannah, L., 2005. *Climate Change and Biodiversity.* Yale University Press, New Haven, CT, p. 418.

Menzel, A., 2000. Trends in phenological phases in Europe between 1951 and 1996. *International Journal of Biometeorology, 44*(2), pp. 76–81.

Menzel, A. and Fabian, P., 1999. Growing season extend in Europe. Nature, *397*, p. 659.

Montagnini, F. and Nair, P.K.R., 2004. Carbon sequestration: an underexploited environmental benefit of agroforestry systems. In Nair, P.K.R., Nair, P.K.R., Rao, M.R. and Buck, L.E. (Eds.), *New Vistas in Agroforestry* (pp. 281–295). Springer, Dordrecht.

Mugloo, J.A., Qaisar, K.N., Mughal, A.H., Khan, P.A. and Mir, A.A., 2017. *Natural Resource Management*. Faculty of Forestry, SKUAST-K, Benhama, Ganderbal (J&K), p. 84.

Murthy, I.K., Tiwari, R. and Ravindranath, N.H., 2011. Climate change and forests in India: adaptation opportunities and challenges. *Mitigation and Adaptation Strategies for Global Change, 16*(2), pp. 161–175.

Nair, P.R., Nair, V.D., Kumar, B.M. and Showalter, J.M., 2010. Carbon sequestration in agroforestry systems. In: Sparks, D.L. (Ed.), *Advances in Agronomy* (Vol. 108, pp. 237–307). Academic Press, Amsterdam.

Paulsen, J., Weber, U.M. and Körner, C., 2000. Tree growth near treeline: abrupt or gradual reduction with altitude? *Arctic, Antarctic, and Alpine Research, 32*(1), pp. 14–20.

Perala, D.A., 1985. Predicting red pine shoot growth using growing degree days. *Forest Science, 31*(4), pp. 913–925.

Petit, J.R., Jouzel, J., Raynaud, D., Barkov, N.I., Barnola, J.M., Basile, I., Bender, M., Chappellaz, J., Davis, M., Delaygue, G. and Delmotte, M., 1999. Climate and atmospheric history of the past 420,000 years from the Vostok ice core, Antarctica. *Nature, 399*(6735), p. 429. 33.

Planning Commision. 2011. Climate Change and 12th Five year Plan, Report of Sub group on Climate Change, Planning Commision, Government of India, p. 97.

Qaisar, K.N., 2010. Annual Progress Report 2009-10. Faculty of Forestry, SKUAST-K, Shalimar, Srinagar.

Qaisar, K.N., Dutt, V., Khan, P.A. and Zafar, S.N., 2016. *Forest Trees Phenology Reference Manual*. Faculty of Forestry, SKUAST-K, Benhama, Ganderbal (J&K), Vol. 20, p. 34.

Ravindranath, N.H. and Sudha, P., 2004. *Joint Forest Management in India: Spread, Performance and Impacts*. Universities Press, Hyderabad, TS.

Ravindranath, N.H., Joshi, N.V., Sukumar, R. and Saxena, A., 2006. Impact of climate change on forests in India. *Current Science, 90*(3), pp. 354–361.

Rounsevell, M.D.A., Berry, P.M. and Harrison, P.A., 2006. Future environmental change impacts on rural land use and biodiversity: a synthesis of the ACCELERATES project. *Environmental Science & Policy, 9*(2), pp. 93–100.

Shah, M., Masoodi, T.H., Wani, P.K.J. and Mir, S., 2015. Vegetation analysis and carbon sequestration potential of salix alba plantations under temperate conditions of Kashmir, India. *Indian Forester, 141*(7), pp. 755–761.

Siegenthaler, U., Stocker, T.F., Monnin, E., Lüthi, D., Schwander, J., Stauffer, B., Raynaud, D., Barnola, J.M., Fischer, H., Masson-Delmotte, V. and Jouzel, J., 2005. Stable carbon cycle–climate relationship during the late Pleistocene. *Science, 310*(5752), pp. 1313–1317.

Singh, K.N., 2014. Climate Change Forecasting and its Impact on Agriculture. Compendium on ICAR Summer School on Temperate Agroforestry for Sustenance and Climate Moderation (5-25th August) Faculty of Forestry, SKUAST-K, Shalimar, pp. 70–79.

Wani, A.A., Joshi, P.K. and Singh, O., 2015. Estimating biomass and carbon mitigation of temperate coniferous forests using spectral modeling and field inventory data. *Ecological Informatics, 25*, pp. 63–70.

Wani, N.R., Qaisar, K.N. and Khan, P.A., 2014a. Biomass and carbon allocation in different components of Ulmus wallichiana (elm): an endangered tree species of Kashmir valley. *International Journal of Pharma and Bio Sciences, 5*(1), pp. 860–872.

Wani, N.R., Qaisar, K.N. and Khan, P.A., 2014b. Biomass, carbon stock and carbon dioxide mitigation potential of Cedrus deodara (deodar) under temperate conditions of Kashmir. *Canadian Journal of Pure and Applied Sciences, 8*(1) pp. 2677–2684.

Wani, N.R., Qaisar, K.N. and Khan, P.A., 2014c. Growth performance, biomass production and carbon stock of 19-year old Fraxinus floribunda (ash tree) plantation in Kashmir valley. *Agriculture & Forestry/Poljoprivreda i Sumarstvo, 60*(1), pp. 125–143.

Worm, B., Barbier, E.B., Beaumont, N., Duffy, J.E., Folke, C., et al. 2006. Impacts of biodiversity loss on ocean ecosystem services. *Science, 314*, pp. 787–790.

Zaffar, S.N., 2013. Carbon Appraisal of Different Land Use Systems of Srinagar District of Kashmir Valley. Ph.D Forestry thesis. Submitted to Faculty of Forestry, SKUAST-K, Wadura, Sopore.

10 Agroforestry Role in Sustainable Management of Degraded Watersheds

S. Sarvade
Jawaharlal Nehru Krishi Vishwa Vidyalaya

Nancy Loria, Indra Singh,
Pankaj Panwar, and Rajesh Kaushal
ICAR-Indian Institute of Soil and Water Conservation

CONTENTS

10.1 INTRODUCTION

Nature has bestowed us with enormous gifts, namely, natural resources which comprises the land, water, and vegetation. India constitutes only 2.4% of the world's land area and the available groundwater resource for irrigation is about $361\,km^3$ (www. nih.ernet.in). Moreover, with 16 major forest types and 251 subtypes, the total forest and tree cover of the country constitutes (79.42 million ha) 24.16% of the country's geographical area (FSI, 2015). These resources have been exhausted by 16.7% of the world's human population (Sarvade, 2014a) and 18% livestock population (Sarvade *et al.*, 2017). The availability of resources for future generation depends on their sustainable management. In the last four decades, the sustainability of natural resources had been threatened due to the misuse and mismanagement of natural resources for human greed. Nevertheless, there is an urgent need to ensure their conservation in the face of changing climate, increased biotic pressure, and declining resource productivity (Bipin *et al.*, 2000; Dalvi *et al.*, 2010; Lenka *et al.*, 2015). The responsibility of care, conservation, and maintenance of basic natural resources such as soil, water, and vegetation lie within mankind. Moreover, the conservation of natural resources is now usually embraced in the broader conception of conserving the earth itself by protecting its capacity for self-renewal. Watershed development aims to balance the conservation, regeneration, and use of land and water resources within a watershed developed by humans. The overall attributes of the watershed development approach, by and large, are threefold, viz, promoting economic development of the rural area, employment generation, and restoring ecological balance.

Watershed is not simply a hydrological unit but also a socio-political-ecological entity which plays a crucial role in determining food, social, and economic security and provides life support services to rural people (Grewal, 2000; Kalra and Mishra, 2006; Madhurima and Banerjee, 2013; Sharda *et al.*, 2006; Lenka *et al.*, 2015; Wani *et al.*, 2008). The watershed is the product of the interactions between land and water, particularly its underlying geology, rainfall patterns, slope, soils, vegetative cover, and land use (Figure 10.1). However, the condition of watershed depends on water quality, water quantity, aquatic habitat, aquatic biota, riparian/wetland vegetation, roads and trails, soils, fire regime or wildfire, forest cover, rangeland vegetation, terrestrial invasive species, and forest health. A large watershed can be managed in plain valley areas or where forest or pasture development is the main objective. In hilly areas or where intensive agriculture development is planned, the small-sized watershed is preferred. The concept of watershed management has evolved to ensure effective use of both natural and social capital.

Evidently, the land, water, and human resources are essential components of watershed development programs. The watershed program is primarily a land-based program, with increased focus on water, and its main objective is to enhance agricultural productivity through increased *in situ* moisture conservation and protective irrigation for socio-economic development of rural people. It has been essential in a country like India where majority of the population depends on agriculture and about 60% of total arable land (142 million ha) in the country is rain-fed. A large portion of the rain-fed areas (65% of arable land) in India is characterized by low productivity, high risk and uncertainty, low level of technological change, and vulnerability to degradation of natural resources (Joshi *et al.*, 2004). Over the years, the

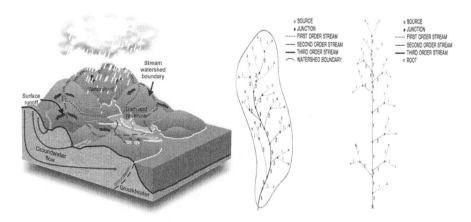

FIGURE 10.1   Watershed area.

sustainable use of land and water has received wider attention among policymakers, administrators, scientists, and researchers.

A watershed approach is a system-based approach that facilitates the holistic development of agriculture, forestry, and allied activities in the proposed watershed. The components of watershed development program would include; (i) soil and land management (ii) water management (iii) crop management (iv) afforestation (v) pasture or fodder development (vi) livestock management (vii) rural energy management (viii) other farm and non-farm activities and (ix) development of community skills and resources. All these components are interdependent and interactive and must, therefore, be considered together (Tideman, 2000; Biswas, 2008; Sucharita, 2008; Varis *et al.*, 2014).

Agroforestry can complement the efforts in watershed management by providing a set of tree-based conservation and production practices for agricultural lands. Agroforestry is the dominant sustainable farming system that is being followed/advocated in upland management (Shukla, 2015; Bhan, 2013; MoA, 1996). An ecologically, socially and economically sustainable system requires deliberate growth and management of trees along with agricultural crops and livestock in systems (Alao and Shuaibu, 2013). Agroforestry has two functions: (a) productive functions, in which the three 'Fs' (fuelwood, fodder, and fruit) are the most important besides construction wood, gums, resins, medicines, fibers, and a host of other economic bases and greater food security (Sarvade *et al.*, 2014); (b) service functions which include shade, reduction in wind speed, control of erosion, and maintenance and improvement of soil fertility (Pandey, 2007). Fundamentally, agroforestry refers to the use of trees in farming systems. The villagers in hilly areas normally own 0.1–1.0 ha of the land and adopt sophisticated land management patterns to get all the benefits from limited land, that is, food crops, fodder, fuelwood, and cash crops. Precisely, the existing agroforestry systems can be broadly classified into three categories, that is, need, economy, and environment-based systems.

Models of agroforestry systems, for upland and lowland conditions with the main emphasis of tree-based agroforestry systems in the upland and agricultural crop-based systems in the lowlands, shows that on cultivated lands, the primary system of land management should be horticulture based with sequential integration of agricultural crops as horti-agriculture, horti-silviculture, and horti-silvi-agriculture system (Ong and Swallow, 2003; Anonymous, 2004; Limnirankul *et al.*, 2015).

10.2   REHABILITATION OF DEGRADED WATERSHEDS

Rehabilitation of the degraded watershed followed a multitier ridge-to-valley sequenced approach. Prior to the start of rehabilitation work, benchmark survey and resource inventory should be done which proves helpful in proper planning and execution of the work. Watershed problem should be identified using participatory rural appraisal (PRA) exercise, as it helps in building rapport with the villagers. Using the basic resource information of the watershed program (with regard to tree and grass species), decisions should be taken considering the area under cropping and new areas which can be brought under cultivation by employing suitable soil conservation techniques, type of tree species to be planted, the location and area where it has to be planted, and the acceptability of these species by the farmers. Conceptually, watershed has three main parts: catchment area, command area, and submergence area. These parts have specific roles to play and are hence managed accordingly. Following are some of the important measures which can be taken for rehabilitation of degraded areas in different parts of the watershed:

10.2.1 CATCHMENT AREA

10.2.1.1 Upper Catchment

A majority of forests and degraded lands are located in the catchment area of watershed. The treatment of these areas is often ignored, as they are typically under the control of the forest department, which does not permit other departments to operate on their lands. The responsibility of implementation of watershed in these areas, therefore, lies with the Forest Departments and the Joint Forest Management Committees (JFMCs).

As per the guidelines, degraded forest lands can be developed in two phases, viz, natural regeneration phase and new plantation phase using soil and water conservation measures.

i. *Soil and water conservation measure:* The soil and water conservation measure aims at checking surface runoff, soil loss, and increase the infiltration of water in the soil profile in such a way that it can be harvested in the reservoirs downstream. Slope stabilization by low-cost gunny bags/ *katta-crate* structure should be used for sloppy areas. First-order gullies/ channels receiving small quantity of runoff should be stabilized either by brushwood check dams or loose stone masonry check dams (Angima *et al.,* 2003; Barungi and Maonga, 2011; Mishra and Rai, 2013; Vanlauwe *et al.,* 2013). Similarly, for larger gullies, masonry check dams are provided with aprons on lower side and their wing wall should be properly embedded in the *Nala* sites. Perennial woody tree and other plant species help to minimize erosion effects through surface litter cover and understory vegetation; increase organic matter and diversity, through continuous degeneration of roots and decomposition of litter; nitrogen fixation and enhancement of soil's physico-chemical properties (Sanchez, 1987; Young, 1989). On silting masonry dams, the area should be planted with species such as *Agave, Ipomea, Vitex, Arundo* and so on. Geotextile membranes should be used in those areas where mass erosion poses a major problem. Trees such as *Leucaena leucocephala, Melia azadirachta, Sesbania grandiflora, S. sesban, Gliricidia spp., Calliandra spp., Grevillea robusta* and so on are employed for efficient soil and water conservation. Drainage line treatment should be done to reduce the runoff velocity within permissible limits for which series of check dams are usually designed (Figures 10.2 and 10.3). Diversion drain should also be provided at appropriate places to protect the downstream area and to discharge the flow safely (Mishra and Rai, 2013).

FIGURE 10.2   Series of check dams for *nala* stabilization.

FIGURE 10.3   Construction of check dams and gully plugs.

ii. *Natural Regeneration:* Securing adequate regeneration from already existing mother trees in the watershed can help in improving the degraded condition of land at a low cost within a reasonable time period (Islam *et al.*, 2016; Camacho *et al.*, 2016). Cultural operations such as weeding, cleaning, cutting, improvement felling, piling of debris, control burning helps in assisting and completing the regeneration process should be given prime importance.

Weeding should be done immediately after seed germination as weeds compete with seedlings for light, nutrients, and moisture. Care should be taken to avoid weeding beyond October as it helps in protecting the dormant seedlings from frost. Approximately, three weedings should be done during the first year, two during the second and one in the third year. Hoeing is preferably done along with weeding while mulching is done along with the last weeding of the year. Once the regeneration has reached sapling stage, cleaning of climbers and inferior growth near the saplings should be done periodically at an interval of 3–5 years.

iii. *Afforestation/Plantation:* For sustainable results, plantation of new species should be taken up only where the first phase of natural regeneration has been successfully completed through social fencing (Sreedevi *et al.*, 2007).

Before the start of any plantation work, a reconnaissance survey should be conducted to know the topographic variations, aspect, slope, and elevation. Information on presence of weeds, rocks, and outcrops drainage pattern should also be collected. The site to be afforested should be made free from the bushes for proper layout. Prior to start of plantation work, the area should be fenced or should be managed undercut and carry system to check the grazing (Dwivedi, 1992; Nair, 1993). Proper soil working is important for the establishment of plantation (Palanisamia and Kumar, 2009). In low rainfall areas, staggered contour trenches should be made to enhance the survival and growth of the seedlings. Trenches should be designed to store 60%–70% runoff and thus the number and size of trenches should be decided in accordance. However, through a rough estimate it is advised to maintain the numbers of trenches equal to the number of trees to be planted. Trenches are usually $3 \times 0.30 \times 0.30$ m (Figure 10.4). However, if trees are to be closely spaced, size of trench should be reduced. The dug-up soil is placed on the downhill slope along the trench. The seeds are either sown or seedlings are planted on berm of the trenches or in diagonally half-filled trenches. The successive lines of contour trenches are kept at 2–4.5 m depending on the slope angle. If soils are shallow, spacing should be ignored and direct sowing of seeds or planting should be done in patches or pits. The dug-up soil is exposed for about 3–4 weeks. Planting should be done after a good rainfall when moisture penetrates to a depth of 30 cm in the profile pit. Care should be taken so that planting operations are carried out within a short time interval. The seedlings should be kept free from weeds. Fire is one of the major factors responsible for large scale mortality in both natural and artificial regenerated areas. It makes the soil dry, accelerates soil erosion and depletes water table, and affects water supply in the lower part of the watershed. For proper rehabilitation, fire needs to be controlled effectively which can be done by making fire lines and by control burning prior to fire season. Care should be taken to cut the grasses before it dries up. The casualties should be beaten up immediately.

FIGURE 10.4   Staggered and continuous contour trenching.

iv. *Energy Plantations:* In India, 55% of the fuelwood comes from the land outside the forest, hence energy plantations are very important. The plantation of fast-growing short rotation tree species both on fertile as well as marginal and wastelands for fuel and energy purpose is called energy plantation (Pandey, 2002). In many studies tree species such as *Eucalyptus teretecornis, Melia azadirachta, Leucaena leucocephala* (Sharma, 1989, Nayak, 1996), *Acacia mollissima, Eucalyptus hybrid* (Verma *et al.,* 2003) have been planted with 4,000–27,000 trees/ha to produce fuelwood and amelioration of degraded lands. The most suitable fuelwood species recommended in the state of Himachal Pradesh are *Salix spp, Acacia mollissima, Melia azadirachta, Eucalyptus hybrid, Populus deltoides, Leucaena leucocephala,* Bamboo *spp., Morus spp, Dalbergia sissoo, Alnus spp., Albizzia lebbeck, Albizia procera* and so on.

v. *Bamboo groves as a component of agriculture holdings (Agri-silviculture):* Along the streams and irrigation/drainage channels, cultivation of bamboo holds a common place in all agricultural practices. Bamboo's are extensively used in building small farmhouses, goat sheds, piggery enclosures, small baskets, and in string making. Large bamboo pipes are used as water conveyers in the farm irrigation/drainage system. Bamboo leaves serve as an excellent winter fodder for goat. Bamboo stumps/culms also protect the water channel from erosion. This system is limited only to the high rainfall areas or where enough water is available to grow bamboo. Patil *et al.* (2004) analyzed the effect of bamboo-based agroforestry system on soil profile and surface soil properties. Soil profile investigations revealed that nutrients were found in an increased proportion in sites having bamboo-based agroforestry. In addition, bamboo-based agroforestry system also increases the biodiversity under its habitat.

vi. *Silvo-pastoral System:* Quli and Siddiqui (1996) reported the excellence of silvo-pastoral system for eco-restoration. The system advocates implication of afforestation on uncultivable wastelands in combination with native grasses. Moreover, suitable trees as per region-specific needs are planted with simultaneous planting of grasses (preferably endemic) on the below-ground area, for stabilizing the soil, and conservation of moisture. The tree planting is preferably done using nursery raised seedlings, however, for grasses, planting is done by broadcast or solid sod method. Plantation on the steep slopes is done after digging contour trenches for water conservation. Plants having good economic potentials and which are good stabilizers like doob grass (*Cynadon dactylon*), khas (*Vetiveria zizanoides*), and agave (*Agave sislana*) are suitable for such problematic areas. After proper site stabilization, direct sowing of *Acacia nilotica, Cassia tora, Acacia catechu, Zyzyphus xylocarpa* and so on are recommended as woody perennial component for silvo-pastoral system (Yadava and Quli, 2007). This system is again classified into four categories, namely, scattered trees and shrubs on pastures, intensive silvo-pastoral (fodder trees with agri. fodder crops), live fence of fodder trees and hedges, and protein bank (cut and carry) (Nair, 1984; Dwivedi, 1992; Nair, 1993; Puri, 2007; Panda, 2013).

vii. *Protein Bank (cut and carry)*: Various multipurpose trees (protein-rich trees) are planted in and around farmlands and rangelands, for cut-and-carry fodder production, to meet the feed requirements of livestock during the fodder-deficit period in winter. Following are the kind of protein bank multipurpose trees planted in and around farmlands:

- For humid and sub-humid areas: *Artocarpus spp., Anogeissus lattifolia, Bombax malabaricum, Cordia dichotoma, Dalbergia jambolana, Samanea spp., Zizyphus* and so on.
- For dry regions: *Acacia nilotica, Ailanthus excelsa, Opuntia, Ficus, Prosopis spp., Rhus* and so on.

viii. *Live Fence of Fodder Trees and Hedges*: Various fodder trees and hedges are planted as live fence to protect the property from stray animals or other biotic influences in this system. The trees generally used are *Sesbania grandiflora, Gliricidia sepium, Erythrina abyssinica, Euphorbia* spp., *Acacia spp.* and so on.

ix. *Scattered Trees and Shrubs on Pastures*: In this system, various tree and shrub species are scattered irregularly or arranged according to some systematic pattern, especially to supplement forage production. The following tree and shrub species are used:

- **For humid and sub-humid regions:** *Derris indica, Emblica officinalis, Psidium guajava, Tamarindus indica.*
- **For dry regions:** *Acacia spp., Prosopis spp. and Tamarindus indica.*

In the Shivaliks of India, the well-known tree species are *Acacia spp., Leucaena leucocephala, Dalbergia sissoo, Melia azadirachta, Eucalyptus hybrid* and *Emblica officinalis*. These trees are grown with local fodder grasses such as *Chrysopogon fulvus* and *Heteropogon contortus*. Bhabbar (*Eulaliopsis binata*), which is a commercial grass, used in the manufacture of paper, has also done very well in association with most of the above tree species. These two-layered agroforestry systems involving trees and grasses have demonstrated their superiority over traditional rain-fed crops on poor quality soils.

Improvement in status of soil organic matter, biological nitrogen fixation, and reduced outflow of nutrients from the system are some of the cardinal aspects (Sarvade et al., 2014b). Kumar (1994) has reviewed the important role of tree species like *Casuarina, Leucaena, Prosopis cineraria, Acacia nilotica* and *Sesbania grandiflora* in soil improvement under agroforestry. Silvopastoral agroforestry system comprising *Prosopis juliflora* and *Leptochloa fusca* grass was found most promising in terms of firewood and forage production and for soil amelioration in sodic soils (Dagar and Gurbachan, 1993). However, there may also be temporary nutrient depletion, adverse chemical/biological effects, and shading effects in such systems (Nair, 1989).

x. *Intensive silvo-pastoral system (fodder trees with agri. fodder crops):* Improved fodder crop species and nutritious and palatable fodder shrub species are preferred alongwith intensive management practices for intensive silvo-pastoral system. Clason and Sharrow (2000), Mahecha *et al.* (2011) and Raj and Jhariya (2016) reported intensive silvo-pasture systems with high-yielding grass and high-density trees/shrubs for quality forage production, for small dairy farms in the humid tropics of India. Intensive silvo-pastoral systems are a good example of sustainable forage production in a natural way. Different strategies and incentives may be used in these systems for their overall improvement (Murgueitio *et al.*, 2013).

xi. *Eucalyptus and Bhabhar Grass:* Several agroforestry systems in the Shivalik foothills were established to study their efficacy in achieving production and conservation goals (Singh *et al.*, 1990; Verma *et al.*, 2017). Furthermore, it has been reported that traditional agriculture systems are transforming in to multifunctional agroforestry in Uttar Pradesh. *Eucalyptus tereticornis* and *bhabhar* grass (*Eulaliopsis binata*) were raised together on a light-textured soil in 1985. The plant population of 2,500 trees/ha was adjusted in a paired row technique in the north-south direction. In the intervening, 2 m space *bhabhar* grass was planted at 50 × 50 cm. The above system not only conserved soil and water but also gave much higher returns than the traditional rain-fed crops. Grewal in 1993 found that under *Eucalyptus hybrid-Eulaliopsis binata* system, soil loss was negligible (0.07 t/ha) followed by *Acacia catechu-Pennisetum purpureum* (Napier grass) (0.24 t/ha) and *Tectona grandis-Leucaena leucocephala-Eulaliopsis binata* system (0.43 t/ha).

xii. *Poplar, Leucaena, and Bhabhar Grass:* In a study conducted by (Singh *et al.*, 1990) on 40 × 40 m plot having 2% slope, sandy loam soil, and a pan of gravel at 1.5 m depth; *Populus deltoides* (G-3) was planted at 4 × 4 m in February 1986, *Leucaena* at 2 × 2 m was added to the system in July the same year and *bhabhar* grass was planted in july at 50 × 50 cm. The three plant species occupied the top, middle, and lower canopy of the composite vegetation system. The above system resulted in a mean runoff of 4.76% and soil loss of 1.6 mt/ha/year against 25%–30% runoff and 5–10 mt/ha/year soil loss from similar slopes under traditional farming systems. The average net returns from grass alone was higher than from field crops. Moreover, additional returns would be obtained from poplar wood as well.

xiii. *Acacia Species and Bhabhar Grass:* Several *Acacia* species were planted on degraded marginal lands (boundary soil with 30%–40% slope) at 3 × 3 m in 1976. *Bhabhar* grass was later introduced in 1982 at a spacing of 75 × 75 cm under these species. After 6 years of study, all these systems demonstrated their superiority over traditional rain-fed crops on poor quality soils not only in terms of higher financial returns but also in terms of better soil and water conservation. Singh *et al.*, 1990; *Grewal et al.*, 1996 reported that after 10 years *bhabhar* grass yield was highest under *Acacia nilotica*. They recommended that eroded slopes can be best managed under *Accacia nilotica* and *bhabhar* grass association. Based on the average yield of grass for 3 years, (Mathur and Joshi, 1975) studies showed that *Dalbergia sissoo* and *Chrysopogon fulvus* were the ideal combination on such degraded land.

xiv. *Agri-silvi-pastoral system:* This system is a slight modification of the silvi-pastoral system in which plantation of some agricultural crops such as groundnut, millets, maize, cucumber, pumpkins, tuber crops and so on came up well in sandy and well-drained soils as recommended by Yadava and Quli, 2007. While selecting the annual/seasonal crops, care should be taken to use the local varieties (as per the region-specific demand), which are suitable for growing on moisture and organic matter deficient soils. Specific rainwater harvesting structures such as staggered contour trenches, bench terraces, retaining walls, brushwood dams, and rock fill dams should be constructed, following standard engineering norms, at appropriate places with quick-growing plantation (shrubs, grasses, and trees) on the upper and lower ends. The entire slope should be revegetated with local variety of grasses by solid sodding, and some fast-growing trees like Subabul (*Leucaena leucocephala*), Khair (*Acacia catechu*), Babool (*Acacia nilotica*) and so on should be planted by broadcasting the seeds during premonsoon showers. The plantation of Agave (*Agave sislana*) along with Khas grass (*Vitveria zizanoides*), should be practiced on high risk zones while in areas with heavy grazing pressures, the nonpalatable plants like *Calotropis, Lantana camara, Ipomea cornea, Vitex negundu* and so on among shrubs and *Eucalyptus*, chakundi (*Senna siamia*), akashi (*Acacia auriculiformis*), and karanj (*Derris indica*) among trees should be planted. In addition, *Acacia nilotica* and *Dendrocalamus strictus* were planted on humps, slope sand gully bottoms of ravinous land to prevent soil erosion and runoff (Singh *et al.*, 2011).

xv. *Woody Hedgerows:* Fast-growing woody hedges and coppicing fodder shrubs and trees are planted in general for browse, mulch, green manure, soil conservation and so on. The woody hedgerows provide a semipermeable membrane to surface movement of water while mulch from the trees reduces the impact of raindrops on the soil and thus minimizes the splash and sheet erosion (Young, 1989). The main aim of this system is production of food/fodder/fuelwood and soil conservation. Contour tree-rows (hedgerows) helps to reduce runoff and soil loss by 40% and 48%, respectively. Thus, such tree-based systems were widely used to reduce soil and water losses in agroecosystems on steep slopes in the western Himalayas. Narain *et al.* (1997) and Kiepe (1995) noticed that around 98% of the soil was conserved with combinations of hedgerows plus mulch compared to control (cropped area). Plantation of hedgerows reduced the soil loss from 5.3 to 1.0 t/ha/year, while the runoff reduced from 12% to 2% in humid tropics having 16% slope and 2,200 mm annual rainfall at Yurimaguas, Peru (Alegre and Fernandes, 1991). In addition, the hedgerows also provide mechanical and biological barrier that helps to minimize soil erosion by reducing runoff (Singh *et al.*, 2000).

10.2.1.2   Second Tier/Intermediate Slopes

In the catchment area which are just above the agricultural lands, the second tier is the intermediate tier on the slopes. The medium upland situations and medium lowland topography which generally constitutes about 5%–8% of the watershed, have better moisture regime and comparatively better soil fertility in the watershed, therefore is more suitable for the cultivation of commercial crops such as vegetables, flowers, pulses, and oilseeds. These medium upland situations if utilized properly shall add significantly to the overall productivity of the system.

The slopes can be used for agriculture after undertaking following mechanical measures such as bench terracing and bunding (Figure 10.5). Stonewalls should be constructed across the hill slopes at predetermined spacing for developing land under cultivation. Diversion drains should be made to divert the runoff water away to protect the downstream area and discharge it safely into a protected waterway. These slopes can also be sustainably used for promoting agri-horticulture, silvi-pasture, and other alternative land-use system.

The terrace risers of agriculture field should be planted with suitable grasses as it helps in binding the soil, checking soil erosion and increasing crop productivity. Grass species such as *Cenchrus ciliaris*, Hybrid Napier, *Panicum maximum*, *Chrysopogon fulvus, Vetivera zizanoides, Eulaliopsis binnata* and so on have been found suitable in the Shivalik foothills and on lower hills. In a study conducted by Ghosh *et al.*, 2011, maize and wheat yield were found to increase by 23%–40% and 10%–20%, respectively when planted with vegetative grass barriers. In addition, grass yield of 6–17 q/ha/year was also observed. Vegetative barriers were also found to reduce the runoff and soil loss by 18%–21% and 23%–68%, respectively on slopes varying from 2%–8%. On steep slopes, trees can also be maintained in the form of hedges for checking soil loss. A study conducted at Dehradun indicated that sediment deposition along hedgerows during a period of 3 years and along trees in 9 years varied from 184 to 256 t/ha which is equivalent to 15–20 mm of soil depth. *Eucalyptus tereticornis* and *Eulaliopsis binata* raised in Shivalik foothills (at 2,500 trees/ha) in paired rows with understory grass planted at 50 × 50 cm spacing, allowed no soil loss with an annual return of about INR 4,000 ha/year from commercial grass alone besides additional returns from Eucalyptus and has proved to be more remunerative than traditional rain-fed crop.

Fruits like citrus, banana, mango, apple, walnut, plum, peach, cherry and so on can be grown on nonarable land with soil moisture conservation practices such as half-moon trenching, mulching and so on. On terrace risers, fruit trees can be grown in combination with vegetables (potato, pea, capsicum, cabbage, cauliflower etc.), spices (ginger, chilies, cardamom, saffron etc.), and flowers (orchids, gladiolus, marigold, chrysanthemum etc.). Runoff water should be utilized to irrigate the fruit

FIGURE 10.5   Terracing to check soil erosion.

trees. Each tree depending upon the species, age, slope, and intensity of rainfall should be provided with micro-catchment for water harvesting. If possible, drip irrigation system should be installed for efficient water use.

Community lands in watersheds should be used for promoting grasslands or silvo-pasture system using appropriate soil and water conservation measures. Light harrowing, contour trenching, and furrowing are very effective methods to enhance soil and moisture conservation and for the establishment of grasses. Small brushwood check dams should be constructed for gully plugging, stabilization, and improvement of gullied rangelands. Poor rangelands should be stocked with high forage yielding perennial grasses by reseeding or slip planting. A suitable tree-grass combination (the silvo-pastoral system) is often desirable, since they mimic natural forest cover to a large extent, besides serving as source material for firewood and fodder, which shall continue to be utilized by rural communities for the years to come. Trees having high fodder values such as *Acacia nilotica, Acacia catechu, Ailanthus excelsa, Anogeissus pendula, Albizzia lebbek, Bauhinia variegata, Butea monosperma, Cordia dichtoma, Celtis australis, Ficus glomerata, Grewia optiva, Kydia calycina, Lannea coromandelica,* and *Morus alba* should be preferred in such degraded areas. The plantation of a mixture of plant species is always better. Pollarding and lopping of the trees should be done during the lean period. The left out branches can be utilized as fuelwood. As community lands in the watershed belongs to all but responsibility of no one, participatory approach is better in selecting, planting, tending, and protecting the trees and grasses. Group of villages, whose livestock utilize the common grazing lands, should jointly form societies to carry out the work efficiently (Dhyani *et al.*, 2007).

i. *Horti-silvicultural system:* Horti-silviculture models are recommended for areas, located mostly on the moderate to lower slopes, with moderate to low erosion and having comparatively low moisture and nutrients (Yadava and Quli, 2007). Such a model is highly profitable in the ecological perspective. The spacing for this model must be kept slightly large (ranging from 5–10 m between the plants), as the horticultural trees have large crown size. The forest trees like Arjun (*Terminalia arjuna*), Karanj (*Derris indica*), Neem (*Azadirechta indica*), Bakain (*Melia azadirachta*), Kend (*Diospyros melanoxylon*), Semal (*Bombax ceiba*), Kusum (*Schleichera oleosa*), Moonga (*Moringa oleifera*), Ber (*Zizyphus mauritiana*), Cashew (*Anacardium occidentale*), Mahua (*Madhuca indica*) and so on are usually recommended. Alder-large cardamom agroforestry system is most prevalent in Sikkim under horti-silviculture and supports conservation of tree biodiversity in the region (Sharma *et al.*, 2000). In some cases, suitable grasses such as Dinanath grass (*Pennisetum pedicellatum*) or Napier grass (*Pennisetum purpureum*) or Anjan grass (*Cenchrus ciliaris*) are introduced in the interspaces of fruit trees under horti-silvi-pastoral system.

ii. *Agri-horticultural system:* In these systems, short-duration arable crops are raised in the interspaces of fruit trees. Although the woody components of the system are fruit trees, agricultural crops such as peas, cabbage, colocacia, turmeric, and pulses are generally grown in the inter-spaces of

Malus domestica (apple) *Prunus domestica* (plum), *P. armeniaca* (apricot) *P. persica* (peach), *P. deueis* (almond) and *Pyrus communis* (pear). Pulses are the important arable crops for this system. Trees are uniformly spaced in the field and average density of trees is generally 5/100 m². In northeast Indian state of Meghalaya, the guava and in Assam lemon-based agri-horticultural agroforestry systems gave 2.96 and 1.98-fold higher net returns respectively, in comparison to farmlands without trees which further helps in checking soil erosion (Bhatt and Mishra, 2003).

iii. *Agri-Horti-Silvicultural system:* Fuel, fodder, and timber trees are retained on the field bunds along with agri-horticultural crops and managed under agri-horti-silvicultural system. Besides providing fruits/fuelwood/fodder trees grown along bunds it also restricts erosion. Some suitable tree species for different agroforestry models are Subabul (*Leucaena* spp.), Casuarinas, *Ailanthus sp*; *Gmelina spp*; Silver Oak (*Grevillea robusta*), Mulberry (*Morus alba*), Kadam (*Anthocephalus cadamba*), *Melia spp*; *Acacia arabica*, *Acacia auriculiformis*, *Acacia mangium*; *Acacia lenticularis*, *Albizzia sp*; *Azadirachta indica*, *Borassus flabelliformis*, *Hovea brasiliensis*, *Prosopis spp*; *Butea monosperma*, *Cedrela toona*, *Tamarindus indica*, *Grewia oppositifolia* and so on and horticultural tree species like Mango, Guava, Coconut, Cashew nut, Citrus, Areca nut, *Artocarpus spp.*, *Zizyphus spp.*; Ashok, Gulmohar (*Delonix regia*), *Cassia fistula* and so on.

Growing legumes or other inter-crops (shade tolerant-ginger and turmeric) in the interspaces between fruit plantations has been an age old practice. Arable crops are grown in interspaces till the trees (fruits or MPT's) develop canopy or bear fruits or there is reduction in the crop yields. Ginger in interspaces of *Ailanthus triphysa* (2,500 trees/ha) helps in getting better rhizome development compared to solo cropping in southern India (Kumar *et al.*, 2001). In some cases, grass component is also mixed within the system thereby being called as Agri-horti-silvi-pastoral agroforestry system. Growing *Alnus, Ficus, Symingtonia, Thysanolanea maxima* (Broom grass) at top, Khasi mandarin (*Citrus reticulate*) at middle and maize, ginger, groundnut, pea and so on at lower slope portion under agri-horti-silvi-pastoral system resulted in higher available nutrients when compared to natural fallow and silvi-pastoral agroforestry system (Majumdar *et al.*, 2002). Thus agri-horti-silvi-pastoral and multi-storeyed systems, with proper soil conservation measures, manures, and fertilization schedule are the most suitable techniques for long term sustainable production in the North-East Himalayan region of India (Majumdar and Venkatesh, 2006).

iv. *Boundary planting:* It includes planting trees all along the boundaries of the fields, farm, or along the margins of footpaths, roads, and canals. It is a simple but effective practice, particularly for small farmers. It is also called four-sided forestry with the object of gaining production from trees whilst having no adverse effect on adjacent crops and possibly a beneficial effect through fertilization by trees or their leaf litter, protection from wind, or aiding soil conservation, that is, watershed protection. Farmers in different parts of the country use diverse kinds of trees for this purpose

(Kumar, 1994). Examples include: *Tectona grandis, Acacia mangium, Casuarina equisetifolia, Artocarpus spp.* and so on.

v. *Vegetative Barriers:* Vegetative barriers are cheap and effective as compared to mechanical measures in mild slopes, in controlling runoff, and soil loss (Narain *et al.*, 1997). Vegetative barrier of *Panicum maximum* has proved effective in controlling erosion in maize cropping system by reducing runoff by 28% and erosion by 48% at Dehradun (Bhardwaj and Khola, 1999). The barriers influenced redistribution of sediment in the catchment resulting in deposition of 30–60 t/ha/year sediments on contours and thus forming bunds/terraces naturally. Meanwhile, Bhardwaj (1999) and Sur and Sandhu (1994) compared different grasses for their efficacy to reduce runoff and soil loss. The soil loss was reduced from 45 to 6.12 t/ha with *Panicum maximum* vegetative barrier on 4% slope in the subhumid region at Dehradun in lower western Uttranchal Himalayas whereas *Saccharum munja* was found effective in the Shivalik's.

10.2.2   COMMAND AREA

These are the plains and the flat areas, where typically, the farmers operate and a large concentration of labour-intensive works are required. The main aim of this area is production of agriculture and livelihood security. The agronomical measures such as contour farming, intercropping, strip cropping, terracing, mixed cropping, land cover or canopy management, mulching, conservation tillage practices, and diversified cropping systems are generally recommended in this part of watershed for maximizing *in-situ* rainwater conservation. Water productivity on farmlands can be increased manifold (2–4 times) by adoption of suitable cropping systems, good variety of seeds, integrated nutrient management (INM), integrated pest management (IPM), weed management, *in situ* and inter-plot water harvesting and so on. Runoff water from stored tanks/ponds in submergence area should be utilized for irrigating the annual crops during moisture stress period.

i. *Agri-silviculture:* Under this system agricultural crops can be grown intensively under protective irrigated conditions while fodder crops, shade-loving crops, and shallow-rooted crops can also be grown economically. Wider spacing is adopted, without sacrificing tree population, for easy cultural operation and to get more sunlight to the intercrop. The performance of the tree crops is better in this system as compared to monoculture system. Microbial biomass carbon which was low in rice-berseem crop ($96.14\,\mathrm{gg^{-1}}$ soil) increased in soils under tree plantation ($109.12\,\mathrm{gg^{-1}}$ soil). Moreover, the soil carbon increased by 11%–52% due to integration of trees and crops leading to improvement of soil fertility of moderately alkaline soils of northern India (Kaur *et al.*, 2000). In agri-silviculture, the optimum density of trees should be 200 plants/ha to minimize the effect of shade and biochemical interactions on crop productivity (Bhatt, 2003). Poplar is a commonly found as intercrop, alongwith a variety of complimentary crops, including wheat, sugarcane, berseem clover, various pulses, and

TABLE 10.1

Average Runoff and Soil Loss under Different Land Use Systems

Treatment	Runoff (%)	Soil Loss (mt/ha)
Chrysopogon fulvus	12.7	8.65
Maize	27.5	28.27
Maize + *Leucaena*	21.4	17.83
Maize + Eucalyptus	20.8	13.51
Leucaena + grass	17.6	10.51
Leucaena	2.4	1.74
Eucalyptus + grass	6.3	3.52
Eucalyptus	2.1	1.20
Cultivated fallow	38.2	56.58

Source: Singh *et al.* (1990).

trees such as teak and mango. Flood irrigation is commonly used to apply water to both crops and trees. *Leucaena leucocephala* when intercropped with agricultural crops and fodder grass, increased the total yield of food grain, fodder, and fuel (Tiwari, 1970). In another study at Dehradun, (Singh *et al.*, 1990) it was observed that incorporation of *Eucalyptus hybrid* and *Leucaena leucocephala* in the agri-system resulted in considerable reduction in runoff and soil loss even though this led to reduction in the yields of wheat and maize crops (Table 10.1). Association of grass with Eucalyptus gave lesser runoff and soil loss as compared to *Leucaena* and grass.

ii. *Horticulture-based Agroforestry Systems:* Dyal *et al.* (1996) have developed a horticulture-based agro forestry systems for small farmers by integrating plants yielding a number of products having synergistic effects on each other. *Subabul* (*Leucaena leucocephala* Lam. De Wit 'K8") was raised for fuel, fodder, and live hedge on the border row, followed by lemon (*Citrus aurantifolia*) "*Baramasi*" on the peripheral row, papaya (*Carica papaya* Linn.) and okra (*Hibiscus esculentus* Linn.) as understory crops in the main plot. During first 5 years (1985–1989) the system was well spread out and provided handsome net returns (INR 15,000–21,000/ha/year) with limited irrigation and INR 10,000–18,000/ha/year for 3 years (1990–1992) under rainfed conditions, compared with INR 4,900–9,800/ha/year with limited irrigation and INR 1,740–4,600/ha/year under rainfed conditions, from agricultural crops (maize-wheat). In the Doon valley of Uttar Pradesh, Singh *et al.* (1990) reported that both *rabi* and *kharif* crops can be grown with peach. Based on 5-year data, they reported that turmeric and ginger perform particularly well.

iii. *Alley farming:* Alley farming has a potential on hills where cultivation is carried on slopes and when hedges are planted on contours. In addition, it can be a successful soil erosion control measure. Studies conducted (Kidd and Pimental, 1992) by alternating three rows of *Leucaena leucocephala* with three rows of maize perpendicular to the slope of 10%–15% revealed that the system can control erosion if all the maize residues and the pruning weighing 2,000 and 2,500 kg, respectively, are left on the surface of land. Erosion rate under these conditions with 1,000 mm of rainfall would be about 1 ton/ha/year (Kidd and Pimental, 1992). A comparable study with maize-*Leucaena* intercrop on steep slope of 44% produced similar results. Plot protected by *Leucaena* hedgerows lost 2-ton/ha/yr topsoil compared with 80 ton/ha/yr for an unprotected plot (Banda *et al.*, 1994). Prabhakar *et al.* (2010) reported that relatively high awareness (87% and above) is noticed in agro-forestry, social forestry, and alley cropping, whereas more than 40% of farmers fully adopted knowledge of farm forestry (in fallow lands), agro-forestry, social forestry, alley cropping, silvi-pasture system silvi-horti-system, and agri-horti-system for their multipurpose uses such as soil and water conservation, crop production etc.

iv. *Multipurpose trees on the cropland:* It is a common practice in the western Himalayan region to cultivate wheat, peas, potato, cauliflower, mustard and so on during winter and maize, tomato, chillies and so on during the summer season, either in monoculture or mixed type, on the permanent terraces prepared across the hill slopes. Fodder, fuel, and timber trees, viz, *Grewia optiva* (bheemal), *Celtis australis* (khirak), *Bauhinia variegata* (kachnar), *Albizzia chinensis* (ohi), *Toona ciliata* (toon), *Morus alba* (toot), *Ulmusla viegata* (meryano) and so on are deliberately left or grown on the bunds of terraces. Agri-silviculture system is prevalent in sub-montane and mid hills, and sub-humid zones of Himachal Pradesh. The MPTs generally follow a random or irregular geometry, except in the case of plantations and orchards, and serve multiple objectives such as providing green manure, fodder, fuel, fruits, nuts, small timber and so on. Competition, however, is a major constrain in this respect and roots of many species grown on the same land management unit, frequently get intermingled and often this overlap of the roots can be extensive, especially in older stands (Jamaludheen *et al.*, 1996; Divakara *et al.*, 2001).

10.2.3 Submergence Area

The lower part of watershed can be utilized for development of small water harvesting structures such as low-cost farm ponds, check dams, percolation tanks, and groundwater recharge measures which can be used for collecting and storing rainfall, runoff, and subsurface flow, to meet increased water demand and overcome moisture stress conditions and to provide ecosystem services like groundwater recharge. The harvested water can be used for drinking, irrigation, livestock, fisheries, groundwater recharge and so on to optimize water productivity. The runoff water could be harvested and stored in suitable storage structures for supplemental irrigations during moisture stress periods for the crops. Samra and Narain (1999) reported that the yield increases up to 200% even with one supplemental irrigation of 5 cm.

Aqua forestry: Various trees and shrubs preferred by fish are planted on the boundary and around fish-ponds in this system (Fanish and Priya, 2013; Verma *et al.*, 2017). Tree leaves are used as forage for fish. The main or primary role of this system is fish production and bund stabilization around fishponds. Planting trees such as *Dalbergia sissoo, Pongamia pinnata, Tectona grandis, Acacia lenticularis, Bombax ceiba* and *Wendlandia exserta* around fishponds augment the fish diet. The trees are managed by the farmers to produce fuel, fodder, timber, and soil conservation (Chaturvedi and Das, 2007). Subhabrata *et al.* (2017) reported that aqua-forestry, an innovation of agroforestry practice, would prove to be a timely accepted best agro-technique for the restoration of the Sundarban Biosphere Reserve. Fish production and bund stabilization around fishponds are the main or primary functions of aquaforestry. In such systems, different trees such as *Moringa oleifera* and variety of shrubs preferred by fish are planted on the boundary and around fishponds (Hemrom and Nema, 2015).

REFERENCES

Alao, J.S. and Shuaibu, R.B., 2013. Agroforestry practices and concepts in sustainable land use systems in Nigeria. *Journal of Horticulture and Forestry*, 5(10), pp. 156–159.

Alegre, J.C. and Fernandes, E.C.M., 1991. Runoff and erosion losses under forest, low input agriculture and alley cropping on slopes (pp. 227–228). TropSoils technical report, 1988–1989.

Angima, S.D., Stott, D.E., O'neill, M.K., Ong, C.K. and Weesies, G.A., 2003. Soil erosion prediction using RUSLE for central Kenyan highland conditions. *Agriculture, Ecosystems & Environment*, 97(1–3), pp. 295–308.

Anonymous. 2004. *Sustainable Farming Systems in Upland Areas*. Asian Productivity Organization, Tokyo, Japan, p. 175.

Banda, A.Z., Maghembe, J.A., Ngugi, D.N. and Chome, V.A., 1994. Effect of intercropping maize and closely spacedLeucaena hedgerows on soil conservation and maize yield on a steep slope at Ntcheu, Malawi. *Agroforestry Systems*, 27(1), pp. 17–22.

Barungi, M. and Maonga, B.B., 2011. Adoption of soil management technologies by smallholder farmers in central and southern Malawi. *Journal of Sustainable Development in Africa*, 13(3), pp. 28–38.

Bhan, S., 2013. Land degradation and integrated watershed management in India. *International Soil and Water Conservation Research*, 1(1), pp. 49–57.

Bhardwaj, S.P., 1999. Vegetative barrier in an effective economic and eco-friendly measure for erosion control on agricultural lands. In: *Abstracts of 8th ISCO Conference, New Delhi*, pp. 203–206.

Bhardwaj, S.P. and Khola, O.P.S., 1999. Biological and mechanical measures for erosion control and crop production on 4 per cent slope. *Annual Report, CSWCRTI, Dehradun*, p. 22.

Bhatt, B.P., 2003. Agroforestry for sustainable mountain development in N.E.H. region. In: Rawat, M.S.S. (ed.), *Central Himalaya: Environment and Development (Potentials, Actions and Challenges)*. Published by Transmedia Publisher, Srinagar Garhwal, Uttarakhand, India, pp. 206–223.

Bhatt, B.P. and Misra, L.K., 2003. Production potential and cost-benefit analysis of agrihorticulture agroforestry systems in Northeast India. *Journal of Sustainable Agriculture*, 22(2), pp. 99–108.

Bipin, B., Rashmi, A., Singh, A.K. and Banerjee, S.K., 2000. Vegetation development in a degraded area under bamboo based agro-forestry system. *Indian Forester, 126*(7), pp. 710–720.

Biswas, A.K., 2008. Integrated water resources management: is it working? *International Journal of Water Resources Development, 24*(1), pp. 5–22.

Camacho, L.D., Gevana, D.T., Carandang, A.P. and Camacho, S.C., 2016. Indigenous knowledge and practices for the sustainable management of Ifugao forests in Cordillera, Philippines. *International Journal of Biodiversity Science, Ecosystem Services & Management, 12*(1–2), pp. 5–13.

Chaturvedi, O.P. and Das, D.K., 2007. Agroforestry systems and practices prevailing in Bihar. *Agroforestry Systems and Practices.* New India Publishing Agency, New Delhi, pp. 277–304.

Clason, T.R. and Sharrow, S.H., 2000. Silvopastoral practices. *North American Agroforestry: An Integrated Science and Practice.* American Society of Agronomy, Madison, WI, pp. 119–147.

Dagar, J.C. and Gurbachan, S., 1993. Afforestation and agroforestry for salt affected soils. *Indian Review of Life Sciences, 13*, pp. 215–240.

Dalvi, V.B., Nagdeve, M.B. and Sethi, L.N., 2010. Rain water harvesting to develop non-arable lands using Continuous Contour Trench (CCT). *Assam University Journal of Science and Technology, 4*(2), pp. 54–57.

Dhyani, S.K., Samra, J.S. and Handa, A.K., 2007. Forestry to support increased agricultural production: focus on employment generation and rural development. *Agricultural Economics Research Review, 20*(2), pp. 179–202.

Divakara, B.N., Kumar, B.M., Balachandran, P.V. and Kamalam, N.V., 2001. Bamboo hedge-row systems in Kerala, India: root distribution and competition with trees for phosphorus. *Agroforestry Systems, 51*(3), pp. 189–200.

Dwivedi, A.P., 1992. *Agroforestry: Principles and Practices.* Oxford & Ibh Publishing Company, New Delhi, p. 365.

Dyal, S.K.N., Grewal, S. S. and Singh, S.C., 1996. An agri-silvi-horticultural system to optimize production and cash returns for Shivalik Foothills. *Indian Journal of Soil Conservation, 24*(2), pp. 150–155.

Fanish, S.A. and Priya, R.S., 2013. Review on benefits of agro forestry system. *International Journal of Education and Research, 1*(1), pp. 1–12.

FSI. 2015. State of Forest Report-2015. Forest Survey of India (Ministry of Environment & Forests), Kaulagarh Road, P.O-IPE, Dehradun -248195.

Ghosh, B.N., Sharma, N.K. and Dadhwal, K.S., 2011. Integrated nutrient management and cropping systems impact on yield, water productivity and net return in valley soils of north-west Himalayas. *Indian Journal of Soil Conservation, 39*(3), pp. 236–242.

Grewal, S.S., 2000. Agroforestry systems for soil and water conservation in Shiwaliks. In: Gill, A.S., Deb Roy, R. and Bisaria, A.K. (eds.), *Agroforestry in 2000 AD for the Semi Arid and arid Tropics.* NRCAF, Jhansi, Uttar Pradesh, India, pp.82–85.

Grewal, S.S., Kehar, S., Juneja, M.L. and Singh, S.C., 1996. Relative growth, fuel-wood yield, litter accumulation and conservation potential of seven Acacia species and an under-storey grass on a sloping bouldery soil. *Indian Journal of Forestry, 19*(2), pp. 174–182.

Hemrom, A. and Nema, S., 2015. A study on traditional agroforestry practices existingat Bastar region of Chhattisgarh. *International Journal of Multidisciplinary Research and Development, 2*(3), pp. 56–64.

Islam, M., Salim, S.H., Kawsar, M.H. and Rahman, M., 2016. The effect of soil moisture content and forest canopy openness on the regeneration of Dipterocarpus turbinatus CF Gaertn. (Dipterocarpaceae) in a protected forest area of Bangladesh. *Tropical Ecology, 57*(3), pp. 455–464.

Jamaludheen, V., Kumar, B.M., Wahid, P.A. and Kamalam, N.V., 1996. Root distribution pattern of the wild jack tree (Artocarpus hirsutus Lamk.) as studied by 32P soil injection method. *Agroforestry Systems*, *35*(3), pp. 329–336.

Joshi, P.K., Pangare, V., Shiferaw, B., Wani, S.P., Bouma, J. and Scott, C. 2004. Socioeconomic and policy research on watershed management in India: Synthesis of past experiences and needs for future research. Global Theme on Agroecosystems Report no. 7. Patancheru 502 324, Andhra Pradesh, India: International Crops Research Institute for the Semi-Arid Tropics. p. 88.

Kalra, B.S. and Mishra, A.K., 2006. Watershed development programmes in India and institutional imperatives. *Journal of Indian School of Political Economy*, *18*(4), pp. 649–668.

Kaur, B., Gupta, S.R. and Singh, G., 2000. Soil carbon, microbial activity and nitrogen availability in agroforestry systems on moderately alkaline soils in northern India. *Applied Soil Ecology*, *15*(3), pp. 283–294.

Kidd, C. and Pimental, D., 1992. *Integrated Resources Management: Agroforestry for Development*. Academic Press, San Diego, CA.

Kiepe, P., 1995. No Runoff, No Soil Loss: Soil and water conservation in hedgerow barrier systems. Doctoral thesis, Agricultural University, Wageningen, Netherlands. Also published in the series Tropical Resource Management Papers. p. 156.

Kumar, B.M., 1994. Agroforestry principles and practices. In: Thampan, P.K. (ed.), *Trees and Tree Farming*. Peekay Tree Research Foundation, Cochin, Kerala, India, pp. 25–64.

Kumar, B.M., Thomas, J. and Fisher, R.F., 2001. Ailanthus triphysa at different density and fertiliser levels in Kerala, India: tree growth, light transmittance and understorey ginger yield. *Agroforestry Systems*, *52*(2), pp. 133–144.

Lenka, N.K., Lenka, S. and Biswas, A.K., 2015. Scientific endeavours for natural resource management in India. *Current Science*, *108*, pp. 39–44.

Limnirankul, B., Onprapai, T. and Gypmantasiri, P., 2015. Building local capacities in natural resources management for food security in the highlands of northern Thailand. *Agriculture and Agricultural Science Procedia*, *5*, pp. 30–37.

Madhurima, C. and Banerjee, A., 2013. Forest degradation and livelihood of local communities in India: a human rights approach. *Journal of Horticulture and Forestry*, *5*(8), pp. 122–129.

Mahecha, L., Murgueitio, M.M., Angulo, J., Olivera, M., Zapata, A., Cuartas, C.A., Naranjo, J.F. and Murgueitio, E., 2011. Desempeño animal y características de la canal de dos grupos raciales de bovinos doble propósito pastoreando en sistemas silvopastoriles intensivos. *Revista Colombiana de Ciencias Pecuarias*, *24*(3), p. 470.

Majumdar, B. and Venkatesh, M.S., 2006. Soil fertility build up through agroforestry practices: a case study. *Agroforestry in north east India: opportunities and challenges*. *ICAR Research Complex for NEH Region, Umiam, Meghalaya, India*, pp. 507–514.

Majumdar, B., Venkatesh, M.S., Satapathy, K.K., Kumar, K. and Patiram, P., 2002. Effect of alternative farming systems to shifting cultivation on soil fertility. *The Indian Journal of Agricultural Sciences*, *72*(2), pp. 122–124.

Mathur, R.S. and Joshi, P.K., 1975. Economic utilization of class V and VI lands for fuel and fodder plantations. Annual Report. CSWCRTI, Dehradun, India, pp. 37–39.

Ministry of Agriculture. 1996. *National Workshop on Himalayan Farming Systems*. Government of India, New Delhi.

Mishra, P.K. and Rai, S.C., 2013. Use of indigenous soil and water conservation practices among farmer's in Sikkim Himalaya. *Indian Journal of Traditional Knowledge*, *12*(3), pp. 454–464.

Murgueitio, E., Chará, J.D., Solarte, A.J., Uribe, F., Zapata, C. and Rivera, J.E., 2013. Agroforestería Pecuaria y Sistemas Silvopastoriles Intensivos (SSPi) para la adaptación ganadera al cambio climático con sostenibilidad. *Revista Colombiana de Ciencias Pecuarias*, *26*, pp. 313–316.

Nair, P.K.R., 1984. Tropical agroforestry systems and practices. In: Furtado, J.I. and Ruddle, K. (Eds.), *Tropical Resource Ecology and Development*. John Wiley, Chichester, England, p. 39.

Nair, P.K.R., 1989. The role of trees in soil productivity and protection. In: Nair, P.K.R. (ed.), *Agroforestry Systems in the Tropics*. Kluver Academic Publishers, Dordrecht, Netherlands/Boston, MA/London, England, p. 664.

Nair, P.K.R., 1993. *An Introduction to Agroforestry*. Kluwer Academic Publishers, Dordrecht, Netherlands, p. 491.

Narain, P., Singh, R.K., Sindhwal, N.S. and Joshie, P., 1997. Agroforestry for soil and water conservation in the western Himalayan Valley Region of India 1. Runoff, soil and nutrient losses. *Agroforestry Systems*, 39(2), pp. 175–189.

Nayak, B.K., 1996. Studies on biomass productivity and nutrient content in *Eucalyptus terreticornis*, Leucaena leucocephala and Melia azadirachta under high density plantation. MSc. Thesis, Deptt. Of Silviculture and Agroforestry, Dr. Y S Parmar University of Horticulture and Forestry, Nauni, Solan (Himachal Pradesh), India, p. 112.

Ong, C.K. and Swallow, B.M., 2003. Water productivity in forestry and agroforestry. In: Kijne, J.W., Barker, R. and Molden, D. (eds.), *Water Productivity in Agriculture: Limits and Opportunities for Improvement*. Comprehensive Assessment of Water Management in Agriculture Series 1. CAB International, Wallingford; UK in association with International Water Management Institute, Colombo, pp. 217–228.

Palanisamia, K. and Kumar, D.S., 2009. Impacts of watershed development programmes: experiences and evidences from Tamil Nadu. *Agricultural Economics Research Review*, 22, pp. 387–396.

Panda, S.C., 2013. *Cropping and Farming Systems*. Agrobios (India), Jodhpur, Rajasthan, India, p. 413.

Pandey, D. 2002. *Fuelwood Studies in India: Myth and Reality*. Centre for International Forestry Research, Jakarta, Indonesia, p. 94.

Pandey, D.N., 2007. Multifunctional agroforestry systems in India. *Current Science*, 92(4), pp. 455–463.

Patil, V.D., Sarnikar, P.N., Adsul, P.B. and Thengal, P.D., 2004. Profile studies, organic matter build-up and nutritional status of soil under bamboo (Dendrocalamus strictus) based agroforestry system. *Journal of Soils and Crops*, 14(1), pp. 31–35.

Prabhakar, K., Latha, K.L. and Rao, A.P., 2010. Watershed programme: adoption of knowledge in farming by farmers. *International NGO Journal*, 5(5), pp. 101–108.

Puri, S., 2007. *Agroforestry: Systems and Practices*. New India Publishing, New Delhi, p. 657.

Quli, S.M.S. and Siddiqui, M.H., 1996. Restoration of economic and environmental viability in rural areas of Chota Nagpur through Silvi-pastoral techniques. In: Abbasi, S.A. and Abbasi, N. (eds.), *Alternative Energy Development and Management*. International Book Distributors, Dehradun, Uttarakhand, India. pp: 201–212.

Raj, A. and Jhariya, M.K. 2016. Joint forest management: a program to conserve forest and environment. *Van Sangyan,* 3(6), pp. 38–42.

Samra, J.S. and Narain, P., 1999. Soil and water conservation. In: *Fifty Years of Natural Resource Management Research*. ICAR, New Delhi, pp. 145–176.

Sanches, P. A. 1987. Soil productivity and sustainable in agroforestry. In *Agroforestry: A decade of development* (Stepiler, H.A. and Nair, P.K.R. Eds.), Int. Conf. For Res. Narobi, Kenya, pp. 205–223.

Sarvade, S., 2014. Agroforestry: refuge for biodiversity conservation. *International Journal of Innovative Research in Science & Engineering*, 2(5), pp. 424–429.

Sarvade, S., Gautam, D.S., Kathal, D. and Tiwari, P., 2017. Waterlogged wasteland treatment through agro-forestry: a review. *Journal of Applied and Natural Science*, 9(1), pp. 44–50.

Sarvade, S., Singh, R., Ghumare, V., Kachawaya, D.S. and Khachi, B., 2014a. Agroforestry: an approach for food security. *Indian Journal of. Ecology*, *41*(1), pp. 95–98.

Sarvade, S., Singh, R., Prasad, H. and Prasad, D., 2014b. Agroforestry practices for improving soil nutrient status. *Popular Kheti*, *2*(1), pp. 60–64.

Sharda, V.N., Sikka, A.K. and Juyal, G.P., 2006. *Participatory integrated watershed management: a field manual*. Central Soil & Water Conservation Research & Training Institute, Dehradun, Uttarakhand, India, p. 366.

Sharma, A.K., 1989. Studies on growth performance and site amelioration under high density plantation. MSc. Thesis, Deptt. Of Silviculture and Agroforestry, Dr. Y. S. Parmar University of Horticulture and Forestry, Nauni, Solan (Himachal Pradesh), India.

Sharma, E., Sharma, R., Singh, K.K. and Sharma, G., 2000. A boon for mountain populations: large cardamom farming in the Sikkim Himalaya. *Mountain Research and Development*, *20*(2), pp. 108–111.

Shukla, H., 2015. Watershed management: its role in environmental planning and management. *IOSR Journal of Environmental Science, Toxicology and Food Technology*, *1*(5), pp. 8–11.

Singh, G., Arora, Y.K., Narain, P., and Grewal, S.S., 1990. *Agroforestry Research in India and Other Countries*. Surya Publications, Dehradun, Uttarakhand, India, p. 189.

Singh, K.A., Satapathy, K.K., Ramchandra, Dutta, K.K. and Singh, J.L., April-June 2000. Bio-terracing in the North Eastern Hills. *ICAR News Letter*, pp. 6–7.

Singh, R.A., Singh, M.K. and Sharma, V.K., 2011. Intervention of forestry and horticulture in watershed management under different climatic conditions. *Asian Journal of Horticulture*, *6*(2), pp. 528–530.

Sreedevi, T.K., Wani, S.P. and Pathak, P., 2007. Harnessing gender power and collective action through integrated watershed management for minimizing land degradation and sustainable development. *Financing Agriculture*, *39*(6), pp. 23–32.

Subhabrata, P., Pratap, K.D., Smritikana, S., Nitai, C.D. and Gokul, B.R., 2017. Inventory studies on various livelihood activities along with agroforestry techniques in the adjoining land of fringe areas for restoration of the Sunderbans under humid climate in India. *Agricultural Research & Technology: Open Access Journal*, *3*(3), pp. 1–3.

Sucharita, S., 2008. Watershed development programmes and rural development: a review of Indian policies. In: Lahiri-Dutt, K. and Wasson, R.J. (eds.), *Water First: Issues and Challenges for Nations and Communities in South Asia*. Sage Publications, New Delhi.

Sur, H.S. and Sandhu, I.S., December 1994. Effect on different grass barriers on runoff, sediment loss and biomass production in foothills of Shiwaliks. *Proceedings of the Abstracts of the 8th International Soil Conservation Organization Conference ISCO Conference, New Delhi, India*, pp. 4–8.

Tideman, E.M., 2000. *Watershed Management: Guidelines for Indian Conditions*. Omega Scientific, New Delhi.

Tiwari, K.M., 1970. Interim results of intercropping of miscellaneous tree species with main crop of tungya plantations to increase the productivity. *Indian Forester*, *96*(9), pp. 650–653.

Vanlauwe, B., Van-Asten, P., and Blomme, G., 2013. *Agro-ecological Intensification of Agricultural Systems in the African Highlands*. Routledge, London, p. 336.

Varis, O., Enckell, K. and Keskinen, M., 2014. Integrated water resources management: horizontal and vertical explorations and the 'water in all policies' approach. *International Journal of Water Resources Development*, *30*(3), pp. 433–444.

Verma, K.S., Mishra, V.K., Kaushal, R. and Sharma, J.K., 2003. Temporal change in the soil properties under high density energy plantation of *Acacia mollissima* and *Eucalyptus hybrid*. In: Verma, K.S., Khuranrand, D.K. and Christerson, L. (eds.), *Short Rotation Forestry for Industrial and Rural Development*. ISTS, Nauni, Solan, India, pp. 99–106.

Verma, P., Bijalwan, A., Dobriyal, M.J., Swamy, S.L. and Thakur, T.K., 2017. A paradigm shift in agroforestry practices in Uttar Pradesh. *Current Science, 112*(3), pp. 509–516.

Wani, S.P., Sreedevi, T.K., Reddy, T.V., Venkateswarlu, B. and Prasad, C.S., 2008. Community watersheds for improved livelihoods through consortium approach in drought prone rain-fed areas. *Journal of Hydrological Research and Development, 23*(1), pp. 55–77.

Yadava, M.S. and Quli, S.M.S., 2007. Agroforestry systems and practices in Jharkhand. In: Puri, S. and Panwar, P. (eds.), *Agroforestry- Systems and Practices.* New India Publishing Agency, New Delhi, pp. 305–318.

Young, A., 1989. *Agroforestry for Soil Conservation.* International Council for Research in Agroforestry. CAB International, Wallingford, Oxford, UK, p. 267.

11 Infiltration Studies of Major Soils under Selected Land Use Practices in Ranikhola Watershed of Sikkim, India

G. T. Patle and D. Jhajharia
Central Agricultural University

Sapam Raju Singh
Indian Institute of Technology

CONTENTS

11.1 INTRODUCTION

Infiltration is the process in which water enters the soil from rainfall, irrigation, and snowmelt and it is useful to estimate the surface runoff, groundwater recharge, evapotranspiration, soil erosion, and transport of chemicals in a surface or subsurface flow (Ahaneku, 2011). The rate at which water infiltrates into the soil is known as the infiltration rate whereas the cumulative infiltration is the total depth of infiltrated water at a time. Infiltration rate varies from soil to soil as well as varies according to the type of land use and land cover. Further, it is influenced by variation in the soil organic matter, porosity, bulk density, specific gravity, unsaturated or saturated hydraulic conductivity, soil water retention characteristics, and the antecedent soil moisture conditions of various layers of the soil profile (Stolte, 2003; Shougrakpam et al., 2010; Ayu et al., 2013; Osuji et al., 2010).

In general, the rate of infiltration is higher for the soils which are under annual vegetation than perennial vegetation (Yusof et al., 2005; Uloma et al., 2013). The process of soil compaction decreases with the rate of infiltration (Gregory et al., 2006). Moreover, it is also influenced by the seasons, that is, the rate of infiltration is higher in summer than in winter (Diamond and Shanley, 1998; Cerdà, 1996). Infiltration rate also decreases with an increase in elapsed time mainly due to increased resistance force developed by the incoming water, and hence, a constant rate of infiltration called basic infiltration rate is attained after a certain time (Mohan and Sangeeta, 2005). Although initially infiltration rate is high, it gradually decreases with time as the soil reaches the saturation point (Dibal et al., 2013). Surface runoff is the excess rainfall from infiltration rate. Various models for analyzing overland flow are based on infiltration rate, and studies have reported that overland flow mass balances are consistently modeled by considering variable infiltration rates corresponding to the depression storage (Rossi and Ares, 2012). Surface runoff is an important parameter for determining the amount of soil loss due to soil erosion of a watershed. A high infiltration rate over a region induces less runoff and sedimentation (Joshi and Tambe, 2010). Many studies have reported that infiltration is of prime importance to irrigation engineering as it influences the application rate of irrigation (Michael, 2005). Considering the importance of infiltration parameter in hilly watershed, basic infiltration rates of different soil under different soil conditions, namely, agriculture and forest land use conditions using double-ring infiltrometer and measured infiltration values from the field studies are compared with calculated values using commonly available infiltration models, viz., Horton's, Philip's, Green-Ampt, and Kostiokov infiltration models (Horton, 1941; Jejurkar and Rajurkar, 2012; Jagdale and Nimbalkar, 2012; Pingping et al., 2013).

11.2 MATERIALS AND METHODS

11.2.1 STUDY AREA: RANIKHOLA WATERSHED

The study area comprises the Ranikhola watershed located in the east district of Sikkim state. The Ranikhola watershed is situated between 27°13' and 27°24' N latitude and 88°29'–88°43' E having an area of 254.5 km² (Figure 11.1). The major

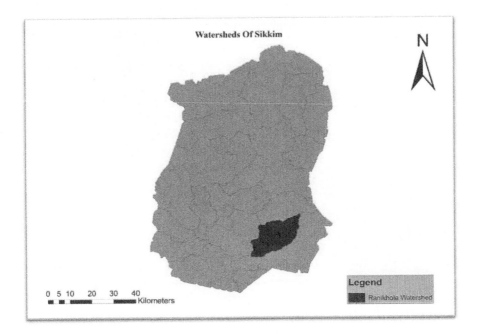

FIGURE 11.1   Study area.

drainage in the study area is provided through the Ranikhola river. The district is a part of the eastern Himalayan region as per the classification of an agroclimatic zone (planning commission). The average annual rainfall of the district varies from 2,525 mm consisting of 135 rainy days in a year. The district receives a major portion of rainfall from the south-west monsoon which contributes 61% of rainfall in the district. As per the land use statistics, the district consists of 10,500 ha area under cultivation, 9,112 ha area under forest, and 3,277 ha area under nonagricultural use. The watershed consists of four major types of soil, namely, fine loamy, coarse loamy, loamy skeletal and fine silty under agriculture, forest, and nonagricultural land use (based on NBSLUP). In this study, only agriculture and forest land use were considered to estimate the infiltration rates.

11.2.2   Study Area: Soil Types of Ranikhola Watershed

The study area consists of four major soil types, viz., coarse loamy, fine loamy, fine silty, and loamy skeletal (Figure 11.2) based on land use and land cover department, Government of India. The soil map was obtained from the Department of Forestry, Government of Sikkim and was digitized to generate the soil map of the Ranikhola watershed. The best-fitted infiltration model was studied for eight infiltration stations under two different land use scenarios and land cover for the above soil types. The considered land use and land cover in this study were agriculture and forest land.

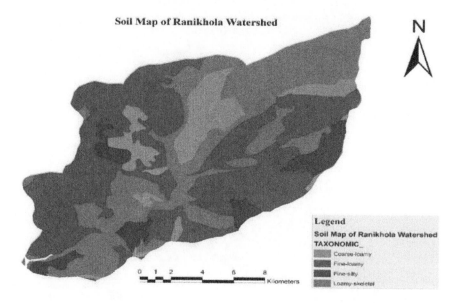

FIGURE 11.2   Soil map of the Ranikhola watershed in the east district of Sikkim.

11.3   METHODOLOGY

11.3.1   DELINEATION OF WATERSHED

The watershed delineation of Sikkim state was done using ArcSwat (version 9.3) from ASTER DEM data which was collected from Land Processes Distributed Active Archive Center-USGS. From this developed macro watershed, a micro watershed, that is, Ranikhola were extracted for this study. The soil map was collected from Department of Forestry, Government of Sikkim and was digitized to find out the soil nature of different places within the watershed.

11.3.2   MEASUREMENT OF INFILTRATION

In this study, the rate of infiltration was measured using double-ring infiltrometer which consists of two concentric rings. The diameter of inner and outer ring were 25 and 35 cm, respectively. Both rings had equal depth of 25 cm each. The longitude, latitude, and altitude of the station were measured using Global Positioning System (GPS-Garmin, model eTrex) to find the locational details of each station. The rate of fall of water was measured using point or hook gauge fixed over the gauging stand over a certain time interval until a basic infiltration rate was observed.

11.3.3   INFILTRATION MODELS

Various infiltration models were used to estimate the infiltration rate of different types of soils under different land use practices. Various parameters of the models were determined graphically by plotting arithmetic graph between observed infiltration data (i.e., infiltration rate/cumulative infiltration) with elapsed time (t).

11.3.3.1 Horton's Infiltration Model

The three parameters model was presented by Horton (1940). The equation presented by Horton to estimate the infiltration rate is in the form of equation 11.1.

$$F_t = f_c + (f_0 - f_c)^{e^{-kt}}$$

(11.1)

Where,

f_t = Infiltration rate at any time t (cm/h),
f_c = Basic/final infiltration rate (cm/h),
f_o = Initial infiltration rate (cm/h), and
k = Decay constant (1/h).

11.3.3.2 Philip's Infiltration Model

The Philip's Two-Term model (1957) relates the accumulated depth of infiltration with time as given in equation 11.2.

$$f_t = \frac{1}{2}st^{-1/2} + K$$

(11.2)

where,

f_t = Infiltration rate at time t (cm/h),
s = Sorptivity, function of soil suction potential,
K = Darcy's hydraulic conductivity (cm/h), and
t = Time (h).

11.3.3.3 Kostiokov Infiltration Model

This model correlates with the accumulated depth of infiltration with time. Mathematically, it is written as equation 11.3.

$$F_t = at^b$$

(11.3)

Where,

F_t = Accumulated depth of infiltration at time t (cm), and
t = Time (h).
a and b are constants which depend on soil and initial soil condition.

11.3.3.4 Green-Ampt Infiltration Model

This model is based on Darcy's law and the equation correlates the infiltration rate with accumulative infiltration depth (1911). Mathematically, it is represented as equation 11.4.

$$f_t = m + (n/F_t)$$

(11.4)

Where,

f_t = Infiltration rate at any time t (cm/h),
F_t = Accumulated depth of infiltration at time t (cm), and
m and n are constant.

11.4 RESULTS AND DISCUSSION

Infiltration rates were measured at various locations situated in the Ranikhola watershed of east district of Sikkim state using double-ring infiltrometer under two different land use conditions for four different types of soil. The longitude, latitude, and altitude of every station were measured using Global Positioning System (GPS-Garmin, model eTrex). Infiltrometer test data from 25 stations were used to develop infiltration map whereas the data from 8 stations were analyzed for estimating the best-suited infiltration model. Table 11.1 shows location of 25 stations. Bulk density varies from 1.1 to 1.9 g/cm³. From Table 11.1, it was also observed that there is large variation in the basic infiltration rates within the watershed and it varies from 0.20 to 45 cm/h. The lowest basic infiltration rate (0.20 cm/h) was observed at station 4 and the highest (45 cm/h) at station 21. It was observed that basic infiltration increased with increase in altitude. Rate of basic infiltration was found more in the forest land followed by agricultural land.

TABLE 11.1

Measured Infiltration Rate and Bulk Density in the Study Area

Location	Longitude (°E)	Latitude (°N)	Altitude (Above MSL), m	Bulk Density (g/cm³)	Basic Infiltration Rate (cm/h)
Station 1	88.5916	27.2879	876	1.9	8.96
Station 2	88.5916	27.2879	877	1.8	1.64
Station 3	88.5934	27.2901	888	1.8	1.4
Station 4	88.5928	27.2901	891	1.9	0.2
Station 5	88.5873	27.2778	931	1.6	0.28
Station 6	88.5872	27.2777	912	1.6	20.4
Station 7	88.6055	27.2798	1083	1.6	6
Station 8	88.6055	27.2800	1059	1.6	6.06
Station 9	88.5616	27.3175	1292	1.4	5.25
Station 10	88.5642	27.2943	1591	1.5	6
Station 11	88.5957	27.2764	795	1.6	6.1
Station 12	88.5957	27.2763	796	1.6	6.1
Station 13	88.5309	27.2543	576	1.2	30.06
Station 14	88.5282	27.2424	643	1.7	4.3
Station 15	88.5888	27.2589	1134	1.2	18
Station 16	88.5797	27.2533	1108	1.1	23.90
Station 17	88.6234	27.3580	1763	1.2	33.6
Station 18	88.6266	27.3631	1874	1.2	36
Station 19	88.7021	27.3803	2100	1.1	4
Station 20	88.4829	27.2176	740	1.8	4
Station 21	88.7321	27.4002	2300	1.1	45
Station 22	88.5951	27.2244	1169	1.6	7.52
Station 23	88.5952	27.2239	1171	1.6	0.64
Station 24	88.6079	27.2125	837	1.7	13.1
Station 25	88.6080	27.2124	838	1.6	6.36

11.4.1 Comparison of Observed and Calculated Infiltration Rates or Accumulated Infiltration

The comparisons of observed and calculated infiltration rates or accumulated infiltration using Horton's, Philip's, Green Ampt's, and Kostikov's equations are represented in Figure 11.3a–h for different soil and land use types. Tables 11.2 and 11.3 show a tabular comparison under loamy skeletal soil.

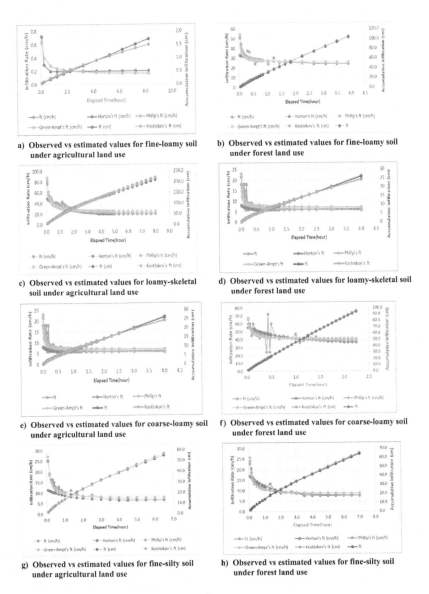

a) Observed vs estimated values for fine-loamy soil under agricultural land use

b) Observed vs estimated values for fine-loamy soil under forest land use

c) Observed vs estimated values for loamy-skeletal soil under agricultural land use

d) Observed vs estimated values for loamy-skeletal soil under forest land use

e) Observed vs estimated values for coarse-loamy soil under agricultural land use

f) Observed vs estimated values for coarse-loamy soil under forest land use

g) Observed vs estimated values for fine-silty soil under agricultural land use

h) Observed vs estimated values for fine-silty soil under forest land use

FIGURE 11.3 Comparison of observed and estimated infiltration rates in the Ranikhola watershed.

TABLE 11.2

Comparison of Observed and Estimated Infiltration Rate in Loamy Skeletal Soil under Agricultural Land Use

Elapsed Time (min)	Observed Infiltration Rate (cm/h)	Observed Accumulated Infiltration (cm)	Infiltration Rate by Horton's Model (cm/h)	Infiltration Rate by Philip's Model (cm/h)	Accumulated Infiltration by Kostiokov Model (cm)	Infiltration Rate by Green-Ampt Model (cm/h)
2	30	1	11.25	27.29	1.00	29.38
4	18	1.6	11.08	20.01	1.60	20.01
6	18.3	2.21	10.92	16.79	2.11	15.70
8	16.2	2.75	10.76	14.87	2.57	13.48
10	9	3.05	10.61	13.56	2.99	12.59
15	16.2	4.4	10.25	11.52	3.94	10.08
20	7.2	5	9.92	10.30	4.80	9.40
25	7.8	5.65	9.61	9.47	5.59	8.82
30	5.4	6.1	9.34	8.86	6.33	8.50
40	6.3	7.15	8.84	8.00	7.70	7.89
50	3	7.65	8.42	7.41	8.96	7.67
60	6.3	8.7	8.07	6.98	10.15	7.27
80	5.25	10.45	7.51	6.37	12.35	6.79
100	9	13.45	7.11	5.96	14.39	6.26
120	8.7	16.35	6.83	5.65	16.29	5.93
150	6	19.35	6.54	5.31	18.97	5.69
180	5.1	21.9	6.37	5.06	21.49	5.54
210	6.1	24.95	6.26	4.87	23.87	5.40
240	6	27.95	6.20	4.71	26.15	5.30
270	6.1	31	6.16	4.58	28.34	5.21
300	6.1	34.05	6.14	4.47	30.45	5.14
Comparison Index						
Coefficient of determination			0.48	0.84	0.99	0.85
Standard error			0.113	0.035	0.002	0.034

Mathematically, the observed and calculated values using various models were compared using the coefficient of determination and standard error values. Table 11.4 shows model parameters for different soil types under agriculture and forest land use for various models.

Coefficient of determination (r^2) and standard errors (e) for different soil types and agricultural and forest land use conditions for Horton's, Philip's, Kostiokov, and Green-Ampt's models are presented in Table 11.5.

For Horton's model, the coefficient of determination and standard error for different soil types and soil conditions were: (a) 0.99 with a minimum standard error of 0.000001 under agriculture and 0.67 with standard error of 0.07 under forest, for fine loamy soil; (b) 0.48 with standard error of 0.11 under agriculture and 0.91 with standard error of 0.01 under forest, for loamy skeletal soil; (c) 0.74 with standard error of 0.04 under

TABLE 11.3
Comparison of Observed and Estimated Infiltration Rate in Loamy Skeletal Soil under Forest Land Use

Elapsed Time (min)	Observed Infiltration Rate (cm/h)	Observed Accumulated Infiltration (cm)	Infiltration Rate by Horton's Model (cm/h)	Infiltration Rate by Philip's Model (cm/h)	Accumulated Infiltration by Kostiokov Model(cm)	Infiltration Rate by Green-Ampt Model (cm/h)
5	6.38	6.38	49.07	82.24	7.04	87.84
10	4.74	11.12	47.29	62.36	11.87	60.02
15	4.27	15.39	45.62	53.55	16.11	49.63
20	4.28	19.67	44.05	48.30	20.02	43.75
25	3.96	23.63	42.59	44.72	23.68	40.20
30	3.09	26.72	41.21	42.07	27.17	38.16
35	3.12	29.84	39.92	40.02	30.51	36.53
40	3.5	33.34	38.70	38.36	33.74	35.07
45	3.46	36.8	37.57	36.99	36.88	33.89
50	3.35	40.15	36.50	35.83	39.92	32.95
55	3.3	43.45	35.50	34.83	42.89	32.16
60	3.06	46.51	34.56	33.96	45.80	31.53
65	3.14	49.65	33.68	33.19	48.65	30.97
70	3.57	53.22	32.86	32.50	51.44	30.40
75	3.32	56.54	32.08	31.89	54.19	29.94
80	3.19	59.73	31.36	31.33	56.89	29.55
85	3	62.73	30.68	30.83	59.55	29.22
90	2.8	65.53	30.04	30.36	62.17	28.93
95	2.77	68.3	29.44	29.94	64.75	28.68
100	2.72	71.02	28.88	29.54	67.30	28.44
105	2.6	73.62	28.35	29.18	69.82	28.24
110	2.42	76.04	27.86	28.83	72.31	28.06
115	2.37	78.41	27.40	28.52	74.78	27.89
120	2.41	80.82	26.96	28.22	77.21	27.73
130	4.87	85.69	26.17	27.67	82.02	27.44
140	4.29	89.98	25.48	27.19	86.73	27.21
150	4.3	94.28	24.87	26.76	91.35	27.00
160	4.2	98.48	24.33	26.36	95.91	26.81
170	4.1	102.58	23.86	26.00	100.39	26.64
180	4.05	106.63	23.44	25.68	104.81	26.48
190	3.96	110.59	23.07	25.37	109.16	26.34
200	3.94	114.53	22.75	25.10	113.47	26.22
210	3.9	118.43	22.47	24.84	117.71	26.10
220	3.75	122.18	22.22	24.60	121.91	25.99
230	3.4	125.58	22.00	24.37	126.07	25.90
240	3.4	128.98	21.81	24.16	130.17	25.81
250	3.4	132.38	21.64	23.96	134.24	25.73
260	3.4	135.78	21.49	23.78	138.27	25.65

(Continued)

TABLE 11.3 (*Continued*)

Comparison of Observed and Estimated Infiltration Rate in Loamy Skeletal Soil under Forest Land Use

Elapsed Time (min)	Observed Infiltration Rate (cm/h)	Observed Accumulated Infiltration (cm)	Infiltration Rate by Horton's Model (cm/h)	Infiltration Rate by Philip's Model (cm/h)	Accumulated Infiltration by Kostiokov Model(cm)	Infiltration Rate by Green-Ampt Model (cm/h)
270	3.4	139.18	21.36	23.60	142.26	25.57
280	3.4	142.58	21.24	23.43	146.21	25.50
290	3.4	145.98	21.14	23.28	150.13	25.43
300	3.4	149.38	21.05	23.13	154.01	25.37
320	6.8	156.18	20.90	22.85	161.68	25.25
340	6.8	162.98	20.79	22.59	169.24	25.13
360	6.8	169.78	20.70	22.36	176.69	25.03
380	6.8	176.58	20.63	22.15	184.04	24.94
400	6.8	183.38	20.58	21.95	191.29	24.85
420	6.8	190.18	20.54	21.77	198.45	24.77
440	6.8	196.98	20.51	21.60	205.53	24.69
460	6.8	203.78	20.48	21.44	212.53	24.62
480	6.8	210.58	20.46	21.29	219.46	24.56
Comparison Index						
Coefficient of determination			0.91	0.91	0.99	0.80
Standard error			0.012	0.013	0.0003	0.029

agriculture and 0.32 with standard error of 0.11 under forest, for coarse loamy soil; (d) 0.86 with standard error of 0.03 under agriculture and under forest, for fine silty soil.

For Philip's model, the coefficient of determination and standard error for different soil types and soil conditions were: (a) 0.86 with standard error of 0.04 under agriculture and 0.89 with standard error of 0.02 under forest, for fine loamy soil; (b) 0.84 with standard error of 0.04 under agriculture and 0.91 with standard error of 0.01 under forest, for loamy skeletal soil; (c) 0.67 with standard error of 0.05 under agriculture and 0.33 with standard error of 0.11 under forest, for coarse loamy soil; (d) 0.97 with standard error of 0.01 under agriculture and 0.95 with standard error of 0.01 under forest, for fine silty soil.

For Kostiokov model, the coefficient of determination and standard error for different soil types and soil conditions were: (a) 0.99 with standard error of 0.003 under agriculture and 0.99 with standard error of 0.00005 under forest, for fine loamy soil; (b) 0.99 with standard error of 0.002 under agriculture and 0.99 with standard error of 0.0003 under forest, for loamy skeletal soil; (c) 0.99 with standard error of 0.0007 under agriculture and 0.99 with standard error of 0.00002 under forest, for coarse loamy soil; (d) 0.99 with standard error of 0.0002 under agriculture and 0.99 with standard error of 0.00001 under forest, for fine silty soil.

For Green-Ampt model, the coefficient of determination and standard error for different soil types and soil conditions were: (a) 0.86 with standard error of 0.04

TABLE 11.4
Model Parameters for Different Soil types under Agriculture and Forest Land Use

Types of Land Use	Horton's Parameters		Philip's Parameters		Kostiokov Parameters		Green-Ampt Parameters	
	K	f_o	S	k	b	a	m	n
Fine Loamy								
Agricultural land	9.892	1.386	0.305	0.105	0.746	0.308	0.146	0.029
Forest land	1.051	33.660	11.057	21.609	0.858	31.741	25.857	49.541
Loamy Skeletal								
Agricultural land	0.995	11.420	9.071	2.445	0.682	10.152	4.402	24.977
Forest land	0.769	59.970	39.192	14.362	0.754	45.801	22.580	416.34
Coarse Loamy								
Agricultural land	5.237	8.180	4.417	4.439	0.766	8.353	6.680	4.892
Forest land	1.351	56.380	11.029	35.343	0.905	45.700	40.315	54.502
Fine Silty								
Agricultural land	0.789	11.847	12.294	4.214	0.743	14.083	6.841	42.632
Forest land	0.752	17.525	10.677	5.606	0.779	14.265	8.011	33.738

TABLE 11.5
Coefficient of Determination and Standard Error Values for Different Models

Types of land use	Horton's Parameters		Philip's Parameters		Kostiokov Parameters		Green-Ampt Parameters	
	r^2	e	r^2	e	r^2	E	r^2	e
Fine Loamy								
Agricultural land	0.99	0.000001	0.86	0.04	0.99	0.003	0.86	0.04
Forest land	0.67	0.07	0.89	0.02	0.99	0.000053	0.83	0.04
Loamy Skeletal								
Agricultural land	0.48	0.11	0.84	0.04	0.99	0.002	0.85	0.03
Forest land	0.91	0.01	0.91	0.01	0.99	0.0003	0.80	0.03
Coarse Loamy								
Agricultural land	0.74	0.04	0.67	0.05	0.99	0.0007	0.52	0.08
Forest land	0.32	0.11	0.33	0.11	0.99	0.00002	0.26	0.12
Fine Silty								
Agricultural land	0.86	0.03	0.97	0.01	0.99	0.0002	0.91	0.02
Forest land	0.86	0.03	0.95	0.01	0.99	0.0001	0.87	0.03

under agriculture and 0.83 with standard error of 0.04 under forest, for fine loamy soil; (b) 0.85 with standard error of 0.03 under agriculture and 0.80 with standard error of 0.03 under forest, for loamy skeletal soil; (c) 0.72 with standard error of 0.08 under agriculture and 0.26 with standard error of 0.12 under forest, for coarse loamy soil; (d) 0.91 with standard error of 0.02 under agriculture and 0.87 with standard error of 0.03 under forest, for fine silty soil.

The above table shows that Horton's model gives a maximum value of the coefficient of determination with minimum standard error for fine loamy under agricultural land use whereas Kostiokov model shows a maximum value of the coefficient of determination with minimum standard error for all soil types and land use except fine loamy under agricultural land use.

11.5   CONCLUSION

The major aim of this study was to identify the best-suited infiltration model for the Ranikhola watershed located in the east district of Sikkim. The infiltration rate of various soils under two land use scenarios was measured using a double-ring infiltrometer test. Commonly available models, namely, Horton, Kostiakov, Philip, and Green-Ampt were used to compare measured and model-predicted infiltration rate during monsoon season. The following major conclusions were drawn from the study:

- Large variation in the basic infiltration rates within the watershed was observed (0.20–45 cm/h) for the Ranikhola watershed. For agricultural land, basic infiltration rate varied from 0.2 cm/h to a maximum of 18 cm/h and for forest land it varied from 20 cm/h to a maximum of 45 cm/h.
- The infiltration rate was highly affected by bulk density. Soils having more bulk density had lower infiltration rates. The maximum bulk density of soil was found to be 1.9 g/cm³ at an altitude of 876 m whereas a minimum of 1.1 was recorded at an altitude of 2,300 m.
- After analysis, it was found that the values of parameters of infiltration models vary from soil to soil and land uses type.
- For fine loamy soil under agriculture, Horton's model is the best-fitted model as it gives a maximum coefficient of determination value of 0.99 and a minimum standard error of 0.000001.
- Except the above condition, Kostiokov model is the best-fitted model giving a maximum coefficient of determination and minimum standard error for all soil types and land use in the study region.

REFERENCES

Ahaneku, I.E., 2011. Infiltration characteristics of two major agricultural soils in north central Nigeria. *Agricultural Science Research Journals*, 1(7), pp. 166–171.
Ayu, I.W., Soemarno, S.P. and Java, I.I., 2013. Assessment of infiltration rate under different drylands types in unter-iwes subdistrict Sumbawa Besar, Indonesia. *Assessment*, 3(10), 71–76.
Cerdà, A., 1996. Seasonal variability of infiltration rates under contrasting slope conditions in southeast Spain. *Geoderma*, 69(3–4), pp. 217–232.
Diamond, J. and Shanley, T., 1998. Infiltration rate assessment of some major soils. *Proc. Agricultural Research forum, Dublin*, pp. 001–013.
Diamond, J. and Shanley, T., 2003. Infiltration rate assessment of some major soils. *Irish Geography*, 36(1), pp. 32–46.

Dibal, J.M., Umara, B.G. and Rimanungra, B., 2013. Water intake characteristics of different soil types in Southern Borno Nigeria. *International Journal of Science Inventions Today*, 2, pp. 502–509.

Eze, E.B., Eni, D.I. and Comfort, O., 2011. Evaluation of the infiltration capacity of soils in Akpabuyo local government area of Cross River, Nigeria. *Journal of Geography and Geology*, 3(1), p. 189.

Green, W.H. and Ampt, G.A., 1911. Studies on soil phyics. *The Journal of Agricultural Science*, 4(1), pp. 1–24.

Gregory, J.H., Dukes, M.D., Jones, P.H. and Miller, G.L., 2006. Effect of urban soil compaction on infiltration rate. *Journal of Soil and Water Conservation*, 61(3), pp. 117–124.

Horton, R.E., 1940. The infiltration-theory of surface-runoff. *Eos, Transactions American Geophysical Union*, 21(2), p. 541.

Horton, R.E., 1941. An approach toward a physical interpretation of infiltration-capacity 1. *Soil Science Society of America Journal*, 5(C), pp. 399–417.

Jagdale, S.D. and Nimbalkar, P.T., 2012. Infiltration studies of different soils under different soil conditions and comparison of infiltration models with field data. *International Journal of Advanced Engineering Technology*, 3(2), pp. 154–157.

Jejurkar, C.L. and Rajurkar, M.P., 2012. Infiltration studies for varying land cover conditions. *International Journal of Computational Engineering Research, ISSN*, 2(1), pp. 2250–3005.

Joshi, V.U. and Tambe, D.T., 2010. Estimation of infiltration rate, run-off and sediment yield under simulated rainfall experiments in upper Pravara Basin, India: Effect of slope angle and grass-cover. *Journal of Earth System Science*, 119(6), pp. 763–773.

Michael, A.M., 2005. *Irrigation Theory and Practices*. Vikas Publishing House Pvt. Ltd, New Delhi, pp. 464–472.

Mohan, S. and Sangeeta, K., 2005. Recharge estimation using infiltration models. *ISH Journal of Hydraulic Engineering*, 11(3), pp. 1–10.

Osuji, G.E., Okon, M.A., Chukwuma, M.C. and Nwarie, I.I., 2010. Infiltration characteristics of soils under selected land use practices in Owerri, Southeastern Nigeria. *World Journal of Agricultural Sciences*, 6(3), pp. 322–326.

Philip, J.R., 1957. The theory of infiltration: 4. Sorptivity and algebraic infiltration equations. *Soil Science*, 84(3), pp. 257–264.

Pingping, H., Xue, S., Li, P. and Zhanbin, L., 2013. Effect of vegetation cover types on soil infiltration under simulating rainfall. *Nature Environment and Pollution Technology*, 12(2), p. 193.

Rossi, M.J. and Ares, J.O., 2012. Depression storage and infiltration effects on overland flow depth-velocity-friction at desert conditions: Field plot results and model. *Hydrology and Earth System Sciences*, 16(9), pp. 3293–3307.

Shougrakpam, S., Sarkar, R. and Dutta, S., 2010. An experimental investigation to characterise soil macroporosity under different land use and land covers of northeast India. *Journal of Earth System Science*, 119(5), pp. 655–674.

Stolte, J., 2003. Effects of Land Use and Infiltration Behaviour on Soil Conservation Strategies. Doctoral Thesis Wageningen University, ISBN 90-5808-854-5.

Uloma, A.R., Onyekachi, C.T., Torti, E.K. and Amos, U., 2013. Infiltration characteristics of soils of some selected schools in Aba, Nigeria. *Archives of Applied Science Research*, 5(3), pp. 11–15.

Yusof, M.F., Zakaria, N.A., Ghani, A.A., Abdullah, R.O.Z.I. and Kiat, C.C., 2005. Infiltration study for urban soil: Case studies–butterworth and engineering campus, Universiti Sains Malaysia. *Proceedings, Congress of the International Association of Hydraulic Research*, 1, pp. 1154–1162.

12 Toward Conservation Agriculture for Improving Soil Biodiversity

Sharmistha Pal
ICAR-Indian Institute of Soil and Water
Conservation, Research Centre

B. B. Basak
Anand Agricultural University

CONTENTS

12.1 INTRODUCTION

Conservation agriculture (CA) is the generic term used to define a set of farming practices designed to enhance the sustainability of food and agriculture production by conserving and protecting the available soil, water, and biological resources such that the need for external inputs can be kept minimal (García-Torres *et al.*, 2003). The Food and Agriculture Organization (FAO; CA website, 2004) has defined CA as a system based on minimal soil disturbance (no-till, minimum tillage) and permanent soil cover (mulch, crop residue) combined with diversified rotations with legumes. Similarly, "conservation tillage" (CT) refers to a set of practices adopted by modern plough-based conventional tillage to enhance water infiltration and reduce erosion risk. CT is commonly applied to no-tillage, direct drilling, or minimum tillage practices when

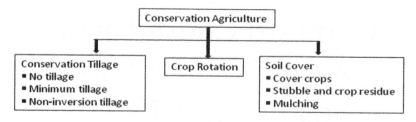

FIGURE 12.1   The three principles of conservation agriculture and the main practices and means needed to achieve each principle.

associated with a cover of crop residues on at least 30% of the soil surface, and when associated with some conservation goals such as the conservation of time, fuel, earthworms, soil water, and nutrients (Baker *et al.*, 2002). CA (Figure 12.1) being referred to as "resource-efficient" agriculture, thus contributes to environmental conservation as well as enhanced and sustained agricultural production (García-Torres *et al.*, 2003).

The negative environmental impacts of intensive agriculture, particularly loss of soil organic matter and contamination of groundwater by agrochemicals, is shifting the focus away from conventional (high-input) agricultural systems to more ecologically sustainable systems. The benefits and costs associated with conservation agriculture are depicted in Table 12.1.

The management of these systems includes the use of reduced tillage, inputs of organic materials, and nutrient cycling strategies based on crop rotations (Pankhurst *et al.*, 1996). In crop rotations under zero tillage, the litter from several crops in the preceding years likely results in a greater variety of substrates than under conventional tillage where the litter does not accumulate. Therefore, conservation tillage and legume-based crop rotations appear to support soil microbial functional diversity (Lupwayi *et al.*, 1998). The interaction of various processes through which conservation tillage can generate environmental benefits is given in Figure 12.2.

TABLE 12.1

Benefits and Costs Associated with Conservation Agriculture

Advantages	Drawbacks
• Reduction in on-farm costs, savings of time	• Purchase of specialized planting equipment
• Increase in soil fertility, moisture retention, resulting in long-term yield increase, greater food security	• Short-term pest problems due to the change in crop management
• Stabilization of soil and protection from erosion	• Acquiring new management skills
• Reduction in toxic contamination of surface water and groundwater	• Development of appropriate technical packages and training programs
• Recharge of aquifers as a result of better infiltration	
• Reduction in air pollution resulting from soil tillage machinery	
• Conservation of terrestrial and soil-based biodiversity	

Source: FAO (2004) Conservation agriculture. website: www.fao.org/ag/ca/index.html.

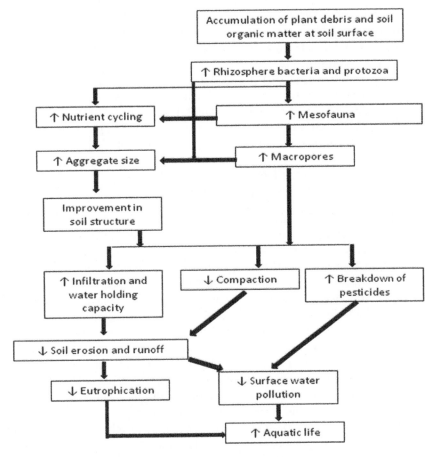

FIGURE 12.2 The interaction of various processes through which conservation tillage can generate environmental benefits. (Modified from Holland, 2004.)

12.2 HOW CONSERVATION TILLAGE AND RESIDUE MANAGEMENT CAN HELP IN IMPROVING SOIL PROPERTIES AND ULTIMATELY SOIL QUALITY

Conservation tillage and residue management can help in influencing some of the soil properties as described below:

- Soil temperature: Surface residues play a significant role in balancing radiant energy and insulation action thereby affecting soil temperature.
- Effects on soil physical properties: Tillage also influences crusting, hydraulic conductivity, and water storage capacity. Soil crusting severely affects germination and emergence of a seedling is caused due to aggregate dispersion and soil particles resorting and rearrangement during rainstorm followed by drying. Conservation tillage and surface residue help in protecting the

dispersion of soil aggregates and in increasing saturated hydraulic conductivity. Better soil organic matter under conservation agricultural practices promotes water-stable aggregates which help in maintaining good infiltration rate, good structure, protection from wind and water erosion.

* Effect on soil organic matter and soil fertility: Conservation agricultural practices help in improving soil organic matter by regular addition of organic wastes and residues, use of green manures, legumes in the rotation, reduced tillage, and recycling of residues. These practices provide a great opportunity in maintaining and restoring soil quality in terms of SOM and N.

12.3   IMPACT OF CONSERVATION TILLAGE (CT) ON SOIL BIOLOGICAL PROPERTIES

Although the microbial biomass is a small but important reservoir of nutrients (C, N, P, and S), many transformations of these nutrients are carried out by them. Numerous studies have shown a decline in SOM and microbial biomass under land disturbance (Sparling, 1997) whereas soil disturbance can cause significant modifications of habitat, affecting the microbial community. The higher level of SOM at the soil surface created using CT encourages a different range of organisms compared to a plough-based system in which residues are buried (Rasmussen and Collins, 1991).

Conventional tillage significantly reduces the diversity of bacteria by reducing both substrate richness and evenness. The conservation tillage and legume-based crop rotations in wheat have been observed to support the diversity of soil microbial communities and may affect the sustainability of agroecosystems (Lupwayi *et al.*, 1998).

From a quantitative perspective, one can postulate that the majority of biodiversity on earth is occurring in soils. There are a multitude of organisms whose ecological requirements are dependent on soils. The number of micro-organisms under a footprint is tremendous. In addition, numerous higher organisms such as arthropods and various worm species inhabit the soil ecosystem (Figure 12.2 and Table 12.2).

TABLE 12.2
Dimensions of Biodiversity under a Foot Print

Taxonomic Group	Number of Individuals	Biomass (g/m^2)
Bacteria	10^{12}–10^{14}	100–700
Fungi	10^9–10^{12}	100–500
Algae	10^6–10^9	20–150
Protozoa	10^7–10^9	6–30
Nematodes	10^4–10^6	5–50
Mites	2×10^2–4×10^3	0.2–4
Springtails	2×10^2–4×10^3	0.2–4
Insect larva	Up to 50	<4.5
Diplopoda	Up to 70	0.5–12.5
Earthworms	Up to 50	30–200

Source: Stahr *et al.* (2008).

12.4 THE ECONOMIC VALUE OF SOIL BIODIVERSITY

Soil biodiversity is economically "priceless" as it provides the base for successful plant life and acts as food source for most human and animal life. Moreover, based on the different services provided, a number of attempts have been made to estimate the economic value of soil biodiversity. Recycling of organic wastes is considered an important aspect of conserving soil biodiversity. Mankind produces more than 38 billion metric tons of organic waste on a global scale annually. In the absence of decomposing/recycling activity by soil micro-organisms, much of the globe's land surface would remain literally covered with organic debris (Table 12.3). The economic value of this service represents approximately 50% of the total benefits of soil biotic activity worldwide (>US $760 billion) (Brussaard *et al.*, 2007).

12.4.1 THE IMPORTANCE OF SOIL BIODIVERSITY

Soil biodiversity is normally indicated as the variability of the living forms such as soil fauna, flora, vertebrates, birds, and mammals within a habitat or a management system of a territory involved in the agricultural activity (Figure 12.3). The importance of soil biodiversity in agriculture has not been given adequate attention because crop productivity has been increased through the use of inorganic fertilizers, pesticides, plant breeding, soil tillage, liming, and irrigation.

12.4.2 MICROORGANISMS

Microbial biomass, diversity, and overall soil biological activity are generally considered higher in soils cultivated using conservation tillage (CT) technique as compared to those receiving deep cultivations (Lupwayi *et al.*, 2001). Long-term trial

TABLE 12.3
The Economic Benefits of Soil Biodiversity

Activity	Role of Soil Biodiversity	Estimated World Economic Benefits ($\times 10^9$ US\$/year)
Waste recycling	Fungi, bacteria, protozoa play a major role in waste recycling	760
Soil formation	Earthworms, ants, termites, fungi facilitate soil formation	25
Nitrogen fixation	Biological nitrogen fixation by bacteria	90
Degradation of chemicals	Soil micro-organisms play a key role in degrading or modifying pollutants	121
Pest control	Soil provides microhabitats for natural enemies of certain animal pest species	160
Pollination	Many pollinators (eg., bumblebees and solitary bees) have a soil-dwelling phase in their life history	200

Source: Gardi and Jefferey, 2009.

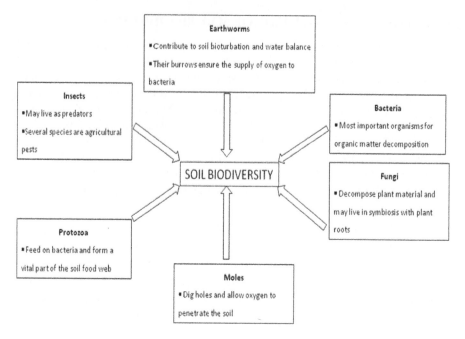

FIGURE 12.3   Functions of soil biodiversity.

reports revealed that CT encourages population growth of rhizosphere bacteria such as *Agrobacterium spp.* and *Pseudomonas spp.* and on a sandy loam soil it increases N_2 fixation and nodulation of pea plants (Höflich *et al.*, 1999). Consequently, the soil organic matter on the surface increases with CT leading to improved aggregate stability. The intensive tillage induces soil compaction, which reduces soil pore space, and alters the exchange and storage of gases, water, and SOM (Brussaard and Van Faassen, 1994). Because free-living nematodes depend on water for movement, they are susceptible to soil structure, aeration and moisture, and soil cultivation. The response to tillage can be variable between different soil functional groups and will depend on other factors such as the cropping and abundance of residues (López-Fando and Bello, 1995). With respect to microfauna, Cochran *et al.* (1994) suggest that management practices that favor bacteria would also be expected to favor protozoa since bacteria are their main food source. The negative effects on microarthropod populations are caused by the physical disturbance of the soil from plough-based tillage. Some individuals may be killed initially by abrasion during tillage operation while others may be killed by being entraped in soil clods post tillage inversion (Wardle, 1995).

12.4.3   MESOFAUNA

Franchini, 1996 reported that a higher abundance of oribatid mites with reduced and no-tillage compared to ploughing and species diversity was also higher in no-tillage compared to ploughing. The soil tillage and compaction act as a major factor determining the vertical distribution of the microarthropods while the reduced tillage facilitates the activity and distribution of microarthropods by promoting a higher density of macropores (Schrader and Lingnau, 1997).

12.4.4 MACROFAUNA

Although there is considerable evidence regarding earthworm populations being directly influenced by tillage method, the degree of impact varies depending on the earthworm species, soil factors and climatic conditions (Chan, 2001). Conservation tillage if combined with the return of crop residues and additional organic manures favors earthworm populations (Kladivko, 2001). It has been reported that when CT was adopted for the integrated system, earthworm biomass (averaged over a 10-years) was 36% higher compared to ploughing (Jordan et al., 2000). Earthworm populations may be especially low in drier climates but adoption of CT has been shown to increase populations substantially. On a long-term basis, CT retains the organic matter on the soil surface, thereby increasing saprophytic and detritus feeding species upon which the predators are dependent.

12.4.5 CONSERVATION AGRICULTURE AND SOIL BIODIVERSITY

It is established that there is a significant and positive correlation between organic carbon content and biodiversity, hence, different management practices that improve soil C status favors soil biodiversity. These management practices include crop rotation, minimum tillage, cover crop and intercropping, which improves soil fertility by increasing soil organic carbon and nutrient recycling due to higher residue retention. Crop rotations also encourage greater biodiversity and posed a positive effect on soil macrofauna. No, and reduced tillage can diminish spring time runoff and soil erosion provided the soil is sufficiently covered (mulch, green manure, catch crops, etc.) and has significant effect on biological activity (Table 12.4).

12.5 SOIL BIODIVERSITY UNDER CONVENTIONAL AND NO TILLAGE SYSTEMS

A number of ecosystem processes are governed by soils and soil biodiversity. Soil organic matter as a key component of soil quality, integrates several soil functions and regulates delivery of many ecosystem services (Palm et al., 2007; Wall et al., 2004).

Although CA practices increase soil organic matter and promote biological properties of soil, they also alter below and above ground species differentially than conventional agricultural practices (McLaughlin and Mineau, 1995). This alteration in species drives a range of responses among different groups of organisms, however, most of them have a greater abundance or biomass and diversity in no-tillage (NT) than in conventional tillage (González-Chávez et al., 2010; Rodríguez et al., 2006).

Soil microbial biomass is an indicator of soil quality due to its role in decomposition, nutrient cycling rates and patterns, formation of SOM, and soil aggregation. According to some authors, fungal communities tend to dominate the soil surface

TABLE 12.4

Summary of Farming Practices on Soil Biological Prosperities

Farming Practices	Soil Biological Characteristic				
	Macro-Fauna Population	Meso-Fauna Population	Micro-Fauna Population	Mycorrhizae Population	Microbial Activity
No tillage	+	+	+	+	+
Reduce tillage	+	+	+	+	+
Cover crops	+	[×]	[×]	[×]	[×]
Crop rotation	+				
Intercropping					
Grassland	+	+	+	+	+
Agro forestry	[×]	[×]	[×]	[×]	[×]
Buffer Terracing	[+]	[+]	[+]	[+]	[+]

Legend:

+: positive observed effect; 0: neutral observed effect

-: negative observed effect; [x]: expected effect

(×): limited (e.g., short term) or indirect effect; empty field: no particulars mentioned

Source: Adapted from Louwagie et al., 2009.

TABLE 12.5

Soil Biological Properties and Residue Retention under Conventional and no Tillage

Soil Biological Properties and Processes	NT Compared to CT	Residue Retention
Soil microbial biomass	↑	↑
Fungal population	↑	↑
Enzymatic activity	↑	↑
Beneficial microbes	↑	↑
Pathogenic micro-fauna	↑	↑
Free-living (beneficial) nematode	NS	↑
Plant parasitic nematode	↓	NS
Earthworms	↑	↑
Arthropod density	↑	↑

Source: Summarized from Verhulst *et al.* (2010).

of NT whereas bacterial communities dominate in conventional tillage systems (Verhulst *et al.*, 2010). This difference results in slower rates of nutrient mineralization and higher nutrient use efficiency with surface residue retention compared to conventional systems. Other long-term studies, however, were not able to identify this shift between fungi and bacteria (Helgason *et al.*, 2009) with minor overall

TABLE 12.6
Possible Effects of Different Farming Practices on Soil Biodiversity

S. No.	Farming Practices	Possible Effects
1.	Minimum use of chemical fertilizer and pesticide	Avoids harmful effects on soil biodiversity resulting from the application of high doses of inorganic fertilizers
2.	Nutrient management through organic sources	Supports a greater abundance of invertebrates that rely on un-degraded plant matter as a food source, *rom.* earthworms (Gerhardt, 1997; Pfiffner and Mäder, 1997), carabids (Kromp, 1999), and more diverse microbial communities (Fraser *et al.*, 1988)
3.	Minimum tillage	Improves soil biodiversity
4.	Crop rotation	Inclusion of legumes in crop rotation enhance soil fertility (e.g., clover in the grass mix); increased crop diversity may benefit a variety of species that require a structurally diverse habitat (Lampkin, 2002; Stoate, 1996)
5.	Intercropping/Mixed farming	An increase in heterogeneity may favor increased invertebrate diversity, e.g., polyphagous predators; significant benefits for biodiversity across a range of taxa
		Increases habitat heterogeneity at multiple spatial and temporal scales (Robinson *et al.*, 2001; Stoate *et al.*, 2001; Vickery *et al.*, 2001; Benton *et al.*, 2003; Sunderland and Samu, 2000)

impacts on decomposition and nutrient availability more likely (Bissett *et al.*, 2013). The impacts of a shift in a microbial composition may become more important on degraded soils (Verhulst *et al.*, 2010).

Fungi, particularly arbuscular mycorrhizal fungi, are important for nutrient acquisition and drought resistance, particularly for low nutrient input systems. They also play a key role in forming stable soil aggregates. Earthworms, on the other hand, play a significant role in maintaining soil structure and nutrient cycling by their movement within the soil, by breaking down litter, and by binding soil particles with their excrement (Table 12.5). The possible effects of different farming practices on soil biodiversity are given in Table 12.6.

12.6 CONCLUSION

Conservation agriculture (CA) refers to a set of agricultural practices encompassing minimum mechanical soil disturbance, diversified crop rotation, and permanent soil cover with crop residues to mitigate soil erosion and improve soil fertility besides performing the basic soil functions. Thus CA aims to conserve, improve, and use resources more efficiently through CA-based technologies.

REFERENCES

Baker, C.J., Saxton, K.E. and Ritchie, W.R., 2002. *No-Tillage Seeding: Science and Practice.* CAB international, Oxon, OX.

Benton, T.G., Vickery, J.A. and Wilson, J.D., 2003. Farmland biodiversity: is habitat heterogeneity the key? *Trends in Ecology & Evolution, 18*(4), pp. 182–188.

Bissett, A., Richardson, A.E., Baker, G., Kirkegaard, J. and Thrall, P.H., 2013. Bacterial community response to tillage and nutrient additions in a long-term wheat cropping experiment. *Soil Biology and Biochemistry, 58*, pp. 281–292.

Brussaard, L. and Van Faassen, H.G., 1994. Effects of compaction on soil biota and soil biological processes. In *Developments in Agricultural Engineering* (Vol. 11, pp. 215–235). Elsevier, Amsterdam, Netherlands.

Brussaard, L., De Ruiter, P.C. and Brown, G.G., 2007. Soil biodiversity for agricultural sustainability. *Agriculture, Ecosystems & Environment, 121*(3), pp. 233–244.

Chan, K.Y., 2001. An overview of some tillage impacts on earthworm population abundance and diversity-implications for functioning in soils. *Soil and Tillage Research, 57*(4), pp. 179–191.

Cochran, V.L., Sparrow, S.D. and Sparrow, E.B., 1994. Residue effects on soil micro- and macroorganisms. *Managing Agricultural Residues*, CRC Press, Boca Raton, FL, pp. 163–184.

FAO. 2004. Conservation agriculture web site. www.fao.org/ag/ca/index.html.

Franchini, P., 1996. Oribatid mites as "indicator" species for estimating the environmental impact of conventional and conservation tillage practices. *Pedobiologia, 40*, pp. 217–225.

Fraser, D.G., Doran, J.W., Sahs, W.W. and Lesoing, G.W., 1988. Soil microbial populations and activities under conventional and organic management. *Journal of Environmental Quality, 17*(4), pp. 585–590.

Garcıa-Torres, L., Benites, J., Martınez-Vilela, A. and Holgado-Cabrera, A., 2003. *Conservation Agriculture: Environment, Farmers Experiences, Innovations, Socio-Economy.* Policy. Kluwer Academic Publishers, Boston, MA.

Gardi, C. and Jefferey, S., 2009. Soil biodiversity. JRC scientific and technical research series.

Gerhardt, R.A., 1997. A comparative analysis of the effects of organic and conventional farming systems on soil structure. *Biological Agriculture & Horticulture, 14*(2), pp. 139–157.

González-Chávez, M.D.C.A., Aitkenhead-Peterson, J.A., Gentry, T.J., Zuberer, D., Hons, F. and Loeppert, R., 2010. Soil microbial community, C, N, and P responses to long-term tillage and crop rotation. *Soil and Tillage Research, 106*(2), pp. 285–293.

Helgason, B.L., Walley, F.L. and Germida, J.J., 2009. Fungal and bacterial abundance in long-term no-till and intensive-till soils of the Northern Great Plains. *Soil Science Society of America Journal, 73*(1), pp. 120–127.

Höflich, G., Tauschke, M., Kühn, G., Werner, K., Frielinghaus, M. and Höhn, W., 1999. Influence of long-term conservation tillage on soil and rhizosphere microorganisms. *Biology and Fertility of Soils, 29*(1), pp. 81–86.

Holland, J.M., 2004. The environmental consequences of adopting conservation tillage in Europe: reviewing the evidence. *Agriculture, Ecosystems & Environment, 103*(1), pp. 1–25.

Jordan, V.W.L., Leake, A.R. and Ogilvy, S., 2000. Agronomic and environmental implications of soil management practices in integrated farming systems. *Aspects of Applied Biology, 62*, pp. 61–66.

Kladivko, E.J., 2001. Tillage systems and soil ecology. *Soil and Tillage Research, 61*(1–2), pp. 61–76.

Kromp, B., 1999. Carabid beetles in sustainable agriculture: a review on pest control efficacy, cultivation impacts and enhancement. *Agriculture, Ecosystems and Environment, 74,* pp. 187–228.

Lampkin, N., 2002. *Organic Farming.* Old Pond, Ipswich, England.

López-Fando, C. and Bello, A., 1995. Variability in soil nematode populations due to tillage and crop rotation in semi-arid Mediterranean agrosystems. *Soil and Tillage Research, 36*(1–2), pp. 59–72.

Lupwayi, N.Z., Arshad, M.A., Rice, W.A. and Clayton, G.W., 2001. Bacterial diversity in water-stable aggregates of soils under conventional and zero tillage management. *Applied Soil Ecology, 16*(3), pp. 251–261.

Lupwayi, N.Z., Rice, W.A. and Clayton, G.W., 1998. Soil microbial diversity and community structure under wheat as influenced by tillage and crop rotation. *Soil Biology and Biochemistry, 30*(13), pp. 1733–1741.

Louwagie, G., Gay, S.H., Sammeth, F., Ratinger, T., Marechal, B., Prosperi, P., Rusco, E., Terres, J.M., van der Velde, M., Baldock, D., Bowyer, C., Cooper, T., Fenn, I., Hagemann, N., Prager, K., Heyn, N. and Schuler, J., 2009. Final report on the project 'Sustainable Agriculture and Soil Conservation (SoCo). Available at http://soco.jrc.ec.europa.eu/documents/EUR-23820, p.172.

McLaughlin, A. and Mineau, P., 1995. The impact of agricultural practices on biodiversity. *Agriculture, Ecosystems & Environment, 55*(3), pp. 201–212.

Palm, C., Sanchez, P., Ahamed, S. and Awiti, A., 2007. Soils: a contemporary perspective. *Annual Review of Environment and Resources, 32,* pp. 99–129.

Pankhurst, C.E., Ophel-Keller, K., Doube, B.M. and Gupta, V.V.S.R., 1996. Biodiversity of soil microbial communities in agricultural systems. *Biodiversity & Conservation, 5*(2), pp. 197–209.

Pfiffner, L. and Mäder, P., 1997. Effects of biodynamic, organic and conventional production systems on earthworm populations. *Biological Agriculture & Horticulture, 15*(1–4), pp. 2–10.

Rasmussen, P.E. and Collins, H.P., 1991. Long-term impacts of tillage, fertilizer, and crop residue on soil organic matter in temperate semiarid regions. In *Advances in Agronomy* (Vol. 45, pp. 93–134). Academic Press.

Robinson, R.A., Wilson, J.D. and Crick, H.Q., 2001. The importance of arable habitat for farmland birds in grassland landscapes. *Journal of Applied Ecology, 38*(5), pp. 1059–1069.

Rodríguez, E., Fernández-Anero, F.J., Ruiz, P. and Campos, M., 2006. Soil arthropod abundance under conventional and no tillage in a Mediterranean climate. *Soil and Tillage Research, 85*(1–2), pp. 229–233.

Schrader, S. and Lingnau, M., 1997. Influence of soil tillage and soil compaction on microarthropods in agricultural land. *Pedobiologia, 41*(1), pp. 202–209.

Sparling, G.P., 1997. Soil microbial biomass, activity and nutrient cycling as indicators of soil health. In: Pankhurst, C., Doube, B.M. and Gupta, V. (Eds.), *Biological Indicators of Soil Health.* CAB International, Wallingford, pp. 97–119.

Stahr, K., Kandeler, E., Herrmann, L. and Streck, T., 2008. *Bodenkunde und Standortlehre* (Vol. 2967). UTB, Stuttgart, Germany.

Stoate, C., 1996. The changing face of lowland farming and wildlife Part 2: 1945–1995. *British Wildlife, 7,* pp. 162–172.

Stoate, C., Boatman, N.D., Borralho, R.J., Carvalho, C.R., De Snoo, G.R. and Eden, P., 2001. Ecological impacts of arable intensification in Europe. *Journal of Environmental Management, 63*(4), pp. 337–365.

Sunderland, K. and Samu, F., 2000. Effects of agricultural diversification on the abundance, distribution, and pest control potential of spiders: a review. *Entomologia Experimentalis et Applicata, 95*(1), pp. 1–13.

Verhulst, N., Govaerts, B., Verachtert, E., Castellanos-Navarrete, A., Mezzalama, M., Wall, P., Deckers, J. and Sayre, K.D., 2010. Conservation agriculture, improving soil quality for sustainable production systems. In *Advances in Soil Science: Food Security and Soil Quality*. CRC Press, Boca Raton, FL, pp. 137–208.

Vickery, J.A., Tallowin, J.R., Feber, R.E., Asteraki, E.J., Atkinson, P.W., Fuller, R.J. and Brown, V.K., 2001. The management of lowland neutral grasslands in Britain: effects of agricultural practices on birds and their food resources. *Journal of Applied Ecology*, *38*(3), pp. 647–664.

Wall, D.H., Bardgett, R.D., Covich, A., and Snelgrove, P.V.R., 2004. The need for understanding how biodiversity and ecosystem functioning affect ecosystem services in soils and sediments. In: Wall, D.H. (Ed.), *Sustaining Biodiversity and Ecosystem Services in Soils and Sediments*. Island Press, Washington, DC.

Wardle, D.A. 1995 Jan 1, Impacts of disturbance on detritus food webs in agro-ecosystems of contrasting tillage and weed management practices. In *Advances in Ecological Research* (Vol. 26, pp. 105–185). Academic Press.

13 Point-Injection Nitrogen Application under Rice Residue Wheat for Resource Conservation

Jagvir Dixit
Sher-e-Kashmir University of Agricultural Sciences
and Technology of Kashmir (SKUAST-K)

J. S. Mahal
Punjab Agricultural University

CONTENTS

13.1 INTRODUCTION

The sustainable food security for the ever-increasing human population along with environmental protection and conservation of natural resources have emerged as the prime concerns for the Indian and global agriculture. Overconsumption/excessive or unnecessary use of resources, inefficient input utilization, over manipulation of soil, and high cost of inorganic fertilizers necessitate judicious use of all the natural resources. The environmental impacts are site-specific and vary with farming systems, technologies used, and other responses/choices of the individual producers. With the recognition of conservation tillage production, apparently long association of traditional fertilizer application practices (broadcasting) with conventional tillage might be ill-suited or impossible. In high-residue zero-till farming, the efficient nitrogen fertilizer management remains a challenge due to slower nitrogen mineralization, greater nitrogen immobilization, and higher de-nitrification and ammonia volatilization losses. The studies suggested greater immobilization or nitrogen losses from surface-applied N in presence of straw than when the straw was burnt before sowing (Singh *et al.*, 2008), which is consistent with the findings of others (Rice and Smith, 1984; Patra *et al.*, 2004). Volatilization of N from urea leaves less N for plant uptake and makes it less efficient than nonurea fertilizers. About 40% of the N fertilizer applied to irrigated wheat is only utilized by the plants due to inefficiency in the application (wrong method or timing of application) and/or the inherent properties of current fertilizer products (Singh *et al.*, 2008). Significant quantities of nitrogen can be lost to the atmosphere from surface-applied urea due to ammonia volatilization. The "bottom line" is that where urea or urea-based fertilizers are surface applied, particularly in the presence of organic residues, crop yields are often reduced (Touchton and Hargrove, 1982; Fox and Hoffman, 1981). The presence of crop residues on the soil surface containing urea can increase the rate of urea hydrolysis thus, increasing the potential for ammonia volatilization in no-till systems (Barreto and Westerman, 1989). A review by Scharf and Alley (1988) found that an average of 25% of the nitrogen applied as urea is lost via ammonia volatilization.

It is a well-known concept that by reducing fertilizer nitrogen contact with the straw mulch, by placing it in the soil surface, can reduce nitrogen immobilization and ammonia volatilization which can increase grain yield, plant N uptake, and nitrogen use efficiency (Blackshaw *et al.*, 2002; Rochette *et al.*, 2009). While information exists on the influence of N fertilization on crop yields and nitrogen use efficiency for conventional till wheat; the effect of point-injection of liquid urea in straw mulched no-till wheat crop under different straw-load conditions on crop yield and N uptake are rarely available. The potential of point-injected nitrogen (liquid urea) in the straw mulched no-till wheat crop is inconclusive, but the method seems to be beneficial compared to surface application (broadcasting) of urea granules. Hence, present study was undertaken to evaluate the effect of point-injected nitrogen (liquid urea) application under straw mulched no-till wheat on crop yields, plant N-uptake, and N retained by mulch under rice-wheat cropping system.

13.2 METHODOLOGY

13.2.1 FIELD EXPERIMENTAL DESIGN

Field experiments were conducted at the research farm of the Department of Farm Machinery and Power Engineering, Punjab Agricultural University, Ludhiana, Punjab, India during *rabi* season of 2012–2013. The experiment was laid out in a factorial randomized block design with three replications. The plot size of each plot was $14 \times 12\,m^2$. The treatment details are given below:

A. Methods of Nitrogen Application (three methods)

M_1: Point-injected N with nitrogen applicator (0 + ½N after 1st irrigation + ½N after 2nd irrigation)

M_2: Broadcasting of N as per package of practice (½N during sowing + ¼N after 1st irrigation + ¼N after 2nd irrigation)

M_3: Broadcasting of N general practice followed by farmers using 'Happy seeder' (0 + ½N before 1st irrigation + ½N before 2nd irrigation)

B. Straw Load (two levels)

L_1: Low straw load (4.6 t/ha)

L_2: High straw load (8.0 t/ha)

All the experimental plots were sown uniformly using 'Happy Seeder' under two straw-load conditions. Wheat (variety PBW-621) was sown on November 9, 2012 and all the plots received a basal dose of 62.5 kg P/ha and 30 kg K/ha prior to sowing. Fertilizer nitrogen @ 125 kg N/ha was applied as per treatments of methods of N application. In all M_1 treatments, nitrogen fertilizer (liquid urea) was point-injected using self-propelled nitrogen applicator (Figure 13.1) in two splits. In M_2 treatment, the recommended dose of urea was broadcasted in three splits after 4–5 days of irrigation while in M_3 treatment, recommended dose of urea was broadcasted in two splits before the application of irrigation. All other practices for growing wheat crop were followed as recommended by Punjab Agricultural University, Ludhiana.

FIGURE 13.1   Point-injected nitrogen (liquid urea) application using-self-propelled nitrogen applicator.

The observations on nitrogen content in plant samples at various stages, decomposition of straw mulch, yield components such as the number of earhead per m row length, earhead length, no. of grains per earhead, 1,000 grain weight, and grain yield were recorded during the study as per standard procedure.

13.2.2 STATISTICAL ANALYSIS

The data were statistically analyzed with the procedure described by Cochran and Cox (1967) and adapted by Cheema and Singh (1991) in statistical package CPCS-1 for significant differences between treatments.

13.3 RESULTS

13.3.1 EFFECT OF METHODS OF NITROGEN APPLICATION ON MULCH DECOMPOSITION

13.3.1.1 Reduction in Straw Mulch Concentration

The straw decomposition rates were slower for point-injected N application (M_1) compared to the broadcasting of fertilizer urea (M_2 and M_3). The average reduction in straw mulch concentration was lowest for M_1 method of N application (5.45%) as compared with 9.28% for M_2 and 10.00% for M_3 method of N application after 60 days of sowing (Figure 13.2). The cumulative reduction in straw mulch concentration was significantly lower (30.25%) in case of M_1 method of nitrogen application compared with that of M_2 and M_3 methods of nitrogen application. The cumulative reduction in straw concentration was almost same for M_2 (40.54%) and M_3 (41.77%) methods of N application after 90 days of sowing. At 120 days after sowing, the cumulative reduction in straw concentration was also significantly lower (62.81%) in case of M_1 method

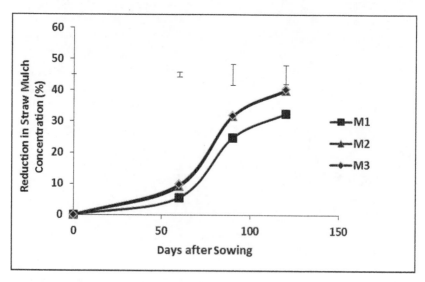

FIGURE 13.2   Percent reduction in straw mulch concentration.

of nitrogen application compared with that of M_2 (80.44%) and M_3 (82.09%) methods of N application. The lower reduction in straw mulch concentration in M_1 method of N application as compared with M_2 and M_3 could be attributed to less contact of nitrogenous fertilizer (urea) to straw; which might be beneficial for moisture conservation and maintaining higher temperature during flowering and anthesis stage of the crop.

13.3.2 RATE OF DECAY OF STANDING STUBBLE

There was the highest reduction in tensile strength (12.7%) in case of M_3 method of N application followed by 12.3% in M_2 and 10.7% in M_1 method of nitrogen application during first 60 days after sowing (Figure 13.3). At 90 and 120 DAS, the highest reduction in tensile strength was also observed in case of M_3 method of N application followed by with M_2 and M_1 methods of N application. The low reduction in tensile strength of rice straw in point-injected N application with developed applicator was due to noncontact of fertilizer urea with stubbles (less loss of nitrogen); which might be beneficial for better carbon sequestration, enhanced crop yield, and nitrogen uptake. Further with the passage of time and irrigation in cold weather stubble residue becomes brittle and slowly dissipates in soil.

13.3.3 EFFECT OF METHODS OF NITROGEN APPLICATION ON NITROGEN UPTAKE

13.3.3.1 Plant N Concentration

The effect of methods of N application on Nconcentration was discernible after crown root initiation (CRI) stage, which was significantly evident at later stages of growth that is, at 75 and 100 days after sowing. The plant N content was highest during CRI

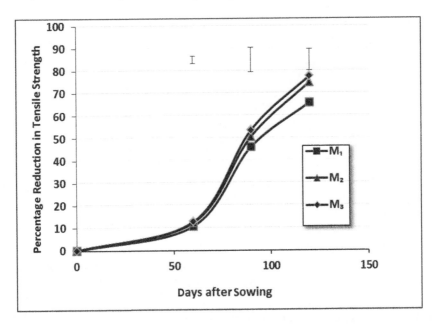

FIGURE 13.3 Percent reduction in tensile strength of standing rice stubbles.

stage and then it declined rapidly up to panicle emergence stage (Figure 13.4). At tillering stage, M_1 method of N application at both straw loads exhibited statistically higher plant N content than that of M_2 and M_3 methods. At crop maturity, M_1 method of N application exhibited highest total N content (2.16%) while M_2 method of N application had the lowest total N content (1.73%). Higher N-content in point-injected nitrogen application wheat may be attributed to more availability of N to the plants through proper and efficient application of urea solution in the soil surface without contact to straw mulch. In case of broadcasting of urea (M_2 and M_3), lower plant nitrogen content was exhibited at high straw-load condition (L_2) than that of low straw-load condition (L_1). These results suggested that straw mulch increased the nitrogen losses in case of broadcasting of urea; which resulted in low availability of N to the plants.

13.3.4 TOTAL PLANT N UPTAKE

N uptake is the interplay of biomass production and N-concentration. Highest total plant N uptake at maturity (121.44 kg/ha) occurred in case of M_1 method of N application at high straw-load while lowest N uptake (76.16 kg/ha) was observed in case of M_2 method of N application at high straw-load condition (Figure 13.5). Total plant N uptake did not differ at different straw loads with M_1 method of N application while plant N uptake differs significantly at different straw loads with M_2 and M_3 methods of N application. Blackshaw et al. (2002) also observed highest N uptake in case of point-injection followed by surface pools and broadcast in the presence of weeds. On the basis of results, point-injection is a proper method of nitrogen application, particularly, for straw-mulched (weed-infested) field.

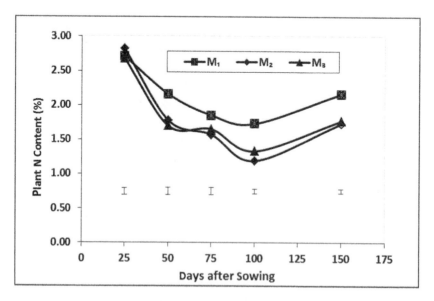

FIGURE 13.4 Effect of methods of N application on plant N content during crop growth stages.

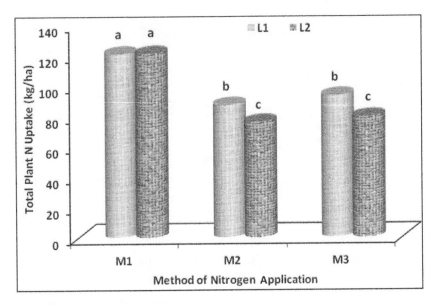

FIGURE 13.5   Total plant N uptake under different straw-load conditions (Note: Common letter does not differ from each other at $P \leq 0.05$.)

13.3.5   Nitrogen Accumulation in Straw Mulch

N accumulation in mulch tended to be similar among treatments at the time of sowing (Figure 13.6). The mean N content in straw mulch was in the range of 0.38%–0.39% for different methods of N application. The N accumulation in straw mulch in the case of M_2 and M_3 methods of N application increased or remained same during the crop growing season whereas it showed a decreasing trend in case of M_1 method of N application. At 70 DAS, mean N accumulation in mulch was significantly lower in case of M_1 than M_3 and M_2 methods of N application. But no significant difference of N accumulation in mulch among M_2 and M_3 methods of N application was observed at this stage. At the time of crop maturity, the N accumulation in straw mulch in case of M_1 method of N application was significantly lower than that of M_2 and M_3 methods. The lower accumulation of N in straw mulch in M_1 was due to noncontact of nitrogen fertilizer with straw mulch during application, which could be coupled with the favorable effect for enhancing the crop yield. The lower N accumulation in straw mulch in case of point-injected N application indicated the reduced N loss under this system of fertilizer urea application, particularly, under straw mulched conditions. Power *et al.* (1986) also found 7%–13% of the labeled N in visible residues in case of no-till with broadcasting of urea.

13.3.6   Grain Yield and Nitrogen Use Efficiency (NUE)

Methods of N application had significant effect on the number of earhead per m row length ($P < 0.05$), length of earhead ($P < 0.05$), grain number per earhead ($P < 0.01$), 1,000-grain weight ($P < 0.05$) and the grain yield ($P < 0.01$). Mean comparisons

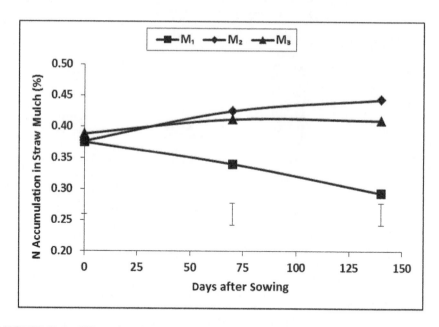

FIGURE 13.6   Effect of methods of N application on N accumulation in the straw mulch.

showed that point-injected nitrogen increased all of the above yield components (Table 13.1). The highest number of earhead per m row length, higher earhead length, higher numbers of grains per earhead resulted for point-injected nitrogen (liquid urea) applied by nitrogen applicator (M_1) at recommended dose than other broadcasting treatments (M_2 and M_3), but there were no significant differences of number of grains per earhead among the treatments M_2 and M_3. Similar trends were obtained for 1,000 grain weight with highest 1,000 grain weight in case of point-injected nitrogen application (M_1). The higher 1,000-grain yield in case of M_1 might be due to higher nitrogen concentration in plant tissue. It can also be observed from the table that there is significantly higher grain yield for point-injected nitrogen application with nitrogen applicator (M_1) than that of broadcasting urea (M_2 & M_3). There were no significant differences in grain yield between M_2 and M_3 treatments of N application. Higher root biomass and more nutrient uptake in point-injected nitrogen application resulted in increased number of earhead per m row length, earhead length, no of grains per earhead, 1,000 grain weight, and increased grain yield in this method. These findings are in line with the work of Kubesova *et al.* (2013), who observed 15.5% statistically significant higher wheat yield by injection method in comparison with the conventional method (broadcasting).

    The nitrogen use efficiency (NUE) in the case of urea solution injected with nitrogen applicator (M_1) was 47.1% more than that of broadcasting of urea (M_2). Reduced wheat NUE in broadcasting of urea under straw mulch conditions was due to lesser availability of N to the plant and lower grain yield where as high wheat NUE in nitrogen application with applicator may be associated with more N uptake and higher grain yield. Based on the result, Kücke and Gref (2006) had strongly recommended injection fertilization in case of minimum or zero tillage cereal crops.

TABLE 13.1

Yield Components for Different Nitrogen Application Methods

	Yield Attributes[a]					
Treatment	No of Earheads per m Row Length	Earhead Length (cm)	Number of Grains per Earhead	1,000 Grain Weight (gm)	NUE (%)	Grain Yield (q/ha)
M₁	97.61a	11.00a	54.42a	42.18a	72.97a	54.23a
M₂	86.56b	10.10b	42.56b	39.38b	45.59b	45.28b
M₃	93.42c	10.33c	44.33b	39.16b	52.73c	46.92b

Note: Common letters in the same column are not significantly different at $P \leq 0.05$.

[a] Average values at two straw-load conditions

13.4 CONCLUSION

The study showed that point-injected nitrogen application with a self-propelled nitrogen (liquid urea) applicator is a better alternative over conventional practice of broadcasting of urea under straw mulched no-till wheat farming. The point-injected nitrogen application was superior to other nitrogen application methods considering crop yield, plant N uptake, N use efficiency, and N accumulation in straw mulch. The lower nitrogen accumulation in straw mulch in case of point-injected nitrogen application indicated the reduced nitrogen loss under this system of fertilizer urea application, particularly, under straw-mulched conditions.

REFERENCES

Barreto, H.J. and Westerman, R.L., 1989. Soil urease activity in winter wheat residue management systems. *Soil Sci. Soc. Am. J.* 53: 1455–1458.

Blackshaw, R.E., Semach, G. and Janzen, H.H., 2002. Fertilizer application method affects nitrogen uptake in weeds and wheat. *Weed Sci.* 50: 634–641.

Cheema, H.S. and Singh, B., 1991. *Software Statistical Package CPCS-1.* Department of Statistics, Punjab Agric University, Ludhiana, India.

Cochran, W.G. and Cox, G.M., 1967. *Experimental Designs.* Asia Publishing House, New Delhi.

Fox, R.H. and Hoffman, L.D., 1981. The effect of N fertilizer source on grain yield, N uptake, soil pH, and lime requirement in no-till corn. *Agron. J.* 73: 891–895.

Kubesova, K., Balik, J., Sedlar, O. and Peklova, L., 2013. The effect of injection application of ammonium fertilizer on the yield of maize. *Sci. Agri. Bohemoslav.* 44(1): 1–5.

Kücke, M. and Gref, J.M., 2006. Experimental results and practical experiences with the fluid fertilizers point injection fertilization in Europe and potentials to optimize fertilization and to minimize environmental pollution. *Proceeding of 18th World Congress Soil Sci.* pp. 170–222. Philadelphia, Pennsylvania, USA.

Patra, A.K., Chhonkar, P.K. and Khan, M.A., 2004. Nitrogen loss and wheat yields in response to zero tillage and sowing time in a semi-arid tropical environment. *J. Agron. Crop Sci.* 190: 324–333.

Power, J.F., Wilhelm, W.W. and Doran, J.W., 1986. Recovery of fertilizer nitrogen by wheat as affected by fallow method. *Soil Sci. Soc. Am. J.* 50(6): 1499–1503.

Rice, C.W. and Smith, M.S., 1984. Denitrification in no-till and plowed soils. *Soil Sci. Soc. Am. J.* 46: 1168–1171.

Rochette, P., Angers, D.A., Chantigny, M.H., MacDonald, J. D., Bissonnette, N. and Bertrand, N., 2009. Ammonia volatilization following surface application of urea to tilled and no-till soils: A laboratory comparison. *Soil and Tillage Res.* 103: 310–315.

Scharf, P.C. and Alley, M.M., 1988. Nitrogen loss pathways and nitrogen loss inhibitors- a review. *Fert J.* 5: 109–125.

Singh, Y., Sidhu, H.S., Singh, M., Humphreys, E., Kukal, S.S. and Brar, N.K., 2008. Straw mulch, irrigation water and fertilizer N management effects on yield, water use and N use efficiency of wheat sown after rice. *Proceeding of Permanent beds and rice residue management for rice-wheat systems in the Indo-Gangetic plain.* pp. 171–181. Centre for International Agricultural Research, Canberra.

Touchton, J.T. and Hargrove, W.L., 1982. Nitrogen sources and methods of application for no-tillage corn production. *Agron. J.* 74: 823–826.

14 The Importance of Water in Relation to Plant Growth

Amit Kumar, M. K. Sharma,
Angrej Ali, and S. A. Banday
Sher-e-Kashmir University of Agricultural Sciences
and Technology of Kashmir (SKUAST-K)

CONTENTS

14.1 INTRODUCTION

With two-thirds of the earth's surface covered by water and the human body consisting of 75% of it, it is evident that water is one of the prime elements responsible for life on earth. Water circulates through the land just as it does through the human body, transporting, dissolving, and replenishing nutrients and organic matter while carrying away waste material. Water on a global scale is plentiful. However, 96.5% of it is oceans, seas and bays, and inland seas, 1.74% is trapped in the glaciers, ice and snow whereas the rest 0.6% is available in groundwater and soil moisture, less than 1% is in the atmosphere, lakes, rivers living plants and animals (USGS, 2012). About 70% of available freshwater is used for agricultural production, 22% for industrial purposes, and 8% for domestic purposes (Fishman et al., 2015). Increasing competition for domestic and industrial purposes is bound to reduce the availability of water for agriculture in future.

Water is also one of the most important inputs in crop production. It influences almost all the biophysiological processes of plants. It also influences the availability and uptake of plant nutrients from the soil and maintains plant temperature within desirable limits through transpiration. Hence, its management in crop production assumes great significance. Plants use large amounts of water during growth with important consequences for agriculture and the distribution of plant communities. The distribution of plants over the earth's surface is controlled by the availability of the water wherever temperature permits growth. Scientific water management aims at the optimal utilization of available water resources in crop production without affecting the soil health. Inadequate supply of water causes reduction in crop growth and yield. Whenever soil water becomes inadequate to support normal plant growth, water is supplied artificially, which is known as irrigation. The basic questions regarding irrigation are: (i) when, (ii) how much, and (iii) how best to irrigate.

Water management is not only important in irrigated agriculture but under rainfed conditions as well. Under rainfed conditions, it chiefly involves moisture conservation and recycling of harvested water. It is needless to say that under semi-arid conditions precipitation has to be conserved as much as possible and is used very efficiently for raising crops. Runoff is inevitable under high rainfall, but this excess water, if harvested, can be used for supplementary irrigation.

The water requirement of crops varies with the soil type (sandy, loamy, or clayey) and climate (tropical, subtropical, and temperate). Rainfall determines the periodicity and amount of irrigated water. Topography determines irrigation practices. The plant age, variety, root system, etc. also significantly influence water requirement.

14.2   ROLE OF WATER IN PLANT GROWTH

Water is the most abundant material in a growing plant. The weight of water contained in a plant is usually four to five times the total weight of dry matter (Narasimhan, 2008). Water is one of the constituents of many complex substances found in the plants, but it is interesting to note that liquid water is never found in a pure state in the environment of living organisms.

The necessity of water is readily apparent and has many functions in the plant as follows:

1. It is a major component of the plant body
2. Water is an essential solvent in which mineral nutrients are dissolved and translocated from the roots to the apex of the plant body. Minerals are also absorbed through water.
3. Numerous metabolic reactions take place in the water medium.
4. It maintains the structure of nucleic acids and proteins by supplying hydrogen bonding.
5. Several processes like photosynthesis use water as a reactant of raw material. Thus, the formation of complex carbohydrates from the simple ones also involves the removal of water while the reverse reaction requires water as a reactant.

6. This essential component is required to maintain the turgidity of the cell. Thus, it helps the cells to retain their tensile strength and provides proper shape to the cells. Turgidity is essential for the opening of flowers; folding of leaves occurs due to the changes in the turgidity of the cells.

7. It even acts as a structural agent. When plant cells contain abundant water, cells are turgid and the plant stands upright and when there is a moisture deficiency, the cells are flaccid and the plant droops and wilts.

8. It is a transportation agent thus, maintains the equilibrium of salts and other dissolved products among various parts of the plant

9. It acts as a temperature regulator in that water vapor given off by leaves produces a cooling effect.

10. It also acts as a temperature buffer since it has an exceptionally high heat capacity for specific heat.

11. Water molecules have the unique property of adhesion and cohesion and thus, these processes keep the water molecule together. This property helps in the upward movement of water in the plant body.

12. The elongation phase of cell growth is mostly dependent on water absorption.

13. Water is also a metabolic end-product of respiration.

14. The plant absorbs water quantity enormously and simultaneously loses significant amounts of water through transpiration.

We think of protoplasm as one of the basic material of life in all organisms whether plant or animal. Protoplasm is the material contained within each living cell. This is an extremely active material, the protein and other substances of which protoplasm is composed of become inactive and lifeless without water. This relationship is similar to protein in a fresh egg white as compared to that in the dried and powdered form.

We may consider the effect of water on protoplasm by using seed as an example. As seeds mature, their water content is reduced. A mature seed actually is living but it is in a relatively inactive state thus, it lives and does not grow. When water enters the seed environment, it is imbibed which helps begin the enzyme activity, seed germination, and growth to produce a living and active plant.

Water outside and within the plant performs an important role in the growth and well-being of the organism. Water dissolves the soil minerals which are essential for plant growth, and the major part of plant nutrients are absorbed into the plant in solution form. Water moving through the soil performs a function in the flushing of pore spaces. As water moves down in the soil, it replaces the gases such as carbon dioxide which occupies pore spaces; as water drains out of soil sufficient air is drawn in thus, it aids in gaseous exchange and supplying oxygen to the roots. It is known that roots best grow in moist soil and numerous observations indicate that roots will not penetrate a completely dry soil. Vapor in the atmosphere, surrounding the aerial parts of plants, play a major role with plants' performance and growth, as transpiration is affected significantly by the humidity and surrounding temperature.

Since water is essential in the performance of several functions of a plant, it is not possible to consider any one of these functions as more essential than the others. Water as a solvent dissolves the minerals in the soil which are absorbed by the plant roots. Then water as a transportation agent moves these dissolved minerals upward through the plant to the stem and leaves where they are used in the synthesis of complex compounds. The synthesized products from plant leaves are moved downward through the plant to the roots, crowns, and stems thus, water serves as a dissolving agent for both simple and complex materials, and as a transportation agent for these same materials. Bypassing the membranes of the cell walls, water helps maintain the dissolved substances in equilibrium.

Water is one of the raw materials through which carbohydrates are synthesized from carbon dioxide and water in the presence of light by the process of photosynthesis. The hydrogen for synthesizes of carbohydrates is contributed by the water. Carbohydrates are represented by the sugars and starches, which are components of other complex compounds, and by cellulose, which is the chief component of the plant skeleton. Other essential but less abundant compounds are also classified as carbohydrates.

Water may be considered as a structural agent since it maintains the turgor of plant cells. As long as the cells are tightly filled with water, the plant is relatively a rigid structure. When they become less turgid owing to the loss of water then the relatively thin walls of the skeletal components of the plant may not be rigid enough to keep it from drooping.

Many observers have found that turf gets trampled in wilted condition and more prone to injury severely than turf with an adequate supply of moisture and in which the plants are turgid. A lack of turgor in plants also reduces the photosynthetic activity because leaves tend to roll or fold following a water loss, thereby reducing light and retarding this important process.

One of the important but rarely maintained functions of water is temperature control. Water is given off by the leaves through the process of transpiration, and the evaporation of this water is accompanied by cooling. This cooling effect helps to maintain a favorable temperature around the leaves of the plant.

Modification of plant in relation to water environment is one of the important factor determining areas of adaptation of those plants. These modifications are also among the chief factors which determine the management requirements of various plants.

Red fescue may be used as an example of a grass that will tolerate relatively droughty conditions. The red fescue leaf has a heavily cutinized lower-leaf surface. The top surface of the leaf is strongly ribbed and the stomates, through which water is lost to the outside air, are located at the bottom of the grooves in the leaf; therefore, when the red fescue leaf begins to transpire rapidly and guard cells lose their turgor, the grooves close and further water loss is retarded.

On the other hand, Kentucky bluegrass is not equipped with the same mechanism, although it may protect itself from water loss it is damaged more severely by drought. The bluegrass leaf folds, because of two rows of thin-walled cells called bulliform cells, along either side of the midrib of the leaf; therefore, the bluegrass plant wilts from water loss and may get permanently damaged.

Zoysia serves as an interesting example of a plant having a defense mechanism against water loss. Members of this genus have leaves that contain numerous parallel rows of thin-walled cells through which water loss occurs (del Amor and Marcelis, 2004; Ferguson, 1959). The fact that these rows of cells are closely spaced causes the leaf to roll tightly in the presence of conditions causing rapid transpiration. The plant thus, defends itself against continuous water loss, but tight rolling of the leaves reduces the surface available for photosynthetic activity and growth is retarded. This is one of the reasons that Zoysia grow better in humid than in dry areas.

The mechanism whereby water loss from plants is prevented are too numerous varied to be discussed here, but the presence and effectiveness of such mechanism have much to do with the range of adaptation of the turf grasses that are used in this country.

Many other modifications affect the ability of the plant to grow willingly in a given set of environmental conditions relative to water. Rice grows in standing water whereas many plants would die because of a lack of oxygen in the root zone. Plants differ in the development of their root systems, some extract a greater amount of the available soil water than others.

The plant growers of any kind will learn widely about his/her particular plant species regarding its water needs, its response to an abundance water deficit in the plant soil, and as humidity in the atmosphere surrounding the plant.

Water is the source of the hydrogen which is combined with carbon and oxygen from air to form carbohydrates by the process of photosynthesis. Photosynthesis is believed to be the underlining process which supports life. Evaporation of water from leaf surface provides refrigeration, a temperature control for the plant.

Water may be considered as structural agent. Plant cells with abundant water are turgid and the plant stands upright. When there is a water deficit, the cells are flaccid and the plant drops, droops, or wilts. This principle can be demonstrated with a toy balloon. Inflated fully, it will support considerable weight while partially inflated it will support practically nothing. Many observers have noted that wilted turf suffers greater damage from traffic than does well-watered turf.

There are significant anatomical and morphological plant modifications that determine a plant's ability to survive in a given environment with respect to water. These same modifications dictate considerably the management practices that must be followed.

The depth and form of the root system play a major role in the plant behavior in abundance-water and water-deficit conditions. . The biochemical and biophysical characterization of plants differ widely; rice can grow in standing water while most plants cannot, and the cactus family members can maintain water in their tissue even in served parts under extremely hot and dry conditions.

Grass leaves display many interesting anatomical differences that are correlated with the environmental conditions in which the grasses are found. The grower should learn as much as possible about the species and its water needs.

14.3   EFFECT OF MOISTURE STRESS ON CROP GROWTH

Plant water stress, oftentimes caused by drought, can have major impacts on the plant growth and development. When it comes to crops, plant water stress can be the cause of low yields and possible crop failure. The effects of plant water stress vary among plant species. Early water-stress symptoms can be critical for maintaining the growth of a crop. The most common symptom of plant water stress is wilt. As the plant undergoes water stress, the water pressure inside the leaves decreases and the plant wilts. Drying to a condition of wilt will reduce the growth of nearly any plant. From an irrigator's perspective, managing water to minimize stress means knowing plant water availability, recognizing symptoms of water stress, and planning ahead.

Plants absorb water through their roots. The amount of force needed for a plant to remove water from the soil is known as the matric potential. During low soil moisture, plants need excess energy to extract water from the soil thus, the matric potential is greater. When the soil is dry and the matric potential is strong, plants show symptoms of stress. This is known as the matric effect.

Due to presence of salts in the roots, it becomes difficult extracting water from the soil for the plants. In general, when the soil solution is more saline than the plant, excess energy is required to absorb water than in nonsaline condition. Plant water stress caused by saline conditions is known as the osmotic effect. Stress from the osmotic effect will cause the same symptoms as stress from the matric effect.

In places with both salinity and drought conditions, plants must overcome matric and osmotic forces to absorb water. If the osmotic effect is strong (high salinity), then the matric force (how dry the soil is) needs to be minimized to reduce the combined effect of matric and osmotic forces. In simple words, more water is needed to minimize the effects of plant water stress in saline soils.

Most amount of water for irrigated crops is extracted from shallow soil depths where the majority of the roots are present. In the first few days following irrigation, 40% or more of plant water comes from the top 12 inches of soil. Plants use roots to absorb water from the soil hence, adequate root density, distribution, and conditions conducive to the root growth are important in aiding water utilization of the plant.

The main consequence of moisture stress is decreased growth and development caused by reduced photosynthesis. Photosynthesis is the process in which plants combine water, carbon dioxide, and light to synthesize carbohydrates for energy. Chemical limitations due to reductions in critical photosynthetic components such as water can negatively impact plant growth.

Low water availability can also cause physical limitations in plants. Stomates are plant cells that control the movement of water, carbon dioxide, and oxygen in and out of the plant. During moisture stress, stomates close to conserve water. This also closes the pathway for the exchange of water, carbon dioxide, and oxygen resulting in decrease in photosynthesis. Leaf growth will be more affected by moisture stress than root growth since roots have the ability to compensate for moisture stress.

Excessive water or moisture level results in a number of component stresses which include:

1. Low soil oxygen causing anaerobic conditions for roots.
2. Phytotoxic accumulation of reduced ions
3. The decrease in aerobic soil moisture
4. Accumulation of phytotoxic by-products of anaerobes
5. Attack by water-borne pathogens
6. Decreased stomatal resistance

14.4   MANAGEMENT OF WATER STRESS

Crop selection can be a key component when dealing with or anticipating moisture stress. Generalizations about plant groups and their behavior under moisture stress may help in deciding about crop selection for drought and saline conditions.

Resistant to moisture stress during vegetative stages, determinate crops are grown for harvest of mature seed and include small grains, cereal crops, peas, beans, and oilseed crops. Determinate crops show a linear relationship between water stress and seed production. These crops are most sensitive to stress during seed formation including heading, flowering, and pollination. Each has a minimum threshold growth and water requirement for the seed production. This process can be interrupted by stress and generally cannot be recovered by stress removals.

Indeterminate crops include tubers and root crops such as potatoes, carrots, and sugar beets. These crops are relatively insensitive to moisture stress in short intervals (4–5 days) throughout the growing season and have no specific critical periods. If an indeterminate crop is subjected to moisture stress, quality will be affected rather than yield. Harvestable yield increases as water use increases. Indeterminate crops are more directly related to climatic demand and cumulative water use during the season than to stress during any particular growth stage.

Forage crops are grown for hay, pasture, and biomass production. In comparison to determinate and indeterminate crops, perennial forages are affected least by moisture stress. Perennials usually have deep well-established roots systems. Forage yields depend on climatic conditions. Moisture stressed forages will have lower yields than normal forages. Annual forages are an effective way to take advantage of early-season moisture and cool temperatures. In general, as water stress is increased, forage nutritional value is increased yet overall yield and harvestable protein are decreased.

14.5   SOIL MOISTURE AVAILABILITY AND UNAVAILABILITY

The soil water availability for plant growth is the amount of water retained in soil between field capacity and the permanent wilting percentage. Since field capacity represents the upper limit of soil water availability and the permanent wilting percentage represents the lower limit, this range has considerable significance in determining the agricultural value of soils. The available water capacity of different soils varies widely as shown in Figure 14.1.

In general, fine-textured soils have a wider range of water between field capacity and permanent wilting than coarse-textured soils. Also, the slope of the curve of water potential over water content in fine-textured soils indicates a more gradual release of the water with decreasing soil water potential. In contrast, sandy soils with their larger proportion of noncapillary pore spacerelease most of their water within a narrow range of potential as the predominance of large pores and release of additional water require very low water potentials.

Data on readily available water must be used cautiously because the availability of water depends on several variables. For example, in any given soil, increased rooting depth in the profile as a whole can compensate for a narrow range of available water in one or more horizons.

Conversely, restricted root distribution combined with a narrow range of available water result in considerable hazard for plant growth from an inadequate water supply, especially, in summers where droughts are frequent. Also, it should be remembered that in many soils the range of water available for survival is substantially

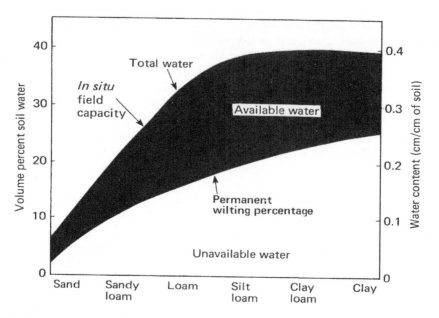

FIGURE 14.1   Relative amounts of available and unavailable water in soils. (Saadi, 2018.)

greater than that available for good growth. Furthermore, within the range of available water, the degree of availability usually tends to decline as soil water content and Ψ soil decline. It should be clear that there is no sharp limit between available and nonavailable water and that the permanent wilting percentage is only a convenient point on a curve of decreasing water potential and decreasing availability.

14.6   WATER BUDGETING

Land and water are the basic needs of agriculture and the economic development of any country. The demand for these resources is continuously increasing. Water is a vital resource of life and human/social development, and therefore, the water-balance studies or water budgeting is an essential exercise to ensure the good crop yields, land use management, and sustainable agricultural development (Figure 14.2).

A sharp increase in water consumption during the last decade, caused by intensive agricultural production and economic development, has changed the viewpoint that water is an unlimited gift of nature. It appeared that availability of freshwater due to complete intersectoral water use is far from unlimited. The ability to meet water needs varies considerably with time and space, and has caused and will continue to cause serious problems impeding the economic and social development in the many regions. The problem of water availability is a major constraint for development in many regions of the world including India. These problems are generated not only by natural factors such as uneven precipitation in space and time but numerous have also been caused by mismanagement, lack of knowledge, and coordination between stakeholders and farmers in irrigation, water delivery, and distribution

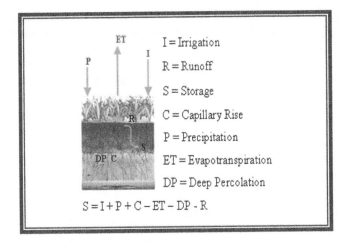

FIGURE 14.2   Water budget diagram.

Water budget is essential for comprehending the status of water availability and use, grasping the significance of the magnitudes of the budget components requires much effort. One way to verify the internal consistency of India's water budget components is to examine the magnitude of evapotranspiration implicit in them and compare the compatibility of following estimate with the estimate from other sources. Such a comparison as presented below shows an apparent discrepancy, in fact, exists between India's water budget and evapotranspiration. Water budget in its elementary form can be represented by the equation:

$$\text{Total rainfall input} = \text{Surface water flows} + \text{Ground water recharge}$$
$$+ \text{Evapotranspiration.}$$

This equation neglects the stream inflows in India from outside its borders. India's average annual rainfall is 1,170 mm and the land area is 3.28 million km² as seen from the country profile. The product of these quantities yields, for India's total rainfall input, 3,838 km³, apparently assuming that the figure has been rounded off to 4,000. Of the total rainfall, 1,869 km³ constitutes average annual potential flow in rivers while 432 km³ is considered to be replenishable groundwater (Table 14.1).

TABLE 14.1
Principal Annual Components of India's Water Budget

Component	Volume (km³)	Precipitation (%)
Precipitation	3,838	100
Potential flow in rivers	1,869	48.7
Natural recharge	432	11.3
Available water	1,869 + 432 = 2,301	60
Evapotranspiration	3,838 − (1,869 + 432) = 1,537	100 − (48.7 + 11.3) = 40.0

These figures are presented in Table 14.1 along with the magnitude of evapotranspiration implicit in them. Out of total rainfall, evapotranspiration is 40%. The remaining 60% constitutes water accessible for human use. If India's evapotranspiration estimates from other sources significantly differ from this estimate of 40%, then such a discrepancy should merit careful evaluation and reconciliation. Water budgeting reflects a balance between the inputs and outputs of water to and from the plant zone. The method is similar to balancing a checkbook. Water budgeting inputs include precipitation, irrigation, dew, and capillary rise from groundwater. The outputs include evapotranspiration, runoff, and deep percolation. Evapotranspiration is the loss of water to the atmosphere by the combined processes of evaporation and transpiration. Transpiration is water transfer to the air through plant tissues.

14.7   ROOTING CHARACTERISTICS

The role of roots in the absorption of water and minerals is well-known but they have other important functions. They are essential for the anchorage of plants in the upright position and some roots store considerable amounts of food. Root converts inorganic nitrogen into organic nitrogen compounds and synthesizes growth regulators such as cytokinins and gibberellins and other compounds such as nicotine. In turn roots are dependent on the shoots for carbohydrates, auxins, and certain vitamins.

The effectiveness of roots as absorbing organs depends on the anatomy of individual roots and on the extent and degree of branching of the root system. During growth and maturation, roots undergo extensive anatomical changes that greatly affect their permeability to water and solutes. Elongating roots usually possess four regions – the root cap, the meristematic region, the region of cell elongation, differentiation and maturation, but these regions are not always clearly delimited. Although the root cap is composed of loosely arranged cells, it is usually well-defined (Figure 14.3).

However, rootcap is absent from certain roots, such as short roots of pine. Since it has no direct connection with the vascular system, it probably has no role in absorption. It is said to be the site of perception of the gravitational stimulus, but this is debatable.

The meristematic region typically consists of numerous small, compactly arranged, and thin-walled cells almost completely filled with cytoplasm. Relatively little water or salt is absorbed through this region, mainly because of the high resistance to movement through the cytoplasm and the lack of a conducting system. Growth in the apical portion of the meristematic region is probably limited by food supply as the phloem is not differentiated in the apex and food moves by diffusion through a thick layer of cells.

It is important to determine how much of the root surface is available for the entrance of water and minerals. Water, obviously, will enter most rapidly through regions offering the lowest resistance to its movement. However, the location of the region of the lowest resistance varies with the species, the age, and rate of growth and sometimes with the magnitude of the tension developed in the water-conducting system. The anatomy of the primary roots suggests that the absorption of water and minerals occurs chiefly in a region a few centimeters behind the root tip, where root hairs are present.

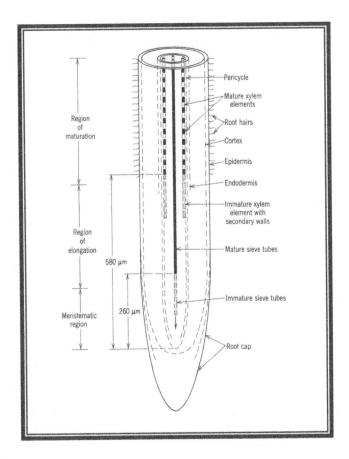

FIGURE 14.3   Root cap details.

During growth and maturation, roots undergo extensive changes in anatomy that significantly modify their permeability to water and solutes. The radial walls of the endodermal cells develop bands of suberized tissue and the epidermal cells and root hairs often collapse and die. It is generally assumed that most amount of water and solutes are absorbed near the tips of roots, but it is more evident that considerable absorption occurs through the older regions, including completely suberized roots.

The depth, the spread of root systems, and the density of root branching affect the success and survival of plants subjected to drought. The type and size of root systems are controlled by heredity and soil conditions. There are also important interactions between root and shoots, roots are dependent on shoots for carbohydrates, growth regulators, and certain vitamins while shoots are dependent on roots for water, minerals, and certain growth regulators.

14.8   MOISTURE EXTRACTION PATTERN

A tendency to over or under irrigate soil results due to the absence of information about the soil-moisture status down the soil profile. Monitoring the soil moisture is the principal fact to develop a management program to optimize plant growth and yield. There are many different methods for determining soil water content on a volume basis or tension basis like gravimetric method, resistance method, and neutron scattering method.

14.8.1 Neutron Scattering Method

The most advance developed technique for determining soil moisture –called as neutron scattering method (McKim et al., 1980; Karmer, 1983 and Narasimhan, 2008). This is used for measurement of moisture *in situ* (in the field). The instrument is known as moisture meter. This is based on the principle that hydrogen atoms (constituent of soil water) have the ability to drastically reduce the speed of the fast-moving neutron. Thus, fast neutron is emitted in the soil where they colloid with proton nuclei of water. Neutrons in this state are called slow neutrons, which can be detected with the help of a scale or meter. The density of slow neutrons so formed around the source nearly proportional to the concentration of hydrogen in the soil that is, proportional to the volume fraction of water present in the soil. These meters give accurate results in mineralizedsoils where water is the main source of combined hydrogen. In organic soils since some of the hydrogens are combined with organic substances, this method may not be suitable. For irrigation, scheduling water is the only form of H^+ that will change from one measurement to another. Therefore any change in counts recorded by NP is due to a change in the moisture with an increase in counts relating to an increase in moisture content. For field use, aluminum tubes are inserted into the soil and stopped to minimize water entry. Readings are taken at depths down the profile (e.g., 20, 30, 40, 50, 60, 80, 100, and 120 cm) with a 16 s count. Three aluminum tubes are average to read one site and utilized for appropriate irrigation management. Measurements are taken two to four times a week and information is downloaded to a personal computer for interpretation.

14.8.2 Advantages of Neutron Scattering Method

1. The measurement is not affected by salt content, soil texture, and structure soil temperature variations.
2. It is sensitive over the entire range of available soil moisture and provides a direct measure of water content by volume with the help of a suitable calibration curve.
3. Generally, the calibration curve supplied with the equipment is adequate. But under certain conditions like change in access tube or unusual neutron absorption, fresh calibration may be required. Accurate calibration curves require the use of a large homogenous body of soil with constant and uniform water content.
4. The field installations involve minimum soil disturbance and repeated readings are possible without disturbing the experimental area. The measurements are rapid and equilibrium time between the soil and the instrument is small.

14.8.3 LIMITATIONS OF NEUTRON SCATTERING METHOD

1. The main disadvantage of the method is with regard to the initial cost of the equipment, which is very high.
2. It is not advisable to measure soil moisture of surface soil (less than 20 cm) with this equipment because of radiations hazards.
3. Because of the radioactive source, the instrument is to be handled carefully.

14.9 CONCLUSION

Plant growth is affected by several factors such as seed variety, amount of water, soil type, amount of light, temperature, humidity, etc. Among these factors, water is essential for all living organisms and plants are no exception. In fact, most of the actively growing plants may contain almost 90% of water. Irrigation is water application to ensure sufficient soil-moisture availability for good plant growth throughout the growing season. Under moisture stress conditions, the vegetative growth, yield, and quality of the fruits are significantly affected, however, the soil and crop environment are also affected by excess water through the depletion of oxygen, which leads to reduced root respiration and other vital plant processes, as well as the production and accumulation of phytotoxic compounds such as ethylene in plant roots and soil. Water budget is essential for understanding the status of water availability and use, grasping the significance of the magnitudes of the budget components requires much effort. An understanding of water budget and underlying hydrologic processes provides a foundation for effective water- resource, environmental planning and management.

REFERENCES

del Amor, F.M. and Marcelis, L.F.M., 2004, November. Regulation of growth and nutrient uptake under different transpiration regimes. *International Symposium on Soilless Culture and Hydroponics 697*, pp. 523–528.

Ferguson, M.H., 1959. The role of water in plant growth. *USGA Journal and Turf Management, 11*(1), pp. 30–32.

Fishman, R., Devineni, N. and Raman, S., 2015. Can improved agricultural water use efficiency save India's groundwater? *Environmental Research Letters, 10*(8), pp. 22–31.

Karmer, J.P., 1983. *Water Relations of Plants.* Academic Press, New York, p. 489.

McKim, H.L., Walsh, J.E. and Arion, D.N., 1980. *Review of Techniques for Measuring Soil Moisture in Situ.* Directorate of Civil Works Office of the Chief of Engineers, Hanover, NH.

Narasimhan, T.N., 2008. A note on India's water budget and evapotranspiration. *Journal of Earth System Science, 117*(3), pp. 237–240.

Saadi, S., 2018. Spatial estimation of actual evapotranspiration and irrigation volumes using water and energy balance models forced by optical remote sensing data (VIS/NIT/TIR). Submitted to Ph.D. University De Toulouse, Institute National Agronimique de Tunisie (INAT).

USGS, 2012. Earth's water distribution 2012 at the Wayback Machine. Ga.Water.USGS.gov. Irrigation in Southern and Eastern Asia in Figures-AQUASTAT Survey 2011.

15 Influence of Deficit Irrigation on Various Phenological Stages of Temperate Fruits

Tsering Dolkar, Amit Kumar, M. K. Sharma, and Angrej Ali
Sher-e-Kashmir University of Agricultural Sciences and Technology of Kashmir (SKUAST-K)

CONTENTS

15.1 INTRODUCTION

Fundamentally, plants require energy (light), water, carbon and mineral nutrients for growth. Abiotic stress can be defined as the negative impact of environmental factors on the organisms in a specific situation. It is a natural phenomenon that occurs in multiple ways and interdependent, and its impact varies across the sectors of agriculture. Unlike a biotic stress that would include living disturbances such as fungi or harmful insects, abiotic stress factors or stressors are naturally occurring, often intangible, factors such as intense sunlight or wind that may cause harm to the plants. Plants are especially dependent on environmental factors, so it is particularly constraining. The most common of the stressors are easy to identify, but there are many other, less recognizable abiotic stress factors that affect environments constantly. The abiotic stresses like temperature (heat, cold chilling/frost), water (drought, flooding/hypoxia), radiation (UV, ionizing radiation), chemicals (mineral/nutrient deficiency/excess, pollutants heavy metals/pesticides, gaseous toxins), mechanical (wind, soil movement, submergence) are responsible for major reduction in agricultural production. The lesser-known stressors generally occur on a smaller scale and so are less noticeable; they include poor edaphic conditions like physical and physico-chemical properties, high radiation, compaction, contamination, rapid dehydration during seed germination, etc.

Climatic variability is the biggest challenging factor that affects agriculture in India and elsewhere (Goyary, 2009). Drought, salinity, extreme temperatures, and oxidative stress are interconnected and affect the water relations of a plant on the cellular as well as whole plant level causing specific as well as unspecific reactions (Beck et al., 2007). This leads to a series of morphological, physiological, biochemical and molecular changes that adversely affect plant growth and productivity.

Worldwide, 6,510 million hectares (mha) of land is under rainfed agriculture of which approximately 60% are in the developing countries (FAO, 2007; Pascual et al., 2009). India ranks first among the dry land agricultural countries in terms of both the extent and value of produce (Abdel-Fattah et al., 2011; Lisar et al., 2012; DAC, 2013). Out of a total 142.1 mha of cultivated area in India, dry land accounts for 91.0 mha and also in the foreseeable future nearly 60% of our population will continue to depend on dryland/rainfed agriculture. Water, an integral part of living systems, is ecologically important because it is a major force in shaping climatic patterns and is biochemically important because it is a necessary component in physiological processes. Water plays a key role in transpiration and photosynthesis and regulates the stomata, and thus is crucial to growth and leaf expansion of plants. Also, it is the primary solvent in physiological processes by which gases, minerals, and other materials enter plant cells and by which these materials are translocated to various parts of the plant.

Too much rain, which can drown the crop, delay harvest and accelerate soil erosion, can be just as serious as too little rain. It is apparent that annual rainfall averages alone are not a dependable gauge of the rainfall in an area, although it gives a good general indication of the amount of moisture available for crop production. The amount of rainfall that ends up stored in the soil for crop use depends on other factors such as water run-off and evaporation from the soil surface. The rate at which crops use water is the highest under hot, dry conditions and the lowest when it is very humid. In contrast, under flood condition, all pores are filled with water; so the oxygen supply is almost completely deprived (waterlogging) and plant roots cannot obtain oxygen for respiration to maintain their activities for nutrient and water uptake. Plants weakened by the lack of oxygen are much more susceptible to diseases caused by soil-borne pathogens. Flood tolerance is dependent upon crop species, prior plant stress (e.g., freezing weather, drought), crop load, air temperatures (warm temperatures are more detrimental), soil type, flooding depth and duration. When plants are unable to absorb enough water to replace that lost by transpiration, water stress develops in the plant system. The results may be wilting, cessation of growth, or even death of the plant or plant parts. In India, irrigation is available for only 40% of the cultivated area and the remaining 60% depends on scanty rains.

As cities grow and populations increase, the problem worsens since needs for water increase in households, industry and agriculture. Climate change has also contributed significantly to the water scarcity problem. Rising temperatures increase the rate of evaporation from land and surface water resources, thus caused reductions in river run-off in several areas. The rise in temperatures has also greatly affected areas that rely on snowmelt and mountain glaciers as a water source. Water scarcity does not only occur in arid and semi-arid areas but also occurs in areas that receive ample rainfall and/or have abundant fresh water resources. How the available water is used, managed and conserved determines if there is enough to meet household, agricultural, industrial and environmental demands. Water is an essential resource to sustain life. It is a principal factor in agricultural production as for the proper development every plant needs an optimum water supply that meets its physiological needs.

Plants are diverse and differ in the minimum amounts of water each plant needs for survival and optimum production. During dry growing periods when there is not enough rainfall to compensate for soil moisture losses through evapotranspiration, application of irrigation water by artificial means is required to maintain proper crop growth and productivity. Irrigated agriculture is the primary user of fresh water resources (Kenny et al., 2009). Irrigation uses take almost 60% of all the world's freshwater withdrawals. It is, therefore, not surprising that irrigated agriculture, especially in arid and semi-arid areas, is facing pressures to reduce its water use in order to also cater for other water uses like power and water needs for growing urban and industrial areas, and the ample water is needed to provide in-stream flows to preserve native fish populations in various regions. Irrigated agriculture is, therefore, forced to operate under conditions of water scarcity. Irrigation, therefore, needs to be managed more efficiently and sustainably, aiming at saving water, maximizing its productivity and reducing non-point sources of pollution of the environment. Deficit irrigation (DI) is profitable when the revenue lost due to yield reduction is less than the savings in costs of production due to applying less than the required water. The impact of water stress on yields and economic returns depend upon the irrigation system, the performance of that system, production costs, and the type of crop. Knowledge of the crop's response to DI, therefore, needs to be known to achieve the desired objectives.

DI can produce significant benefits under favorable circumstances. Benefits of DI are summarized as: (i) reduction in applied water; (ii) reduction in water losses due to deep percolation and runoff leading to reduction in non-point sources of pollution and increase in irrigation efficiency; (iii) reduction in costs of production and (iv) increase in net farm income. The awareness of the growing impact of environmental stress has led to worldwide efforts in adapting horticultural production to adverse environmental conditions focusing on mitigating quantitative yield losses (Godfray et al., 2010).

15.1.1 DROUGHT

Water stress refers to the situation where cells and tissues are less than truly turgid. It occurs whenever the loss of water in transpiration exceeds the rate of absorption. When plants are unable to absorb enough water to replace that lost by transpiration, water stress develops in the plant system. The results may be wilting, reduction in photosynthesis, disturbances in physiological processes, cessation of growth, or even death of the plant or plant parts. Water stress practically affects every aspect of plant growth, modifying the anatomy, morphology, physiology, and biochemistry which are related to a decrease in turgor, water potential and osmotic potential. The occurrence of drought conditions during the production of fruit and vegetable crops is becoming more frequent with climate change patterns. Much work has been devoted to the understanding of drought effects on production and productivity of horticultural crops. The existing literature provides some insight which may lead to better understanding and perhaps also encourages future research. Water stress during the production phase of some fruits and vegetables may affect their physiology and morphology in such a manner as to influence susceptibility to weight loss in storage. There have been both positive effects reported for field water deficits (stress) in tree fruits and root vegetables. Size of fruit is important since larger fruit has lower surface area to volume ratios, which confers lower relative water loss. Another negative effect associated with water deficits is the case of root vegetables, such as carrot, where pre-harvest water stress (watering to 25%–75% of soil water field capacity) can weaken the cells, resulting in higher membrane leakage (i.e., cell damage) and consequently greater weight loss in storage. Timing of a water stress event can also be very important in determining response to postharvest abiotic stress response. One example is that 'Kensington' mango fruit (*Mangifera indica* L.) will be significantly more susceptible to postharvest chilling injury with exposure to water stress during the cell expansion phase of growth as opposed to being exposed to the stress during cell division or at a time near to harvest maturity (Lechaudel and Joas, 2007). Therefore, it is critical to avoid water stress until the fruit has reached maximum size to minimize the incidence of chilling-induced injury in storage.

15.1.2 The Concept of Deficit Irrigation

DI is a watering strategy that can be applied by different types of irrigation application methods, during the drought-sensitive growth stages of a crop and reduces the irrigation application. DI consists of deliberate and systematic under-irrigation of crops. In other words, the amount of water applied is lower than that needed to satisfy the full crop water requirements. The correct application of DI requires a thorough understanding of the yield response to water (crop sensitivity to drought stress) and the economic impact of reductions in the harvest. It is well known that reductions in the water applied usually lowers evapotranspiration (ET) and crop growth rates by limiting their principal component, transpiration (T) and as a consequence carbon assimilation. For this reason, it is of great interest to know the maximal reduction in ET compatible with obtaining benefits similar or even higher to those obtained when crop evapotranspiration (ETc) is fully satisfied in mature fruit trees. In young plantations, on the other hand, the main objective is to maximize growth so that trees can mature as fast as possible which implies the avoidance of even mild water deficits. The potential benefits of DI, therefore, will come from:

i. increased water use efficiency (WUE),
ii. reduced irrigation and production costs, and
iii. the opportunity cost of water.

Crop water productivity (WP) is a key term in the evaluation of DI strategies and was defined by Geerts and Raes (2009) as the ratio of the mass of marketable yield (Ya) to the volume of water consumed by the crop (ETa):

$$WP\left(kg\,m^{-3}\right) = Ya/Eta$$

Water restriction is limited to drought-tolerant phenological stages, often the vegetative stages and the late ripening period. In regions where water resources are restrictive, it can be more profitable for a farmer to maximize crop WP instead of maximizing the harvest per unit land. The saved water can be used for other purposes or to irrigate extra units of land. DI maximizes WP, which is the main limiting factor. In other words, DI aims at stabilizing yields and at obtaining maximum crop WP rather than maximum yields. DI is sometimes referred to as incomplete supplemental irrigation or regulated DI.

15.1.3 Regulated Deficit Irrigation

Regulated deficit irrigation (RDI) is generally defined as an irrigation practice whereby a crop is irrigated with an amount of water below the full requirement for optimal plant growth; this is to reduce the amount of water used for irrigating crops, improve the response of plants to the certain degree of water deficit in a positive manner, and reduce irrigation amounts or increase the crop's WUE. RDI is primarily about restricting irrigation between fruit set and veraison to control shoot growth to influence the quality of the crop. RDI has been adapted successfully for tree crops and mainly for grape production (Girona et al., 2006). Among herbaceous crops, RDI has been applied to sugar beet, cotton and tomatoes. This practice uses water stress to control vegetative and reproductive growth, and it generally imposes water deficits during crop growing phases that are not yield reducing. Precision irrigation strategies, e.g., micro-irrigation, are paramount for a successful application of RDI, as well as timing control and soil water level monitoring. The relevant factors (both positive and negative) affecting the choice to use RDI are

1. RDI admits furrow irrigation,
2. control of fruit size and quality can be achieved,
3. vegetative growth can be controlled,
4. RDI causes potential yield losses,
5. positive effects of RDI mainly recorded on grape and wine quality,
6. marginal water savings, and
7. soil water monitoring is recommended

15.1.4 Partial Root-Zone Drying

Another approach to developing practical solutions to manipulate vegetative and reproductive crop growth is partial rootzone drying (PRD) also known as partial root-zone irrigation. PRD is a modified form of DI, which involves irrigating only one part of the root zone in each irrigation event, leaving another part to dry to certain soil water content before rewetting by shifting irrigation to the dry side; therefore, PRD is a novel irrigation strategy since half of the roots are placed in drying soil and the other half is growing in irrigated soil (Ahmadi et al., 2010).

PRD is a new irrigation technique that improves the WUE (by up to 50%). The technique was developed based on knowledge of the mechanisms controlling transpiration and requires that approximately half of the root system is always maintained in a dry or drying state while the remainder of the root system is irrigated (Figure 15.1). These applications are possible only as a consequence of a better understanding of physiological responses to water deficit and the widespread use of precision irrigation strategies, e.g., drip and other forms of microirrigation that

FIGURE 15.1　Partial root-zone drying using two above-ground drip lines in a vineyard.

enable the precise control of water application rate and timing. In PRD, a percentage of crop evapotranspiration is applied to alternate plant sides, allowing part of the root system to be in contact with wet soil all the time. In the literature, PRD irrigation has been shown to increase WUE and decrease vegetative vigor without significantly reducing crop yield (Liu et al., 2006). The beneficial effects of PRD are hypothesized due to a reduction in stomatal conductance and growth by chemical signals, possibly abscisic acid (ABA) synthesized by the roots and transported to the leaves in the transpiration stream. The main factors that may affect the choice of PRD are the following:

1. Drip irrigation is preferred within PRD; alternate row furrow irrigation is possible.
2. No effects on fruit size.
3. Vegetative growth can be controlled.
4. Positive effects on irrigated crop quality.
5. Significant water savings.
6. Significant cost increase for doubling laterals in cases where it is not necessary for technical reasons.
7. Soil water monitoring is recommended.

15.1.5   Water Stress – Why and How?

Plants experience water stress either when the water supply to their roots becomes limiting or when the transpiration rate becomes intense. Water stress is primarily caused by the water deficit, i.e., drought or high soil salinity. In case of high soil salinity and also in other conditions like flooding and low soil temperature, water exists in soil solution but plants cannot uptake it – a situation commonly known as 'physiological drought.' Drought occurs in many parts of the world every year, frequently experienced in the field-grown plants under arid and semi-arid climates. Regions with adequate but non-uniform precipitation also experience water limiting environments.

15.1.6   Effects of Water Stress on Plants

15.1.6.1   Photosynthesis and Respiration

Photosynthesis is particularly sensitive to the effects of water deficiency. Plants' resistance to water deficiency yields metabolic changes along with functional and structural rearrangements of photosynthesizing apparatus. Photosynthesis of higher plants decreases with the reduction in the relative water content (RWC) and leaf water potential. Lower faced with the scarcity of water resources, water stress is the single most critical threat to world food security. It was the catalyst of the great famines of the past. Because the world's water supply is limiting, future food demand for rapidly increasing population pressures is likely to further aggravate the effects of water stress. The severity of water stress is unpredictable as it depends on many factors such as occurrence and distribution of rainfall, evaporative demands and moisture storing capacity of soils.

Investigations carried out in the past provide considerable insights into the mechanism of drought tolerance in plants at a molecular level. Three main mechanisms reduce crop yield by soil water deficit:

    i. reduced canopy absorption of photosynthetically active radiation,
    ii. decreased radiation-use efficiency, and
    iii. reduced harvest index.

Water stress induces several changes in various physiological, biochemical and molecular components of photosynthesis. Water stress can influence photosynthesis either through pathway regulation by stomatal closure and decreasing flow of CO_2 into mesophyll tissue or by directly impairing metabolic activities. The main metabolic changes are decline in the regeneration of ribulose bisphosphate (RuBP) and ribulose 1, 5-bisphosphate carboxylase/oxygenase (Rubisco) protein content, decreased Rubisco activity, impairment of ATP synthesis and photophosphorylation or decreased inorganic phosphorus. In general, during the initial onset of water stress decreased conductance through stomata is the primary cause of the decline in photosynthesis. At later stages with increasing severity, drought stress causes tissue dehydration, leading to metabolic impairment. In contrast, there is evidence in some species that non-stomatal inhibition (metabolic activities) may occur first, causing a temporary increase in internal CO_2 concentration (Ci), which causes stomata to close. Recent studies suggest that both diffusive limitations through stomatal closure and non-stomatal limitation (such as oxidative damage to chloroplast) are responsible for the decline in photosynthesis under water stress.

The photosynthesis apparatus, photosystem II (PSII), plays a key role in the response of leaf photosynthesis to environmental stresses. Photosystem II is relatively more tolerant to water stress than heat stress. Drought stress disturbs the balance between the productions of reactive oxygen species that induces oxidative stress. Upon reduction in the amount of available water, plants close their stomata (via ABA signaling), which decreases the CO_2 influx. Reduction in CO_2 not only reduces the carboxylation directly but also directs more electrons to form reactive oxygen species. Severe water stress conditions limit photosynthesis due to a decrease in the activities of ribulose-1, 5-bisphosphate carboxylase/oxygenase (Rubisco), phosphoenolpyruvate carboxylase (PEP Case), NADP-malic enzyme (NADP-ME), fructose-1, 6-bisphosphatase (FBPase) and pyruvate orthophosphate dikinase (PPDK). Reduced tissue water contents also increase the activity of Rubisco binding inhibitors. Moreover, non-cyclic electron transport is down-regulated to match the reduced requirements of NADPH production and thus reduces the ATP synthesis. ROS: Reactive oxygen species (Figure 15.2).

Plant responds to osmotic stress caused by drought is at the morphological, anatomical, cellular, and molecular levels (Pessarakli, 2011). These changes include developmental changes such as a life cycle, inhibition of shoot growth and enhancement of root growth, adjustment in ion transport through uptake, extrusion and sequestration of ions and metabolic changes such as carbon metabolism, the synthesis of compatible solutes. Some of these signals are triggered by the primary osmotic stress signals whereas others may result from secondary stresses caused by primary

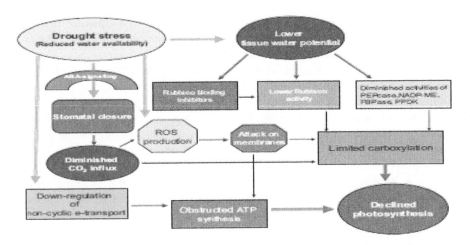

FIGURE 15.2 Reduction in photosynthesis under water stress.

signals. These secondary signals may be phytohormones, mainly ABA, reactive oxygen species (ROS) and intercellular second messengers such as phospholipids. In general, plant responses are of three kinds: maintenance of homeostasis, detoxification of harmful elements, and recovery of growth. ABA signaling pathway plays a vital role in plant stress responses as evidenced by the fact that many of drought inducible genes studied to date are also induced by ABA. Similarly, drought tolerance is dependent upon crop species, prior plant stress, and crop load. In addition, water deficit alters the cell wall nonenzymatically, for example, by the interaction of pectate and calcium (Boyer, 2009). Furthermore, water conductance to the expanding cells is affected by aquaporin activity and xylem embolism (Nardini et al., 2011). The initial growth inhibition by water deficit occurs before any inhibition of photosynthesis or respiration (Hummel et al., 2010).

15.1.6.2 Morphological, Anatomical and Cytological Changes

In the majority of the plant species, water stress is linked to changes in leaf anatomy and ultrastructure. Shrinkage in the size of leaves, decrease in the number of stomata, thickening of leaf cell walls, cutinization of leaf surface, underdevelopment of the conductive system, increase in the number of large vessels, submersion of stomata in succulent plants and xerophytes, formation of tube leaves in cereals and induction of early senescence are the other reported morphological changes.

The root-to-shoot ratio increases under water-stress conditions to facilitate water absorption and to maintain osmotic pressure, although the root dry weight and length decrease. A higher root-to-shoot ratio under the water stress conditions has been linked to the ABA content of roots and shoots. Water stress is linked to a decrease in stem length in plants and up to 25% decrease in plant height in the citrus seedling. Decreased leaf growth, total leaf area and leaf-area plasticity were observed under the water stress conditions in many plant species such as peanut. Although water saving is the important outcome of lower leaf area it causes reduced crop yield through a reduction in photosynthesis. A decrease in plant biomass consequences from the water deficit in crop plants is mainly due to low photosynthesis and plant growth and leaf senescence during the stress conditions.

Plant acclimation to drought stress that tries to eliminate excessive water loss lies in the osmotic adjustment, i.e., a decrease of cell water potential in order to diminish the difference in water potential between the plant cell and the ambient soil. This decline in osmotic potential as a response to water deficit can be achieved by solute accumulation within the plant cells or by a decreased cell volume leading to an increased concentration of osmotic solutes as water leaves from the vacuole. These phenomena are described as osmoregulation and osmotic adjustment. Osmo regulation has been defined as the regulation of osmotic potential within a cell by the addition or removal of solutes from the solution until the intracellular osmotic potential is approximately equal to the potential of the medium surrounding the cell. The osmotic adjustment refers to the lowering of the water potential due to the net accumulation of solutes in response to water deficits. Osmotic adjustment is an important mechanism in drought tolerance because it enables

   a. continuation of cell expansion,
   b. stomatal and photosynthetic adjustments,
   c. better plant growth and
   d. yield production.

The degree of the osmoregulatory process is affected by the rate of stress, stress preconditioning, the organ type, and the genetic variation between and within species. Osmotic adjustment is associated with accumulation of low molecular soluble metabolites collectively called compatible solutes like low-molecular saccharides, monosaccharides: glucose, fructose; disaccharides: sucrose, oligosaccharides like raffinose, stachyose, and verbascose organic acids and sugar alcohols, mannitol, pinitol, sorbitol, nitrogen-containing compounds such as amino acids, amides such as glutamine and asparagine, quaternary ammonium compounds called betaines: alanine betaine, glycine betaine, imino acid proline; polyamines spermine, spermidine, putrescine and with relatively high-molecular hydrophilic proteins inside the cells. Among hydrophilic proteins, several LEA proteins including dehydrins accumulate to the relatively high extents in various plant parts during the process of osmotic adjustment. Although proline role in plant osmo-tolerance remains controversial, proline is thought to contribute to osmotic adjustment, detoxification of ROS and protection of membrane integrity. Proline accumulation is believed to play adaptive roles in plant stress tolerance and has been proposed to act as a compatible osmolyte and to be the way for store carbon and nitrogen (Carillo et al., 2008). Proline has also been proposed to function as a molecular chaperone stabilizing the structure of proteins, and proline accumulation can provide a way to buffer cytosolic pH and to balance cell redox status.

During drought stress, protoplast volume shrinkage by water loss leads to loss of turgor, osmotic stress and a potential change of membrane potentials. Upon severe loss of water from the cells, membrane disintegration and abolition of metabolic processes occur. Moisture stress can also inhibit the production and accumulation of lycopene in tomato. Moisture stress has also been shown to induce phenolic accumulation through up-regulation of phenylalanine ammonia lyase (PAL). This up-regulation of PAL was associated with wound-induced ethylene production. Some of the metabolic shifts are mediated by stress response messengers (e.g., phenolics, suberin and isocoumarin accumulation) and others are a direct consequence of cellular disruption that occurs during wounding or bruising (e.g., methanethiol, allyl isothiocyanate and dimethyl sulfide accumulations). Enzymes such as polygalacturonase and pectinesterase may increase in activity leading to loss of cell wall structure and concomitant increases in soluble sugars. This may explain at least a component of the loss of firmness that has been observed with carrots as they lose water. This hypothesis is borne out by results of work with cucumbers where water stress resulted in up-regulation of polygalacturonase activity, suggesting that water loss itself was not the only factor in causing softening of stressed fruit. Another aspect of water stress is induction ethylene production, which may explain why water stress leads to accelerated ripening in bananas and accelerated senescence in bell peppers.

15.1.6.3 Water and Oxidative Stress in Plants

Exposure of plants to unfavorable environmental conditions such as alteration of temperature, high light intensity, water availability, air pollutants or salt-stress can increase the production of ROS. This phenomenon is called oxidative stress and is known as one of the major causes of plant damage as a result of environmental stresses. ROS include hydrogen peroxide, hydroxyl radicals and Superoxide anions. ROS are usually generated by normal cellular activities such as photorespiration and β-oxidation of fatty acids, but their levels increase when plants are exposed to biotic or abiotic stress conditions. Superoxide radical is regularly synthesized in the chloroplast and mitochondria, though some quantity is also reported to be produced in microbodies. Hydroxyl radical can damage, thus fatally affecting plant metabolism and ultimately growth and yield. The capacity to scavenge ROS and to reduce their damaging effects on macromolecules such as protein, DNA, lipids, chlorophyll and other important macromolecules appears to represent an important stress-tolerance. Increase in activities of antioxidant enzymes such as SOD, APX, CAT and GR under abiotic stresses and also in tolerant species/varieties have also been reported by various workers. Several reports have shown that over-expression of Superoxide dismutases leads to increased tolerance to abiotic stresses such as low temperature and water stress.

Many ROS, particularly hydrogen peroxide, behave as signalling agents to trigger biochemical changes at the gene expression level (Jaspers and Kangasjärvi, 2010). In general, abiotic stressors will induce perturbations in the fruit or vegetable cellular homeostasis which will then result in the increased generation of ROS in the apoplast, mitochondria, peroxisomes, cytoplasm, chloroplasts and endoplasmic reticulum (Jaspers and Kangasjärvi, 2010). The ability of the cell to initially cope will depend largely on the endogenous free radical scavenging capacity (Mittler, 2006). When free radical generation exceeds the endogenous scavenging capacity, the ROS interact with sensors, for which the full nature is not currently understood, that will initiate mitogen activated-protein kinase (MAPK) cascade reactions and also directly up-regulate transcription factors and calcium/calmodulin kinases (Mittler, 2006;

Jaspers and Kangasjärvi, 2010). The MAPK cascade reaction will activate various transcription factors that enable *de novo* production of ROS, ROS scavenging systems, accumulation of heat shock proteins, and modulate NADPH supply in the cell. Some of the MAPK cascade paths have also been shown to be linked specifically to ethylene production (Jaspers and Kangasjärvi, 2010), which is probably why ethylene production seems to be intrinsic to most stress responses. However, not all stressors produce identical response pathways, and so there is still a lot of work to be done in the mapping of stress response networks (Jaspers and Kangasjärvi, 2010). Water stress will lead to accelerated softening, and that response has been associated with the induction of ethylene production in response to water stress.

Oxidative stress which frequently accompanies many abiotic stresses like high temperature, salinity or water stress causes a serious secondary effect on cells. Oxidative stress is accompanied by the formation of ROSs such as O_2^-, $^1O_2$, H_2O, and OH^-. ROSs damage membranes and macromolecules affect cellular metabolism and play a crucial role in causing cellular damage under drought stress.

Water creates an imbalance between light capture and its utilization, which inhibits the photosynthesis in leaves. In this process, imbalance between the generation and utilization of electrons is created. Dissipation of excess light energy in photosynthetic apparatus results in the generation of ROS. Denaturation of functional and structural macromolecules is the well-known results of ROS production in cells. DNA nicking, amino acids, protein and photosynthetic pigments oxidation, and lipid peroxidation are the reported effects of ROS. As a consequence, cells activate some responses such as an increase in the expression of genes for antioxidant functions and production of stress proteins, up-regulation of anti-oxidants systems, including antioxidant enzymes and accumulation of compatible solutes. All these responses increase scavenging capacity against ROSs.

15.1.6.4   Nutrient Availability, Uptake and Metabolism

At the molecular or cellular level, the photosynthetic capacity of plants is closely associated with leaf N. Water stress can decrease activities of N assimilatory enzymes. The two enzymes involved in assimilating intracellular ammonium into organic compounds are nitrate reductase and glutamine synthetase. These changes in enzyme activities could be a result of changes in amino acid composition as altered by water stress. In plants, the carbon and nitrogen assimilation are coupled in plant metabolism and the limitation of photosynthesis and growth by the interaction of stress factors, such as water stress might be associated with an alteration of nitrogen levels and availability. Although nutrient and water absorption processes are independent processes, the need for water for absorption and transport makes them highly dependent on each other. Most nutrients are absorbed by plant roots as ions, and water is the medium of transport. Under fully irrigated conditions when soil water potential is high, the absorption and transport of water and nutrient are higher. Water stress decreases nutrient transport by diffusion and mass flow to the root surfaces and nutrient absorption by roots, which is influenced by water potential. Under water stress, roots are unable to take up nutrients from the soil because of the lack of activity of fine roots, water movement, and ionic diffusion of nutrients. Water stress influences nutrient uptake not only via effects of nutrient availability at the rhizoplane but also by altering the nutrient capability of mycorrhizal or non-mycorrhizal roots (Rennenberg et al., 2006).

Increased soil temperatures improve microbial mineralization of N and P, increasing its resupply to plants. Increased nutrient uptake capacities with increasing temperatures have been observed for NH_4^+, NO_3^-, PO_4^-, and K^+. Increased temperatures can cause increased nitrification and denitrification resulting in loss of N. In contrast, water stress decreases microbial activity which leads to lower nutrient availability. Since nutrient uptake by mycorrhizal or non-mycorrhizal plant roots is mainly an active transport process, it is likely that all energy-consuming enzymatic processes are highly temperature dependent.

Therefore, the reduction in nitrogen accumulation is not due to specific effects of water stress on transport proteins or accumulation mechanisms; rather, the changes in nitrogen use and flow result in conditions that inhibit nitrogen accumulation kinetics. Generally, higher N levels in the leaves enhance photosynthesis and delay leaf senescence and water stress results in a decrease in leaf N content. The absorption and assimilation of nutrients occur normally under optimum temperature conditions and any changes below or above optimum can adversely affect these processes.

15.2   EFFECTS OF WATER STRESS ON PLANTS

The effects of drought range from morphological to molecular levels and are evident at all phenological stages of plant growth at whatever stage the water deficit takes place. An account of various drought stress effects and their extent is elaborated below.

15.2.1   SEED QUALITY

There are three important aspects of seed quality:

    I. size of individual seed,
   II. composition or nutritional quality of seed, and
  III. The ability of seed to germinate and grow.

Growth environment plays an important role in all three aspects of seed quality. The impacts of drought and heat stress on seed size have been discussed in the earlier section. Water stress can have a profound impact on seed quality of fruit, mainly because of their impact on nutrient uptake assimilate supply, partitioning, and remobilization of nutrients. In addition to decreasing seed size and seed composition, water stress can decrease the viability of the harvested seeds. This was mainly attributed to insufficient calcium level that resulted from impaired calcium uptake under water stress conditions.

15.2.2   SOIL–PLANT WATER RELATIONS

Water stress causes changes in both soil and plant water potentials. Under water stress conditions, soil water potential decreases however; decreased soil water potential does not always lead to water stress because the stress response is more dependent on the plant response. Therefore, measuring soil water potential may not be a good indicator of plant drought. High soil temperature generally causes increased evaporation and decreases soil water potential. The manifestation of water in the plant often lags behind the soil water potential. Leaf water potential is often considered as a reliable parameter for quantifying plant water stress. Leaf RWC is a better indicator of water stress than plant water potential. Plant water potential varies diurnally in response to transpiration; plant water potentials are greater during daytime than night time. There is a slight lag as water absorption responds slower than water loss from plant cells. When soil water potential is high, plant water potential approaches soil water potential at night when stomata are closed. As the soil dries under water stress, hydraulic conductivity of soil decreases and the rate of water movement toward root and absorption become slow to completely replace the water lost from the plant during the daytime because of transpiration. Thus, drought results in lower plant water potential. The effects of drought on leaf water potential are progressive rather than immediate. The changes in the plant water potential can be attributed to change in osmotic pressure or osmotic component of the water potential.

When leaf water potential is low, it causes the stomata to close, which causes decreased transpiration which in turn leads to increased water potentials. However, if the drought persists, the water potential will continue to decrease and reach a zero turgor (plant water potential = plant osmotic potential = soil water potential). This point is often referred to as permanent wilting point and is a function of osmotic potential of the plant. As mentioned above, the leaf water potential is influenced by transpiration and is also dependent on vapor pressure deficit (difference in vapor pressure from leaf and ambient air), which is affected by air and leaf temperatures.

15.2.3 CROP GROWTH AND YIELD

The first and foremost effect of drought is impaired germination and poor stand establishment. Water stress has been reported to severely reduce germination and seedling stand (Kaya et al., 2006). Growth is accomplished through cell division, cell enlargement and differentiation, and involves genetic, physiological, ecological and morphological events and their complex interactions. The quality and quantity of plant growth depend on these events, which are affected by water deficit. Cell growth is one of the most drought-sensitive physiological processes due to the reduction in turgor pressure (Figure 15.3).

Under severe water deficiency, cell elongation of higher plants can be inhibited by interruption of water flow from the xylem to the surrounding elongating cells. Impaired mitosis, cell elongation and expansion result in reduced plant height, leaf area and crop growth under water stress (Hussain et al., 2008). Under water stress conditions, cell elongation in higher plants is inhibited by reduced turgor pressure. Reduced water uptake results in a decrease in tissue water contents. As a result, turgor is lost. Likewise, water stress also trims down the photoassimilation and metabolites required for cell division. As a consequence, impaired mitosis, cell elongation and expansion result in reduced growth.

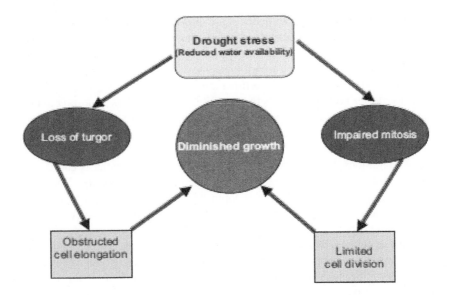

FIGURE 15.3 Description of possible mechanisms of growth reduction under water stress.

15.2.4 GROWTH PROCESSES

The general effects of mild water stress on leaves are a reduction in leaf numbers, rate of expansion, and final leaf size. Under severe stress, the rate of leaf elongation decreases and leaf growth can cease. Water stress can also influence total leaf area through its effect on initiation of new leaves, which is decreased under drought stress. Continued water stress can accelerate leaf senescence and lead to the death of leaf tissue, resulting in leaf drop, particularly old and mature leaves. Re-watering plants after a relatively short period of stress (3–5 days) do not completely eliminate the effects of water stress on the senescence process. Decreased leaf senescence under water stress is often termed as a tolerance mechanism, particularly to post-flowering water stress that occurs during fruit-filling stages. In contrast, loss of leaf area can serve as a drought-avoidance mechanism as a reduction in leaf area can help limit further water loss. Many yield-determining physiological processes in plants respond to water stress. Yield integrates many of these physiological processes in a complex way. Thus, it is difficult to interpret how plants accumulate, combine and display the ever-changing and indefinite physiological processes over the entire life cycle of crops. For water stress, severity, duration and timing of stress, as well as responses of plants after stress removal, and interaction between stress and other factors are extremely important.

15.2.5 REPRODUCTIVE PROCESSES

The success of reproduction is determined largely by the environmental conditions prevailing during the growing season. Among the various environmental factors, drought and heat stress have a direct and major influence on reproduction.

Reproduction is highly phasic, with each phase showing susceptibility to water and heat stress. Early reproductive processes particularly those of micro and megasporogensis, pollen and stigma viability, anthesis, pollination, pollen tube growth, fertilization, and early embryo development are all highly susceptible to water stress. Failure of any of these processes decreases fertilization or increases early embryo abortion, leading to a lower number of seeds thus limiting crop yield. It is important to understand that whether these processes under stress conditions are controlled by changes in carbon or nitrogen or the stress factors have a direct influence on the reproductive processes. In addition, crop developmental stages are differentially sensitive to stress conditions. Water stress just before anthesis and at anthesis caused a significant increase in floral abortion and lower seed numbers in peanut. Water stress during the early stages of embryo development increased the rate of abortion. It is important to know if the abortion is caused directly by decreased water potential in the floral tissues (pollen or ovary) or is a result of decreased carbohydrate or nitrogen flux supply or if it is related to the whole plant signaling system involving hormones particularly ABA. The response of floral parts might be different than those of developing embryos because additional connections (e.g., placenta or chalaza) involved linking the embryo inside the ear or pod. Under water stress, even though the leaf water potential was decreased, the embryos did not respond in a similar fashion and had normal water potentials. Water stress imposed at flowering can also decrease photosynthetic rates and, thus, decrease the number of photosynthates allocated to floral organs, causing increased abortion.

However, the demand for photosynthates by the small embryo is low, particularly during the very early stages of development, and the sink strength of these is much lower than in other tissues (such as vegetative tissues) to experience a shortage of photosynthates. Water stress later during the reproductive development (after fertilization) decreases seed size rather than seed number. Seed size is the final component of yield. Seed size is largely dependent on the availability of photosynthetic reserves that are either currently available or those that can be moved from other parts into the seeds. Seed size is mainly decreased by the reductions in assimilate and nitrogen supplies either through decreased photosynthetic rates or because of a decrease in photosynthetic leaf area observed under drought stress. In addition, drought can also directly shorten the seed-filling duration, resulting in smaller seed size and yield. Some studies suggest faster grain filling and enhanced mobilization of stored carbohydrates can minimize the effects of drought on yield.

15.2.6 Partitioning, Yield and Yield Components

Under mild water stress, the pattern of resource allocation generally favors root growth rather than shoot growth. Severe stress conditions often decrease root growth. Timing of drought stress also has a great influence on the partitioning of carbohydrates and nitrogen. If drought stress occurs during early vegetative growth stages, there is a shift of partitioning toward roots rather than shoots, increasing the root-to-shoot ratio. This increase is mainly due to decreased shoot weight rather than increased root weight. Root mass rarely increases under stress, whereas root length and root volume often increase in response to mild stress. Water stress occurs during the reproductive phase; there is no influence on the root-to-shoot ratio, but flowering and seed-set are decreased. If drought stress occurs after flowering, there is generally increased partitioning of resources toward seed filling. Yield is mainly a function of various components that can be broadly divided into the number of plants (germination), dry matter production (growth, potential reproductive sites), seed numbers (reproductive processes and seed-set), and seed size (product of seed-filling rate and seed-filling duration).

15.2.7   EFFECT OF MOISTURE STRESS ON FRUIT GROWTH AND QUALITY

There have been both positive and negative effects reported for field water deficits (stress) in tree fruits and vegetables. Size of fruit is important since larger fruit has a lower surface area to volume ratios, which confers lower relative water loss. Timing of a water stress event can also be very important in determining response to post-harvest abiotic stress response. Carbon metabolism and the levels of specific sugars are severely affected by abiotic stress. Under stress, a decrease in sucrose, starch and soluble sugars content was reported. The shift of metabolism toward sucrose might occur because the starch synthesis and degradation are more affected than sucrose synthesis. Trehalose is thought to protect biomolecules from environmental stress, as suggested by its reversible water-absorption capacity to protect biological molecules from desiccation-induced damage. Mannitol is another sugar alcohol that accumulates upon salt and water stress and can, thus, alleviate abiotic stress. Stress can also result in the losses of nutrient constituents in the fruit or vegetable, with vitamin C loss being the most sensitive indicator of stress exposure (Ioannidi et al., 2009). Any water deficit during the growth of banana would retard its growth and the effects may sometimes be evident only several months after the drought. The soil moisture deficit stress during the vegetative stage of banana causes a plant to extend its life cycle, poor bunch formation, lesser number of fingers and small-sized fingers (Singh et al., 2010).

Water stress in a certain period of crop growth imparts beneficial effects to plant growth and development. In the case of peaches, it has been shown that lower levels of irrigation result in the higher density of fruit surface trichomes and consequent lower weight losses in storage. In addition, two studies have shown that DI of apples and pears could reduce water loss of these fruit in subsequent storage (Lopez et al., 2011), and this was attributed to a reduction in skin permeance of the deficit irrigated fruit.

Hayward observed a significant loss in fruit weight, especially in plants exposed to stress in early summer (fruit set period). In contrast, an increase in the total soluble solids (TSS) occurred. Kiwifruit harvested from vines exposed to a less severe drought stress were unaffected in size and fruit firmness was retained for 30 days longer in comparison to control. Bordonaba and Terry (2010) testing strawberry (cvs. Elsanta, Sonata, Symphony, Florence and Christine) in response to water deficits obtained promising results. Berry size was equivalent (Florence and Christine) or smaller (Sonata and Symphony) than control plants. Considering that the main components of red color in strawberries are anthocyanins is plausible to presume that these fruit have lower contents of this secondary metabolite. However, in a previous study, Terry et al. (2007) observed the same reduction in red color, but anthocianins measurements pointed to a higher content of this metabolite.

DI has effects on fruit maturation and ripening depending on the timing of application in apple. All DI treatments increased fruit TSS and firmness regardless of maturity but had little or no effect on titratable acidity. Fruit thinning has been proposed as a feasible strategy to compensate for the loss in fruit size caused by water stress. Regarding quality parameters, deficit irrigated plants exhibited higher contents of TSS than fully irrigated plants. Soluble solids content and coloration of peach fruit increase when RDI was applied during production. RDI caused fruit peel stress lowering the content of vitamin C and carotenoids while increasing the phenolic content, mainly anthocyanins and procyanidins in peach (Buendía et al., 2008). The effects of RDI and crop load on Japanese plum (*Prunus salicina*) cv. Black-Gold were investigated by Intrigliolo and Castel (2010). RDI strategy increased the efficiency of water usage, with 30% of water savings, having minimal effect on crop yield and fruit growth. The combination of medium crop load and RDI shifted fruit mass distribution toward the low-value categories. Working with grape (cv. Rizamat), table type, Du et al. (2008) reported in alternate drip irrigation condition; the photosynthetic rate was similar to control whilst the transpiration rate kept in the same level. Stressed grapes presented higher concentrations of both ascorbic acid and TSS, and lower titrated acidity, culminating in healthier and sweeter grapes. In a similar study, Santos et al. (2007) compared the effects of partial root zone drying irrigation system (50% ETc irrigating one side at a time) with the conventional DI system (50% ETc applied on both sides), full irrigation system (100% ETc applied on both sides) and non-irrigated vines. Plants submitted to partial drying regime showed a decreased vegetative growth, expressed by the smaller values of leaf layer number, percentage of water shoots, shoot weight, pruning weight and total leaf area.

15.2.8 EFFECT OF WATER STRESS ON TEMPERATE FRUITS

15.2.8.1 Pome Fruits

Pear tree response to DI was studied in a mature commercial orchard (*Pyrus communis* L. cv. 'Blanquilla') in Lleida. DI during both Stages I or II of fruit growth affected fruit production by increasing fruit numbers but decreasing fruit size, whereas over-irrigation strongly reduced fruit numbers. Optimal fruit production occurred between these extremes. In addition, fully irrigated trees achieved the highest accumulated trunk growth and the largest fruit size. In potted pear trees, DI (15% of the control) also led to a smaller fruit size at harvest than in fully irrigated trees; despite fruit osmotic adjustment and the slightly higher tree water status in DI, when full irrigation was resumed during Stage II fruit development, the fruit growth rate remained lower in DI trees than in the control trees.

15.2.8.2 Prunus Species

In most of the research on plant responses to DI in *Prunus* sp., water restriction were applied during Stages I and II of fruit growth (initial growth and pit hardening, respectively) as well as during the postharvest period, whereas full irrigation was applied during the critical period, namely rapid fruit growth (Stage III). Overall, DI can be used successfully on peach trees. Results indicated that DI at 35% of Etc during Stage II (pit hardening) and/or during the postharvest allowed water saving of up to 22% in shallow soils and 35% in deep soils without affecting yield or final fruit size. However, the carry-over effect of DI affected yield through reductions in tree size after 3 years. Implications from these studies pointed to that water deficit during the postharvest should be managed carefully to avoid reductions in bloom and fruit load.

Peaches intended for industrial use can be managed under RDI conditions (irrigated at 40% of the control during Sage II or at 70% during Stage III) without affecting the grower's profit and even increasing the sugar content, as indicated in a long-term trial in a low water holding capacity soil in Lleida.

A reduction of up to 25% in irrigation water was obtained when the DI strategy based on DI during Stage II of peach fruit growth was applied to a medium maturing peach cultivar 'Babygold' in Murcia, with no differences in final yield. In an early maturing peach cultivar, 'Flordastar' DI strategy based on irrigation at 100% of ETc only during Stage III of peach fruit growth and 25% during the rest of the growing season led to lower yields compared with full irrigation in mature trees (Abrisqueta et al., 2009), although it performed well during young stages of tree growth (Alarcón et al., 2006). The results indicated that severe water deficits applied during the postharvest period (longer in the early maturing varieties) limited vegetative growth and the yield of mature peach trees. The experiment was performed in stony, shallow clay-loam textured soil under Mediterranean conditions in Murcia. Fruits from the DI treatment showed a lower content of vitamin C and carotenoids, while the phenolics content (mainly anthocyanins and procyanidins) increased (Buendía et al., 2008).

Pérez-Pastor et al. (2007) evaluated postharvest fruit quality of apricot (*Prunus armeniaca* L. cv. Búlida) harvested from trees exposed to three different treatments: control treatment (100% of evapotranspiration); RDI, which consists in fully irrigation during critical periods; and 50% water regime compared to control. At harvest was not observed differences in weight, equatorial diameter and firmness of the fruit among the different treatments.

In addition, fruit from water-stressed plants had higher values of TSS, titratable acidity (TA) and $h°$ value (skin color). During storage, stressed plants had lower decreases in fruit color (skin and pulp) compared to fully irrigated plants. During the first 20 days of storage, stressed plants conserved higher values of TSS and TA, after that the differences disappeared between treatments. During a simulated retail sale, fungi of the genera *Rhizopus, Monilinia, Penicillium, Alternaria, Botrytis* and *Cladosporium* caused fruit losses. Interestingly, a lower fungal attack was observed in stressed fruit. This fact was due to a thicker cuticle and to the absence of micro-crackings. They found that some qualitative characteristics such as the level of soluble solids, fruit taste and the color of the fruit are enhanced.

15.2.8.3 Kiwifruit

Plants were submitted to a stress regime composed of three different treatments: Control (water received according to culture demands), water stress in early summer and water stress in late summer. They observed a significant loss in fruit weight, especially in plants exposed to stress in early summer (fruit set period). In contrast, an increase in the TSS occurred. Differences in firmness and performance during storage were not detected for stressed kiwifruit. However, drought stress was unaffected in size and fruit firmness was retained for 30 days longer in comparison to control (fruit harvested from fully irrigated vines).

15.3　FACTORS AFFECTING PLANT SURVIVAL

The ability of landscape plants to survive water stress depends on many factors including

- severity and timing of water stress,
- species of plant,
- soil conditions, and
- additional stresses.

15.3.1　WATER STRESS SEVERITY

The length and severity of the drought are perhaps the most important factors influencing plant survival.

Drought in early spring when water for growth is critical has the greatest impact on plant health and survival.

15.3.2　PLANT SPECIES

Some plants are inherently more tolerant of water stress. Water stress tolerance may be attributed to anatomical structures such as an aggressive, deep root system or thickened, waxy leaves. Drought tolerance may also be attributed to physiological responses within the plant.

15.3.3　SOIL CONDITIONS

Soil type, organic matter content, fertility levels and other soil factors affect water stress tolerance. Plants growing in sandy soils, which have low moisture-holding capacities, are most sensitive to water stress. On heavy clay soils or those that are compacted, root growth is restricted, which can predispose plants to drought damage. Loam soils with at least 5% organic matter are conducive to root development and water retention. Nutrient deficiencies can intensify the effects of water stress. A deficiency of nitrogen or micronutrients can further impede photosynthate production. Phosphorus deficiency can restrict root growth. A deficiency of potassium can interfere with the normal functioning of the stomata's that affect internal water relations.

15.3.4　ADDITIONAL STRESSES

Landscape plants are subjected to a wide range of stresses that can intensify the effects of drought. Root damage from construction, transplanting, soil compaction and pavement over the root system has profound effects on plant survival.

Landscape plants must compete with turf or other ground covers for water and nutrients, which also intensifies the effects of drought. Other stresses that weaken landscape plants include old age, defoliation from pests, bark wounds, reflective heat from pavement and buildings and chemicals such as air pollutants, herbicides, and deicing salts.

15.4   PREVENTIVE AND REMEDIAL TREATMENTS

Water needs for irrigation can be met, in part, by practicing uniformity of water application- precise irrigation with micro-irrigation that delivers water from piped mainlines and laterals directly to the root zone frequently and in small amounts, and at rates matched to crop needs. This irrigation strategy has shown to be the best method for saline waters. However, such precise irrigation systems are expensive, but benefits include reduction of hidden costs of water wastage and land degradation, and the environmental costs of drainage and land reclamation. The net benefits of microirrigation improve markedly when such advantages are taken into account. A tax on groundwater withdrawals in a region where demand exceeds the natural rate of recharge will have a similar impact on the relative cost of microirrigation. Thus, there is a need of widespread adoption of policies that motivate farmers to reduce off-farm impacts and encourage entrepreneurs to develop low-cost microirrigation systems that are financially compatible with a wide range of crops and production environments. It has been found that up to 81% water saving was observed in lemon compared to flood irrigation with the over 35% increase in yield. Similarly, banana, grapes and pomegranate recorded 45% saving in water using drip irrigation. Wastewater can also be used for horticultural plantations in between 2 and 3 irrigation with normal water. As agriculture is the major industry that consumes maximum water judicious use of wastewater or saline water is an alternative strategy to bring down the water consumption and to alleviate moisture stress during drought period.

Most large landscape plants require 1 in. of water per week during the growing season. This is equivalent to approximately 750 gallons of water/1,000 square feet beneath the crown. For new transplants, root damaged trees or plants growing in sandy soil, water should be provided at least twice a week. Water should be concentrated on the root ball of new plantings. On established plantings in clay or loam soils, the recommended quantity of water should be supplied at least once each week. Drip irrigation systems or soaker hoses usually are most efficient since they irrigate only the root zone and minimize runoff.

The correct application of DI for a certain crop:

- maximizes the productivity of water, generally with adequate harvest quality;
- allows economic planning and stable income due to a stabilization of the harvest in comparison with rainfed cultivation; and
- control over the sowing date and length of the growing period independent from the onset of the rainy season and, therefore, improves agricultural planning.

15.4.1 CONSTRAINTS

A number of constraints apply to DI:

- Exact knowledge of the crop response to water stress is imperative.
- There should be sufficient flexibility in access to water during periods of high demand (drought-sensitive stages of a crop).
- A minimum quantity of water should be guaranteed for the crop, below which DI has no significant beneficial effect.
- An individual farmer should consider the benefit for the total water users community (extra land can be irrigated with the saved water) when he faces a below-maximum yield.
- Because irrigation is applied more efficiently, the risk for soil salinization is higher under DI as compared to full irrigation.

15.5 CONCLUSION

Abiotic stresses are significant determinants of quality and nutritional value of fruits and vegetables during growth, harvest, handling, storage and distribution to the consumer. Crop management can have a significant influence on susceptibility to stress. The most important criterion of adaptation in a natural ecosystem is survival in space and time. An important aspect of the breeding and selection approach is that there must be stressors applied in reproducible ways to allow the breeder to identify expression of stress resistance since that characteristic is adaptive, rather than constitutive, in nature. High yield potential has been the most important criterion for man from the domestication during the cultivation, but the mechanism of the adaptation and relationships of plant production is still a dilemma! It needs clarification, but to try them with only one targeted factor is almost impossible. If these points can be adequately answered (in terms of plant domestication, cultivation and agronomic respects), many affected and hidden/invisible obstacle(s) easily can be removed or overcome. Both an early crop cultivar and earlier sowing may be reducing the irrigation requirements and water amount. The increased WUE caused by increasing CO_2 will compensate only partially for the negative effects of increasing water limitation. Farming practices like 'conservation tillage and zero-tillage,' 'optimum land use' practices for the agricultural application lead to decomposition of organic materials and more carbon stored in the soil. Agricultural practices are among the biggest water-consuming activities, considering that alternatives to reduce water use in agricultural practices is of special interest in the present moment. Irrigation practices often aim at the total replacement of evapotranspiration in order to obtain the maximum yield. The use of regulated water stress (RDI) is a feasible strategy to enhance the accumulation of health-promoting compounds in food.

REFERENCES

Abdel-Fattah, G.M., El-Haddad, S.A., Hafez, E.E. and Rashad, Y.M., 2011. Induction of defense responses in common bean plants by arbuscular mycorrhizal fungi. *Microbiological Research, 166*(4), pp. 268–281.

Abrisqueta, I., Tapia, L.M., Conejero, W., Sáncheztoribio, M.I., Abrisqueta, J.M., Vera, J. and Ruizsánchez, M.C., 2009. Riego deficitario en melocotonero extratemprano. In Actas XXVII Congreso Nacional de Riegos, Murcia (pp. 23–24).

Ahmadi, S.H., Andersen, M.N. and Plauborg, F., 2010. Effects of irrigation strategies and soils on field grown potatoes: Gas exchange and xylem. *Agriculture Water Management, 97,* pp. 1486–1494.

Alarcón, J.J., Torrecillas, A., Sánchez-Blanco, M.J., Abrisqueta, J.M., Vera, J., Pedrero, F., Magaña, I., García-Orellana, Y., Ortuño, M.F., Nicolás, E. and Conejero, W., 2006. Estrategias de riego deficitario en melocotonero temprano. *Vida Rural, 225,* pp. 28–32.

Beck, E.H., Fettig, S., Knake, C., Hartig, K. and Bhattarai, T., 2007. Specific and unspecific responses of plants to cold and drought stress. *Journal of Biosciences, 32*(3), pp. 501–510.

Bordonaba, J.G. and Terry, L.A., 2010. Manipulating the taste-related composition of strawberry fruits (Fragaria × ananassa) from different cultivars using deficit irrigation. *Food Chemistry, 122*(4), pp. 1020–1026.

Boyer, J.S., 2009. Evans review: Cell wall biosynthesis and the molecular mechanism of plant enlargement. *Functional Plant Biology, 36*(5), pp. 383–394.

Buendía, B., Allende, A., Nicolás, E., Alarcón, J.J. and Gil, M.I., 2008. Effect of regulated deficit irrigation and crop load on the antioxidant compounds of peaches. *Journal of Agricultural and Food Chemistry, 56*(10), pp. 3601–3608.

Carillo, P., Mastrolonardo, G., Nacca, F., Parisi, D., Verlotta, A. and Fuggi, A., 2008. Nitrogen metabolism in durum wheat under salinity: Accumulation of proline and glycine betaine. *Functional Plant Biology, 35*(5), pp. 412–426.

DAC. 2013. Department of Agriculture and Cooperation, Agricultural statistics at a glance 2012. http://agricoop.nic.in/agristatistics.htm. Accessed 3rd March 2013.

dos Santos, T.P., Lopes, C.M., Rodrigues, M.L., de Souza, C.R., Ricardo-da-Silva, J.M., Maroco, J.P., Pereira, J.S. and Chaves, M.M., 2007. Effects of deficit irrigation strategies on cluster microclimate for improving fruit composition of Moscatel field-grown grapevines. *Scientia Horticulturae, 112*(3), pp. 321–330.

Du, T., Kang, S., Zhang, J., Li, F. and Yan, B., 2008. Water use efficiency and fruit quality of table grape under alternate partial root-zone drip irrigation. *Agricultural Water Management, 95*(6), pp. 659–668.

FAO. 2007. Mapping Biophysical Factors that Influence Agricultural Production and Rural Vulnerability. FAO, Rome. www.fao.org/docrep/010/a1075e/a1075e00.htm. Accessed 26th February 2013.

Fereres, E. and Soriano, M.A., 2006. Deficit irrigation for reducing agricultural water use. *Journal of Experimental Botany, 58*(2), pp. 147–159.

Geerts, S. and Raes, D., 2009. Deficit irrigation as an on-farm strategy to maximize crop water productivity in dry areas. *Agricultural Water Management, 96*(9), pp. 1275–1284.

Girona, J., Mata, M., Del Campo, J., Arbonés, A., Bartra, E. and Marsal, J., 2006. The use of midday leaf water potential for scheduling deficit irrigation in vineyards. *Irrigation Science, 24*(2), pp. 115–127.

Godfray, H.C.J., Beddington, J.R., Crute, I.R., Haddad, L., Lawrence, D., Muir, J.F., Pretty, J., Robinson, S., Thomas, S.M. and Toulmin, C., 2010. Food security: The challenge of feeding 9 billion people. *Science, 327*(5967), pp. 812–818.

Goyary, D., 2009, March. Transgenic crops, and their scope for abiotic stress environment of high altitude: Biochemical and physiological perspectives. *DRDO Science Spectrum,* pp. 195–201.

Hummel, I., Pantin, F., Sulpice, R., Piques, M., Rolland, G., Dauzat, M., Christophe, A., Pervent, M., Bouteillé, M., Stitt, M. and Gibon, Y., 2010. Arabidopsis thaliana plants acclimate to water deficit at low cost through changes of C usage; an integrated perspective using growth, metabolite, enzyme and gene expression analysis. *Plant Physiology, 154*(1), pp. 357–372.

Hussain, M., Malik, M.A., Farooq, M., Ashraf, M.Y. and Cheema, M.A., 2008. Improving drought tolerance by exogenous application of glycinebetaine and salicylic acid in sunflower. *Journal of Agronomy and Crop Science*, *194*(3), pp. 193–199.

Intrigliolo, D.S. and Castel, J.R., 2010. Response of plum trees to deficit irrigation under two crop levels: Tree growth, yield and fruit quality. *Irrigation Science*, *28*(6), pp. 525–534.

Ioannidi, E., Kalamaki, M.S., Engineer, C., Pateraki, I., Alexandrou, D., Mellidou, I., Giovannonni, J. and Kanellis, A.K., 2009. Expression profiling of ascorbic acid-related genes during tomato fruit development and ripening and in response to stress conditions. *Journal of Experimental Botany*, *60*(2), pp. 663–678.

Jaspers, P. and Kangasjärvi, J., 2010. Reactive oxygen species in abiotic stress signaling. *Physiologia Plantarum*, *138*(4), pp. 405–413.

Kaya, M.D., Okçu, G., Atak, M., Cıkılı, Y. and Kolsarıcı, O., 2006. Seed treatments to overcome salt and drought stress during germination in sunflower (Helianthus annuus L.). *European Journal of Agronomy*, *24*(4), pp. 291–295.

Kenny, J.F., Barber, N.L., Hutson, S.S., Linsey, K.S., Lovelace, J.K. and Maupin, M.A., 2009. Estimated use of water in the United States in 2005 *(No. 1344)*. US Geological Survey.

Lechaudel, M. and Joas, J., 2007. An overview of preharvest factors influencing mango fruit growth, quality and postharvest behaviour. *Brazilian Journal of Plant Physiology*, *19*(4), pp. 287–298.

Lisar, S.Y., Motafakkerazad, R., Hossain, M.M. and Rahman, I.M., 2012. Water stress in plants: Causes, effects and responses. *Water Stress*, *7*, pp. 1–14.

Liu, F., Shahnazari, A., Andersen, M.N., Jacobsen, S.E. and Jensen, C.R., 2006. Effects of deficit irrigation (DI) and partial root drying (PRD) on gas exchange, biomass partitioning, and water use efficiency in potato. *Scientia Horticulturae*, *109*(2), pp. 113–117.

Lopez, G., Larrigaudière, C., Girona, J., Behboudian, M.H. and Marsal, J., 2011. Fruit thinning in 'Conference'pear grown under deficit irrigation: Implications for fruit quality at harvest and after cold storage. *Scientia Horticulturae*, *129*(1), pp. 64–70.

Mittler, R., 2006. Abiotic stress, the field environment and stress combination. *Trends in Plant Science*, *11*(1), pp. 15–19.

Nardini, A., Gullo, M.A.L. and Salleo, S., 2011. Refilling embolized xylem conduits: Is it a matter of phloem unloading? *Plant Science*, *180*(4), pp. 604–611.

Pascual, M., Villar, J.M., Domingo, X. and Rufat, J., 2009. Water productivity of peach for processing in a soil with low available water holding capacity. In VI International Symposium on Irrigation of Horticultural Crops (Vol. 889, pp. 189–195).

Pérez-Pastor, A., Ruiz-Sánchez, M.C., Martínez, J.A., Nortes, P.A., Artés, F. and Domingo, R., 2007. Effect of deficit irrigation on apricot fruit quality at harvest and during storage. *Journal of the Science of Food and Agriculture*, *87*(13), pp. 2409–2415.

Pessarakli, M., 2011. Saltgrass, a high salt and drought tolerant species for sustainable agriculture in desert regions. *International Journal of Water Resources and Arid Environments*, *1*, pp. 55–64.

Rennenberg, H., Loreto, F., Polle, A., Brilli, F., Fares, S., Beniwal, R.S. and Gessler, A., 2006. Physiological responses of forest trees to heat and drought. *Plant Biology*, *8*(05), pp. 556–571.

Singh, H.P., Singh, J.P. and Lal, S.S., 2010. *Challenges of Climate Change–Indian Horticulture*. Westville Publishing House, New Delhi.

Terry, L.A., Chope, G.A. and Bordonaba, J.G., 2007. Effect of water deficit irrigation and inoculation with Botrytis cinerea on strawberry (Fragaria x ananassa) fruit quality. *Journal of Agricultural and Food Chemistry*, *55*(26), pp. 10812–10819.

Vandeleur, R.K., Mayo, G., Shelden, M.C., Gilliham, M., Kaiser, B.N. and Tyerman, S.D., 2009. The role of plasma membrane intrinsic protein aquaporins in water transport through roots: Diurnal and drought stress responses reveal different strategies between isohydric and anisohydric cultivars of grapevine. *Plant Physiology*, *149*(1), pp. 445–460.

16 High-Pressure Processing of Seafoods

Ranjna Sirohi, N. C. Shahi, and Anupama Singh
G. B. Pant University of Agriculture and Technology

Rishi Richa
Sher-e-Kashmir University of Agricultural Sciences
and Technology of Kashmir (SKUAST-K)

CONTENTS

16.1 INTRODUCTION

High-pressure processing (HPP) is a complex technology based on a simple scientific fact. When high pressure of >100 MPa is applied on any microorganisms, they are deactivated. This is the most effective technology for food materials containing moisture as moisture can act as a medium for the uniform transmission of pressure. Hence, world food market primarily and increasingly uses this technology for food preservation as it retains fresh food characteristics, flavor, and nutrients and extends shelf life. HPP offers a better alternative to thermal and chemical preservation of foodstuffs. Technologically advanced countries have already adopted HPP technology in the last few decades for food processing including fish products.

The first HPP study was conducted by Hite in 1899 at the West Virginia University in USA for the preservation of milk, meat, and fruit juices. He reported that a pressure of 650 MPa applied for 10 min destroyed microorganisms and extended the shelf life of commodities. By 1990, HPP-processed jam was the earliest commercial product, which went on sale in Japan. Subsequently, a range of food items were processed worldwide by HPP, which exposed the potential applications of high pressure in food industries. A number of universities, research institutions, and governmental research departments actively conducting studies on high-pressure technologies using laboratory-scale high-pressure equipment with the goal of helping the food sector to establish a common technical standard for the preservation of foods. Because HPP technology is a novel preservation technology for the food industry, in recent years, the gradual widespread adoption of HPP equipment has become a key factor driving the development of this technology.

In HPP, an isostatic pressure is applied to entire food molecules, that is, the same pressure is equally distributed to all parts of a food product (Patterson et al., 2006). HPP approach is based on four principles: (i) Le Chatelier principle: according to this, any phenomenon (phase transition, chemical reactivity, change in molecular configuration, and chemical reaction) accompanied by a reduction in volume will be enhanced by pressure and vice versa. One would expect that temperature would have an antagonistic effect because increasing temperature results in volume increase; (ii) Food products are compressed by uniform pressure from every direction following which they return to their original state upon pressure release. In this case, the food products are compressed independently of the product size and geometry because the transmission of pressure to the core is not mass but by time dependent; (iii) Electrostriction: according to this principle, pressure leads to increased ionization because water molecules arrange themselves more compactly around electric charges, which results in more or less negative and reversible charges depending on the chemical nature of the buffer and bio-molecular reactions; (iv) Energy input during pressure processing is very small compared with that of thermal processes; therefore, no chemical reactions are involved in HPP. Usually, hydrostatic pressure is accompanied by a moderate increase in temperature by adiabatic heating. The amount of heat produced depends on the composition of the food product being processed. Every food component has a specific heat of compression, and hydrostatic pressure is generated by the increase in free energy in the system. This can be achieved by the physical compression as a mechanical volume reduction during the pressure treatment in the closed system. HPP utilizes elevated pressure typically in the range of 100–1,000 MPa with or without addition of external heat.

HPP generally uses batch equipment; however, semi-continuous systems can also be used. The HPP apparatus is usually made up of high-strength steel alloys with high oxidization resistance and breakage toughness. The basic structure of this machine consists of a pressure module or a vessel, a process control system, power backup, and a mechanical unit, whereas the pump and vessels are housed. The pressure vessel should have a frame or yoke design to create pressure up to 600 MPa. The product taken for high-pressure treatment is packed in flexible or semi-flexible pouches made of plastic or other polymeric materials mostly. The packed product is then placed inside the vessels. The vessel is completely filled with pressure transmitting medium such as water, oil, or other organic solvents. Water is frequently used pressure-transmitting fluid in industrial-scale HPP equipments as it has low compressibility than other fluids. Once, the vessels' door is closed, the hydraulic fluid inside is pressurized using a pressure intensity pipe placed outside the vessels. The pressure is uniformly transmitted to the whole surface of pouch carrying the food item through the transmitting fluid. Once the pressure reaches the desired

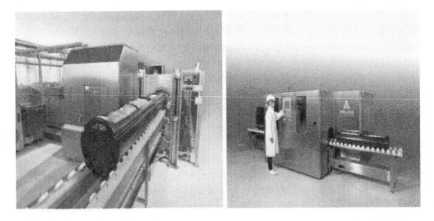

FIGURE 16.1 Commercial-scale high-pressure equipment system (horizontal configuration).

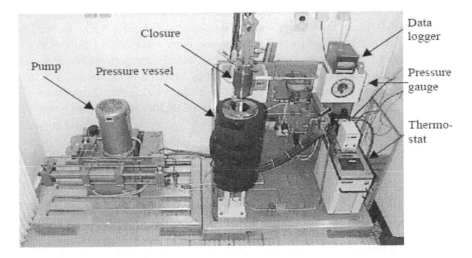

FIGURE 16.2 Schematic diagram of high-pressure equipment.

level, the product is maintained at that pressure for few minutes after which the vessel is depressurized and the product is unloaded from the sample loading basket. Programmable options are available in these machines to control the temperature, phase, and duration of pressurization. More precise results are obtained when temperature is used as an additional variable during pressure treatment. Using an electronic monitoring system, the data of the entire process can be monitored and analyzed, and the outcome can be determined (Figures 16.1 and 16.2).

16.1.1 MECHANISM OF MICROBIAL DESTRUCTION

The mode of action of microorganism completely depends on the applied pressure level. High pressure inactivates most of the spoilage and pathogenic bacteria present in food. Resistance of microorganism to pressure varies and depends on the applied pressure, temperature, duration of the process, and nature of the food material. Gram-positive bacteria are more baro-resistant than gram-negative bacteria, yeast, and molds. This is due to the presence of techoic acid, which is a bacterial polysaccharide. On the other hand, spores are more heat- resistant because they contain calcium-rich dipicolinic acid, which protects against excessive ionization. Heat-resistant microorganisms are usually pressure resistant and may be considered equally stable under pressure. Cells in exponential phase are also more pressure resistant compared with the cells in stationary phase. Yeasts, molds, coliforms, and psychrotrophs are inactivated more rapidly than resistant bacteria. Vegetative bacterial cells are inactivated at pressure in the range of 400–600 MPa. Hydrostatic pressure can induce tetraploidy in *Saccharomyces cerevisiae* (Hamada et al., 1992), showing ultra-high pressure can interfere with the replication of DNA. The hydrostatic pressure between 30 and 50 MPa can influence the gene expression and protein synthesis. HP induces changes in cell structure and its internal organelles including cell lengthening, cell wall contraction, pore formation, cell membrane separation from the cell wall, and gas vacuoles compression. In addition, changes in the distributions of DNA and ribosome, and ribosome destruction have been observed in HP-treated cells. Even though nucleic acids are more HP resistant than proteins, condensation of nuclear material has been observed at very high pressures. There is also evidence that HP can cause degradation of bacterial DNA due to the action of endonucleases, which are not normally in contact with DNA (Murchie et al., 2005). Studies have shown that pressure in the range of 300–600 MPa can inactivate many fungi and vegetative bacteria (Smelt, 1998). The nuclear membrane of yeast is affected at the pressure of approximately 100 MPa. At the pressure more than 400–600 MPa, mitochondria and cytoplasm can be altered.

Inactivation of virus depends on the denaturation of capsid protein, which is essential for host cell attachment. At high pressure, the cell membrane of microorganism is affected and the most of membrane is interrupted leading to expulsion of intracellular material due to the expansion of lipid bi-layer. This ultimately results in the destruction of cell membrane and collapse of cell integrity. Therefore, the cells are unable to reproduce and control the transport of water and ions across the membrane leading to inactivation of viruses.

16.2 APPLICATIONS OF HPP TO EXTEND THE SHELF LIFE OF SEAFOODS

HPP has the potential to increase the shelf life and improve the safety of seafoods. Generally, seafoods are spoiled by gram-negative bacteria, which are relatively pressure sensitive, as discussed in the earlier section of the chapter. High pressure may, therefore, prove to be an efficient and processing technology for such products. Thus, high pressure-treated seafood has higher proportions of gram-positive bacteria, particularly lactic acid bacteria, due to greater susceptibility of gram-negative species to high pressure. Although lactic acid bacteria may not be eliminated by high pressure, their numbers in seafood can be reduced and their growth delayed. Extensions in shelf life of high pressure-treated seafood may also reflect that off-odors associated with spoilage due to lactic acid bacteria are generally less objectionable than those produced by usual spoilage bacteria. It has also been proposed that the inhibition of other spoilage microorganisms by lactic acid bacteria may improve the preservation of foods. In addition, endogenous enzymes are also implicated in the spoilage of seafood, and the inactivation of these enzymes by high pressure would decrease the deterioration of seafood and may extend shelf life. The shelf life of high-pressure-processed food will be more or less equal to that of thermally processed food. Many research institutes now endeavor to standardize the process parameters for a variety of seafoods for high commercial potential through specific trials of this technology. One major application of HPP worldwide is shellfish processing. Shellfish being a filter feeder might carry a heavy load of harmful microbes from the water. The edible portion of shellfish is detached from the shell using HPP to facilitate raw consumption. Typical examples of HPP aquatic food products such as fishes, crabs, prawns, and edible oysters are discussed. Table 16.1 shows the effect of HPP on the visual and textural characteristics of seafoods.

16.2.1 Effect of HPP on Fishes

Fish meat contains a high nutritional and biological value, especially in terms of proteins and lipids. Fatty fish are rich in omega-3, which promotes neural health, decreases the risk of cardiac instances, and acts as an anti-depressant. However, the preservation and supply of fish is difficult due to its highly perishable nature. Fish is prone to microbial and oxidative degradation due to high water activity and a neutral pH. Conventional methods of preservation, such as heat treatment, lead to changes in texture, flavor, and appearance. Freezing and refrigeration extend the shelf life but produce undesirable characteristics including cholesterol oxidation leading to off-flavor. HPP of fish has been practiced since 1990 and is known to affect the proteins, lipids, moisture, pH, and organoleptic properties of fish differently according to the pressure applied. For instance, the spatial conformation of the protein myosin found in fish changes at pressure in the range of 100–200 Mpa, whereas actin and sacroplasmic protein remain relatively stable. Pressurized fish meat exhibits lower drip loss due to more protein denaturation than that of thermal-processed one. Myosin denaturation has also been related to decrease in textural properties such as adhesiveness,

TABLE 16.1
HPP Shucking Effect on Oyster

207 Mpa	242 MPa	242 MPa	242 MPa	276 Mpa	276 MPa	311 MPa
2 min	0 min	1 min	2 min	0 min	1 min	0 min
6% (−)	56% (−)	4% (−)	6% (−)	4% (−)	4% (0)	100% (+)
39% (0)	44% (0)	27% (0)	6% (0)	16% (0)	96% (+)	
55% (+)		69% (+)	88% (+)	80% (+)		

(−): No release of adductor muscle after HPP treatment
(0): Partial release of adductor muscle after HPP treatment
(+): Full release of adductor muscle after HPP treatment

gumminess, and cohesiveness as opposed to actin denaturation (~200 MPa), which shows increase in the same textural characteristics. Fishes are rich in amino acids such as lysine, arginine, and proline. When high pressure is applied to fish meat, apart from lipid oxidation, amino acids are also affected because of the formation of free radicals from amino acid side chains. The protein oxidation is usually detected by measuring the decrease in thiol content and the formation of carbonyls. In tilapia, high pressure in the range of 50–200 MPa for 0–60 min showed increase in free radical formation leading to protein oxidation. Similar results were reported for threadfin bream at 200–600 MPa for 10–50 min.

Fish is rich in poly-unsaturated fatty acids. These lipids are not directly affected by pressure and can remain stable even at pressure in the range of of 600–800 MPa. However, the lipids present in fish may get oxidized because of the combined effect of oxygen, metal ions (especially iron), proteins, and enzymes which may be altered during HPP. Reports indicate that high pressure may modify the melting temperature of fats up to 10°C/100 MPa. This change is induced because of modifications in lipid acyl groups resulting in lateral shrinkage and phase transformation from liquid-crystal to gel. In this regards, the studies conducted on mackerel, shrimp, turbot, and salmon showed no effect of high pressure on lipid hydrolysis. However, increase in free fatty acid (FFA) concentration was reported for horse mackerel and carp in pressure in the range of 150–450 MPa for 0–30 min. Because the oxidation of lipids is an induced phenomenon, the unfolding of myofibrillar proteins along with the disruption of protein-FFA interaction contributes to lipid oxidation.

Color is another major factor that is affected during HPP of fish. CIELAB system of color expression is generally adapted with L* representing brightness from 0 (black) to 100 (white), a* represents greenness (−a) or redness (+a), and b* represents blueness (−b) or yellowness (+b). When high pressure is applied to fish, the L* parameter increases suggesting a general light gray color along with an increased b* (toward yellow) and a decreased a* (loss of red). There are many factors that contribute to the color of fish during HPP such as pigments, state of proteins, muscle hydration, and lipid and protein oxidation. Literature suggests that even a pressure in the range of 200–400 MPa/10 min may be sufficient for significant perceivable change ($\Delta E > 3.0$) in color. The change in color can be explained by three mechanisms: (i) protein denaturation, (ii) release of iron metal ions, and (iii) oxidation of iron ($Fe^{+2} \rightarrow Fe^{+3}$). Apart from these mechanisms, an indirect effect on pigment structure may also be closely related to color change.

16.2.2   EFFECT OF HPP ON CRABS AND PRAWN

Crabmeat contains 76%–80% moisture, 19%–14% protein, 0.5%–0.8% lipids, and fewer amounts of carbohydrates, and it is rich in calcium, phosphorus, magnesium, sodium, potassium, manganese, zinc, and iron. Mostly thermal treatment is used to facilitate the removal of crab meat from the shell. Thermal treatment produces the primary flavor compounds of cooked crab meat and a considerable moisture loss is associated with the cooking, which contributes to a significant decrease in the yield of extracted crab meat (Ward et al., 1983). They also reported that the yield of crab meat after cooking is around 10%–15%, which is a very sensitive problem for crab processors. To remediate the problem, new processing techniques are required to improve the yield and quality of extracted crab meat. HPP is used as a non-thermal technology in this regard and its application in this sector is continuously growing.

HPP causes conformational changes in proteins, as partial protein unfolding promoting covalent (inter and intra molecular bonds) and non-covalent (hydrogen, hydrophobic or ionic bonds) interactions during pressurization and upon release of pressure (Huppertz et al., 2004). Partial conformational changes of proteins can modify several functional properties including water holding capacity, gelification, emulsification, and foaming (Cando et al., 2014). These modifications of the functional properties of proteins could result in important positive changes in the texture of some food products (Ramírez et al., 2011). Martínez et al. (2017) reported that cooked crab meat had higher moisture content (80%) than raw crab meat (78%), which indicates that in crab meat, mainly myofibrillar proteins are hydrated during the thermal treatment. Karthikeyan et al. (2006) reported that the higher hydration of myofibrillar proteins could be influenced by the higher concentration of myofibrillar and some polar residues available in cooked crab meat. Similar findings have been reported for other species as brown crab (Barrento et al., 2010), Chinese mitten crab (Chen et al., 2007), and blue crab (Gökoðlu and Yerlikaya, 2003; Küçükgülmez et al., 2006). Raw crab meat contains more protein in comparison to cooked meat, which could be attributed to some water-soluble proteins washing out at the time of boiling (Niamnuy et al., 2008). Martenez and others reported that as pressure increases, the extraction of crab meat was increased (Martínez et al., 2017). They obtained 15.12% extraction of crab meat through thermal methods as contrary to 20.99%, 23.78%, and 27.31% extraction after HPP at 100, 300, and 600 Mpa, respectively. Water drain during the thermal cooking process was reported as a possible cause for the extraction rate to decrease (Hong et al., 1993). The high-pressure process allowed the whole meat being easily detached from the shell and resulted in easier hand-picking than thermal-treated samples at low pressure (100 MPa). Low pressure-processed crab meat showed a fresh-like appearance, almost like raw crab meat. However, at high pressure (600 MPa), samples showed whitish cooked-like appearance (Martínez et al., 2017), which was considered as a positive result. In addition, the high-pressure-processed crab meat was slightly more voluminous and juicy than the cooked sample. After pressurization, the water-binding capacity of crab meat was also significantly increased. Increased binding capability may be due to the promotion of cross-link interactions by HPP through hydrogen bonds and hydrophobic interactions, which can retain water molecules (Uresti et al., 2005).

Prawn (*Fenneropenaeus indicus*) is one of the major seafood item and it has good price and demand in the global market. However, prawns are highly perishable due to high water activity, amino acid content, pH, bacteria, and autolytic enzymes. The content of tri-methylamine (TMA), total volatile base nitrogen (TVB-N), and FFAs, texture, color, and microbial load indicate the level of the spoilage of prawn. TMA and non-protein nitrogenous content in prawns are converted to TVB-N by enzymatic degradation (Montogomery et al. 1970). Both TMA and TVB-N have been used to assess the spoilage of prawns (Gou et al., 2010). It has been reported that HPP can retard the formation of TMA by inactivating TMAOase and inhibiting microbial growth. Similarly, reduction of TVBN content in head dock muscle after HP treatment was reported by Karim et al. (2011).

Color and appearance of prawn have a great role for consumer acceptability. Color changes of seafood after pressurization are due to the denaturation of myofibrillar and sarcoplasmic proteins, denaturation of myosin (Szczesniak, 2002), the formation of metmyoglobin, oxidation of hemoprotein, and stabilization of ferrous nitroso-myoglobin. According to Marshall et al. (2005), cooked appearance of seafood was accentuated with pressure level. Authors also found an increased lightness and yellowness and decreased redness after HP treatment.

16.2.3 EFFECT OF HPP ON OYSTERS

Curiosity in high pressure for oyster processing was influenced by the coming of *Vibrio vulnificus* as a human pathogen, which is found in oyster. *Vibrio spp.* are sensitive to high pressure; Motivatit Seafood Inc. reported that *V. vulnificus* was significantly reduced in oyster at 260 MPa high pressure for 3 min at ambient temperature. Similarly, *Vibrio parahaemolyticus* was reduced in oysters at 345 MPa for 90 s (Calik et al., 2002). HP treatment also had a positive effect on the appearance of oysters. Oysters are more voluminous following treatment and are more acceptable than untreated oysters. Texture is also an important attribute determining the perceived quality of seafood. High pressure-treated oysters were slightly juicier than untreated oysters due to the increase in moisture content of pressurized oysters (Cruz Romero et al., 2007). Although high pressure-treated oysters retain a raw taste, Hoover et al. reported enrichment in flavor, attributed to the uptake of salty liquor in which the oysters were treated. High pressure has an important additional advantage for oyster processors that during treatment the adductor muscle of oysters detaches from the shell, shucking the oyster. He et al. (2002) reported that treatment at 241 MPa for 2 min causes detachment of adductor muscle in 88% of oysters while treatment at 310 MPa, with intermediate pressure release, results in 100% efficiency of shucking (see Table 16.1). An optimum pressure for oyster shucking, which releases a high percentage of adductor muscles causing minimum other changes, may vary with oyster species and growing conditions. Initially, the opening of oysters by HPP was considered a significant drawback, as a closed shell indicates a live oyster and so is a criterion for judging oyster freshness. Though the fitting of oysters with heat-shrinkable plastic bands before treatment holds the shells together and reduces the loss of interval valve fluid, oyster treated in this way do not gape but can be shucked with minimal effort and skill and have proven an attractive alternative to traditional live oysters. It has been seen that HPP could increase the yield by around 25%.

The color, texture, taste, appearance, and smell of raw oysters are of vital concern to the consumers and industry. Nowadays, there is one perception by the industry that pressure above 300 MPa result in undesirable changes to oyster quality. It is somehow true that the degree to which HPP changes a raw oyster's characteristics is a function of the pressure applied and the temperature at which pressure is applied. When oyster was treated at 600 MPa at room temperature, some whitening and blanching took place but it was minimized when 600 MPa was applied at 5°C.

Overall appearance of the oysters and clams was much better when shucked by HPP, as shown in Figure 16.3. This differed from a hand-shucked oyster which could often be sliced by the shucking knife. Some researchers reported that HPP can induce some firmness or chewiness in seafood, but this change is relatively subtle and may be desirable, because firmness can be considered as an attribute of freshness. Juiciness and flavor were also enhanced by HPP, because oysters took up liquid from the peripheral liquor within the shell. This results in an increase of yield because the shucked oyster became more voluminous due to absorption of liquor fluid.

Oysters have higher carbohydrate content in the form of glycogen compared with other marine products. The spoilage of oysters is often a fermentative process characterized by a gradual decrease in pH. Jay (1996) reported that decrease in pH is considered as a good indicator of spoilage in oysters and other shellfish. A pH of 6.2–5.9 represented good quality while a pH of 5.2 and below exhibited sour and putrid characteristics and were unacceptable.

Cook (1991) reported that oysters with liquor above pH 6.0 are classified as good and those below 5.0 are considered to be in an advanced stage of decomposition. He et al. (2002) reported that HPP-treated oysters showed a gradual decrease in pH less than one unit after 4 weeks of storage. The pH of the control dropped to 5.1 after 16 days of storage while the pH of pressure-treated samples remained above 5.8. It can be concluded that HPP is very effective for the removal of oyster meats from the shell and producing a high-quality product of oyster. The highest pressures are the most effective in shucking and decreased the microbial concentration prolonging the shelf life of the product.

16.3 MISCELLANEOUS APPLICATION

16.3.1 IMPROVEMENT OF YIELD IN CRUSTACEANS

HPP enables seafood processors to obtain meat from smaller parts such as the legs or antennae of the crustaceans where meat is most difficult to obtain. This technology also retains natural flavors, provide shelf life extension, keeping freshness, maintaining higher sensorial qualities, functional properties, and improving food safety. Raghubeer (2007) reported that all meat was released from the shell including leg muscle after being subjected to pressure between 250 and 500 MPa in HPP vessels.

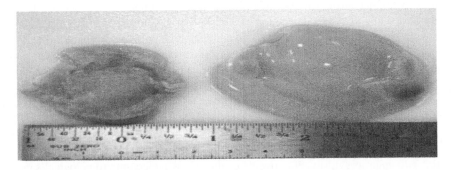

FIGURE 16.3 A hand-shucked clam (left) is compared to a HPP-treated clam (right). (Lefebvre and Robertson, 2010.)

In addition, the average total weight percentage recovered in traditional cooked Maine lobster was 25% of total body weight compared with the HPP shucked lobster, which average recovered weight percentage was 43%. A more significant increase in yield was seen in soft-shelled (recent molts) animals with a 45% recovery compared to 22% from cooking. The HPP conditions were changed to improve the textural quality of the meat from the soft-shelled lobster, which is generally less desirable when processed by cooking. Similarly, for crabs (Blue, Dungeness, Alaskan King, and Golden), average weight recovered is 19% of total body by the traditional cooking methods, on the contrary, HPP increases the percentage to an average of 35%.

16.3.2 Meat Separation (Shucking)

HPP is becoming well established as a processing technology for the shucking of shellfish and crustaceans. The commercial success of oyster shucking by HPP is well known with a number of companies in the US and elsewhere implementing the technology over the last 5 years. The same efficacy in shucking shellfish meat from their shells is seen in mussels, clams, scallops, crabs, and lobsters.

In addition to the destruction of pathogenic microorganisms, HPP shucking of shellfish and crustaceans offers the processor following advantages.

- Limited/or no need for shucking labor
- 100% meat recovery
- No physical damage of meat from shucking knives
- Increased yield weight from hydration of raw protein
- Improvement of product quality
- New markets for raw lobster and crab meats, particularly the sushi industry

Treatment at pressure above 300 MPa can also give seafood cloudiness appearance similar to that obtained by very light cooking (Hoover et al., 1989). Cod and mackerel muscle become opaque, and have higher L-values (an index of visual lightness) following HP treatment. Similar results have been reported for salmon, sheephead, bluefish, hake, carp, plaice, pollack, trout, and turbot. As the changes in appearance of HP-treated products are similar to those produced by heating, they may be of more importance for foods normally eaten raw. In addition, although HP-processed foods may have a slightly cooked appearance, consumers need to make a clear distinction between these and thermal processed. In addition, high pressure can induce the gelation of sarcoplasmic proteins, usually it is lost during the traditional production of surimi, providing the opportunity for their addition in the preparation of surimi. Other potential uses of HP technology, which may be applicable to the seafood industry, include pressure-shift freezing, HP thawing, and non-frozen storage of food.

16.4 CONCLUSION

HPP kills microorganisms effectively, resulting in enhanced microbiological safety and shelf life of seafood. Further studies related to use of HPP with combination of hurdle technology to improve the quality and shelf life at higher temperatures need to be carried out. Processing parameters including time, temperature, and pressure levels need to be optimized with respect to the quality of the final product and the consecutive uses for the pressurized seafood.

HPP offers opportunities for food manufacturers to develop new foodstuff with extended shelf life, maintained organoleptic properties and nutritional values. These processing characteristics cannot be achieved using thermal pasteurization technology, and thus, the emerging HPP technology could meet the consumer demand for safe, wholesome, new foods containing fewer additives in efficient manner.

REFERENCES

Barrento, S., Marques, A., Teixeira, B., Mendes, R., Bandarra, N., Vaz-Pires, P. and Nunes, M.L., 2010. Chemical composition, cholesterol, fatty acid and amino acid in two populations of brown crab Cancer pagurus: Ecological and human health implications. *Journal of Food Composition and Analysis, 23*(7), pp. 716–725.

Calik, H., Morrissey, M.T., Reno, P.W. and An, H., 2002. Effect of high-pressure processing on Vibrio parahaemolyticus strains in pure culture and Pacific oysters. *Journal of Food Science, 67*(4), pp. 1506–1510.

Cando, D., Moreno, H.M., Tovar, C.A., Herranz, B. and Borderias, A.J., 2014. Effect of high pressure and/or temperature over gelation of isolated hake myofibrils. *Food and Bioprocess Technology, 7*(11), pp. 3197–3207.

Chen, D.W., Zhang, M. and Shrestha, S., 2007. Compositional characteristics and nutritional quality of Chinese mitten crab (Eriocheir sinensis). *Food Chemistry, 103*(4), pp. 1343–1349.

Cook, D.W., 1991. Microbiology of bivalve molluscan shellfish. In: Ward, D.R. and Hackney, C. (Eds.), *Microbiology of Marine Food Products*. Springer, Boston, MA, pp. 19–39.

Cruz-Romero, M., Kelly, A.L. and Kerry, J.P., 2007. Effects of high-pressure and heat treatments on physical and biochemical characteristics of oysters (*Crassostrea gigas*). Innovative Food Science & Emerging Technologies, *8*, pp. 30–38.

Gökoðlu, N. and Yerlikaya, P., 2003. Determinaton of proximate composition and mineral contents of blue crab (Callinectes sapidus) and swim crab (Portunus pelagicus) caught off the Gulf of Antalya. *Food Chemistry, 80*(4), pp. 495–498.

Gou, J., Xu, H., Choi, G.P., Lee, H.Y. and Ahn, J., 2010. Application of high pressure processing for extending the shelf-life of sliced raw squid. *Food Science and Biotechnology, 19*(4), pp. 923–927.

Hamada, K., Nakatomi, Y. and Shimada, S., 1992. Direct induction of tetraploids or homozygous diploids in the industrial yeast Saccharomyces cerevisiae by hydrostatic pressure. *Current Genetics, 22*(5), pp. 371–376.

Hong, G.P., Flick, G.J. and Knobl, G.M., 1993. Development of a prediction computer model for blue crab meat yield based on processing and biological variables. *Journal of Aquatic Food Product Technology, 1*(3–4), pp. 109–132.

Huppertz, T., Fox, P.F. and Kelly, A.L., 2004. High pressure-induced denaturation of α-lactalbumin and β-lactoglobulin in bovine milk and whey: a possible mechanism. *Journal of Dairy Research, 71*(4), pp. 489–495.

He, H., Adams, R.M., Farkas, D.F. and Morrissey, M.T., 2002. Use of high-pressure processing for oyster shucking and shelf-life extension. Journal of Food Science, *67*, pp. 640–645.

Hoover, D.G., Metrick, C., Papineau, A.M., Farkas, D.F. and Knorr, D., 1989. Biological effects of high hydrostatic pressure on food microorganisms. Food Technology, *43*, pp. 99–107.

Jay, J.J., 1996. *Modern Food Microbiology*, 5th ed. New York: Chapman & Hall. p. 127.

Karim, N.U., Kennedy, T., Linton, M., Watson, S., Gault, N. and Patterson, M.F., 2011. Effect of high pressure processing on the quality of herring (Clupea harengus) and haddock (Melanogrammus aeglefinus) stored on ice. *Food Control, 22*(3–4), pp. 476–484.

Karthikeyan, M., Dileep, A.A. and Shamasundar, B.A., 2006. Effect of water washing on the functional and rheological properties of proteins from threadfin bream (Nemipterus japonicus) meat. *International Journal of Food Science & Technology, 41*(9), pp. 1002–1010.

Küçükgülmez, A., Celik, M., Yanar, Y., Ersoy, B. and Çikrikçi, M., 2006. Proximate composition and mineral contents of the blue crab (Callinectes sapidus) breast meat, claw meat and hepatopancreas. *International Journal of Food Science & Technology, 41*(9), pp. 1023–1026.

Lefebvre, K.A. and Robertson, A., 2010. Domoic acid and human exposure risks: A review. *Toxicon, 56*(2), pp. 218–230.

Marshall, M.R., Kristinsson, H. and Balaban, M.O., 2005. Effect of high pressure treatment on Omega-3 fatty acids in fish muscle. *Report. University of Florida.*

Martínez, M.A., Velazquez, G., Cando, D., Núñez-Flores, R., Borderías, A.J. and Moreno, H.M., 2017. Effects of high pressure processing on protein fractions of blue crab (Callinectes sapidus) meat. *Innovative Food Science & Emerging Technologies, 41*, pp. 323–329.

Montogomery, W.A., Sidhu, G.S. and Vale, G.I., 1970. The Australian prawn industry. *CSIRO Food Preservation Quarterly, 30*(2), pp. 21–27.

Murchie, L.W., Cruz-Romero, M., Kerry, J.P., Linton, M., Patterson, M.F., Smiddy, M. and Kelly, A.L., 2005. High pressure processing of shellfish: A review of microbiological and other quality aspects. *Innovative Food Science & Emerging Technologies, 6*(3), pp. 257–270.

Niamnuy, C., Devahastin, S. and Soponronnarit, S., 2008. Changes in protein compositions and their effects on physical changes of shrimp during boiling in salt solution. *Food Chemistry, 108*(1), pp. 165–175.

Patterson, M.F., Ledward, D.A., and Rogers, N., 2006. High pressure processing. In: Brennam, J.G., editor. *Food Processing Handbook* (pp. 173–200). Wiley-VCH, Weinheins, Germany.

Raghubeer, E.V., 2007. High hydrostatic pressure processing of seafood. Avure Technologies. Technical note "Seafood white paper". http://www.avure. com/archive/documents/Food-products/seafood-white-paper.pdf

Ramírez, J.A., Uresti, R.M., Velazquez, G. and Vázquez, M., 2011. Food hydrocolloids as additives to improve the mechanical and functional properties of fish products: A review. *Food Hydrocolloids, 25*(8), pp. 1842–1852.

Smelt, J.P.P.M., 1998. Recent advances in the microbiology of high pressure processing. Trends in Food Science &Technology, *9*, pp. 152–158.

Szczesniak, A.S., 2002. Texture is a sensory property. *Food Quality and Preference, 13*(4), pp. 215–225.

Uresti, R.M., Velazquez, G., Vázquez, M., Ramírez, J.A. and Torres, J.A., 2005. Effect of sugars and polyols on the functional and mechanical properties of pressure-treated arrowtooth flounder (Atheresthes stomias) proteins. *Food Hydrocolloids, 19*(6), pp. 964–973.

Ward, D.R., Nickelson, R., Finne, G.U.N.N.A.R. and Hopson, D.J., 1983. Processing technologies and their effects on microbiological properties, thermal processing efficiency, and yield of blue crab. *Marine Fisheries Review, 45*(7/8/9), pp. 38–43.

17 Irrigation Scheduling under Deficit Irrigation

Sushmita M. Dadhich and Hemant Dadhich
Sher-e-Kashmir University of Agricultural Sciences
and Technology of Jammu (SKUAST-J)

N. K. Garg
Indian Institute of Technology

CONTENTS

17.1 INTRODUCTION

Water is precious as it is widely used in many segments such as agricultural, industrial, household, recreational, and environmental. The accessibility of fresh water is one of the utmost emerging concerns. The world population raised by threefold in the last century from 2.2 billion to more than 5 billion and it is projected to reach 8.9 billion by 2,050. However, water availability decreased from 6,363 cubic meter per capita to 2,333 cubic meter per capita. Countries that are in semiarid and arid regions are often in a critical situation concerning fresh water availability (Calvache et al., 1997; Cresswell and Paydar, 1996; Department of Economic and Social Affairs Population Division, 2017).

The agricultural sector is the largest end user of water in the world. On a water use basis, 80%–90% of water is consumed in agriculture sector. However, due to increasing alternative water demands and water scarcity, greater attention is being given to water management in agriculture sector. Irrigation will play a key role in achieving the rates of growth in agricultural production that are needed to make available food to the world's increasing population (Blank, 1975; Wichelns, 2004; Cakir and Ebi, 2010). The pressure has been most severe in developing nations, where population is increasing rapidly but water resources are often scarce and many irrigation systems are primitive. Economic pressure on farms, increasing competition of water, and adverse environmental impacts of irrigation will motivate to optimal utilization of water resources. This new approach could be described simply as "optimization" (English et al., 1990; Zhang and Oweis, 1999; Perry, 1999; Gupta et al., 2001; Dhawan, 2017). There have been efforts how to produce more food with less water (deficit irrigation). Therefore, various innovations are in pipeline in irrigation methods, agricultural water management, crop types, and water monitoring. Improvements in irrigation water management (IWM) can help to maintain the long-term feasibility of the irrigated agricultural sector. Irrigation management means the use of optimal water for irrigation purpose over an irrigation season or over a number of seasons and the selection of cropping pattern. Improved water management may also reduce the expenditures of energy, chemicals, and labor inputs, while enhancing revenues through higher crop yields and improved crop quality. Therefore, IWM has become an important aspect for arid and semiarid regions of the world.

17.2 IRRIGATION WATER MANAGEMENT (IWM)

In arid and semiarid regions, increasing municipal and industrial demands for water are compelling major changes in irrigation management and scheduling in order to increase the water use efficiency (WUE) that is allocated to agriculture. IWM is the planning of timing and regulating irrigation water application to satisfy the crop water requirement without wasting water, soil, and plant nutrients and degrading the soil resources. The aim of irrigation planning is to maximize the net farm income, which is an appropriate goal in all projects and the irrigation should be directed toward increasing the production by meeting the crop water requirement (Garg and Ali, 1998; Paul et al., 2000; Wang et al., 2009; Garg and Dadhich, 2014). Water should be delivered to the crop to have sufficient soil-water storage, be non-erosive, have minimal waste, and be non-degrading to water quality. Irrigations remove the planned depleted soil moisture used by the crop at a fixed-time interval. Effective rainfall during the growing season should be taken into consideration. There are various terminologies used in IWM as follows:

17.2.1 Total Available Water (TAW)

The total available water (TAW) in the root zone is the difference between the water content at field capacity and wilting point. TAW refers to the capacity of a soil to keep water available to plants. Field capacity is the amount of water wherein excess water is drained out and the rate of downward water movement substantially decreases.

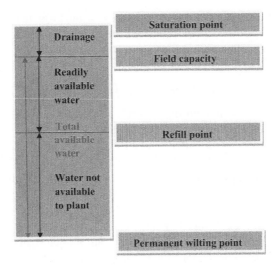

FIGURE 17.1 Water availability in the soil.

In the absence of water supply, the water available in the root zone decreases as water uptake by the crop for evapotranspiration (Figure 17.1). As water uptake is a continuous process, the available water is detained by the soil particles with larger force, lowering its potential energy and making it more difficult for the plant to extract it. Eventually, a point is reached where the crop can no longer extract the remaining water. The water uptake by the crops becomes zero when wilting point is reached. The wilting point is the water content at which plants will wilt permanently.

17.2.2 READILY AVAILABLE WATER (RAW)

The fraction of TAW that a crop can remove from the root zone without suffering to the water stress and waterlogging condition is the readily available soil water. Plant roots continue to take water from the soil, but this water is readily available and the crop finds not much difficult to extract it. It is a thumb rule that the readily available water (RAW) is half of the total available water.

17.2.3 EFFECTIVE RAINFALL (ER)

The key source of water for agricultural crop production in most parts of the world is rainfall. All the rainwater, however, cannot be useful for crop production. In simplest terms, effective rainfall means useful or utilizable rainfall in agriculture. The crop water requirement point of view, Dastane (1974) had defined effective rainfall as "that portion of the total annual or seasonal rainfall which is useful directly and/ or indirectly for meeting the crop water needs in crop production at the site where it falls but without pumping." Therefore, it includes water intercepted by living or dry vegetation and lost by evapotranspiration during crop growth. Consequently, ineffective rainfall is that portion which is lost by surface runoff, deep percolation losses, and moisture remaining in the soil after harvest of the crop and not useful for the succeeding crop. This concept of effective rainfall is suggested for use in planning and operation of irrigation projects.

17.2.4 POTENTIAL EVAPOTRANSPIRATION (ET_m)

It is the evapotranspiration from disease-free, well-fertilized crops, grown in large fields, under optimum soil water conditions and achieving full production under the given climatic conditions (Allen et al., 1998). It can be calculated by multiplying reference evapotranspiration, ET_0, by a crop coefficient, k_c. The concept of the reference evapotranspiration was introduced to study the evaporative demand of the atmosphere independently of crop type, crop development, and management practices. The evapotranspiration from an extensive surface of green grass of uniform height (0.12 m), actively growing, completely shading the ground with an albedo of 0.23 and not short of water is called reference crop evapotranspiration (Allen et al., 1998). The crop coefficient, K_c, is basically the ratio of the potential ET_m to the reference ET_0, and it represents a combination of the effects of four characteristics that are:

    i. Crop height
    ii. Albedo (reflectance) of the crop-soil surface.
    iii. Canopy resistance.
    iv. Evaporation from soil, especially exposed soil.

17.3 DEFICIT IRRIGATION

IWM decisions are often made without considering limited irrigation water availability. The region where water supply is limiting, deficit irrigation is often practiced. Deficit irrigation occurs when irrigation water is insufficient to fully satisfy the soil water efficiency in the entire root zone and subsequently full crop water requirement cannot be met for part of the growing season (Doorenbos and Kassam, 1979). Water deficit in crops has an effect on crop evapotranspiration and crop yield. Water stress in crop can be quantified by the rate of actual evapotranspiration (ET_a) in relation to the rate of potential evapotranspiration (ET_m). If water supply is not sufficient, the crop water use and yield relations, which include the effect of both timing and amount of irrigation water, are called crop water production functions. These relationships are complex as they must include the effect of crop water stress in different crop growth stages. In the literature, the most widely used relations are proposed by Doorenbos and Kassam (1979). The effect of water deficit on crop growth and crop yield can be quantified by empirically derived values known as yield response factor, K_y, (Doorenbos and Kassam, 1979). Greater values of K_y indicate that the yield will be more affected by water deficit and vice-versa. The effect of water stress on crop growth and yield depends on the magnitude and time of the occurrence of water stress. It also depends on crop species and variety. Different crops resist in a different way to water shortages. Some may have a major yield reduction when it is subject to water deficit, while others may have a minor yield reduction. One of the important irrigation strategies to obtain "more crop per drop" is deficit irrigation

(English, 1990; English and Raja, 1996; Oweis, 1997; Ali, 1995; Ali et al., 2007; Hamdy et al., 2003; Smith, 2004; Zhang et al., 2004; Jalota et al., 2006 and Ali and Talukder, 2008). Water deficits in crops, and the resulting water stress on the plant, have an effect on crop evapotranspiration and crop yield. Water stress in the plant can be quantified by the rate of actual evapotranspiration (ET_a) in relation to the rate of potential (maximum) evapotranspiration (ET_m).

Water deficit in the crop affects crop growth and yield which varies with the crop species and crop growth period. In order to quantify the effect of water stress, Doorenbos and Kassam (1979) derived a relationship between relative yield reduction and relative evapotranspiration reduction as follows:

$$\left(1 - \frac{y_a}{y_m}\right) = k_y \left(1 - \frac{ET_a}{ET_m}\right)$$
(17.1)

where,
y_a = actual crop yield
y_m = maximum crop yield
ET_a = actual evapotranspiration
ET_m = potential evapotranspiration
K_y = crop yield response factor.

The values of K_y are crop specific and may vary over the growing season. The K_y values for most of the crops are derived on the basis of assumption that the relationship between relative yield and relative evapotranspiration is linear and is valid for water deficits up to about 50% (Doorenbos and Kassam, 1979). The values of K_y are based on an analysis of experimental field data covering a wide range of growing conditions (Doorenbos and Kassam, 1979).

Following different terms and terminologies are used in deficit irrigation:

17.3.1 Maximum Yield (Y_m)

The maximum yield of the crop is defined as the harvested yield of a high producing variety, well-adapted to the given growing environment, including the time available to reach maturity, under conditions where water, nutrients, and pests and diseases do not limit the yield. The maximum yield varies from place to place as it relates to weather but is independent of soil, which is assumed that soil is physically and chemically favorable for crop growth. The climate factors that influence potential yield are solar energy, surrounding CO_2 concentration, temperature, and photosynthesis process. The maximum yield is the yield of crops where irrigation, the amount and distribution of rainfall, ensures that water deficits do not constrain yield.

17.3.2 Actual Yield (Y_a)

When water supply does not meet crop water requirements, actual evapotranspiration will fall below ET_m. Under this condition, water stress will develop in the plant which adversely affects crop growth and ultimately crop yield. The effect of water stress on growth and yield depends on the crop species and the variety on the one hand and the magnitude and the time of occurrence of water deficit on the other.

17.3.3 Actual Evapotranspiration

Actual evapotranspiration is the field condition evapotranspiration, which is different from standard condition evapotranspiration (ET_m). In field condition evapotranspiration, correction is required in ET_m due to soil water shortage, soil salinity, pests, diseases, and the presence of hard or impenetrable soil layer in the root zone, which may result in scanty plant growth and lower evapotranspiration. The effects of soil water stress are described by multiplying the crop coefficient, K_c by the water stress coefficient, K_s. ET_a in water stress condition can be estimated as follows:

$$ET_a = k_c * k_s * ET_0 \qquad (17.2)$$

17.3.4 Water Stress Coefficient (K_s)

The effects of soil water stress on crop ET are described by reducing the value for the crop coefficient. This is accomplished by multiplying the crop coefficient by the water stress coefficient, K_s. Crop coefficient is the ratio of potential evapotranspiration to the reference evapotranspiration (ET_0). Water content in the root zone can also be expressed by root zone depletion, D_r, that is, water shortage as compared to field capacity. At field capacity, the root zone depletion is zero ($D_r = 0$). When soil water is extracted by evapotranspiration, the depletion increases and stress will be induced when D_r Reaches below the RAW. After the root zone depletion exceeds RAW, the root zone depletion is high enough to limit evapotranspiration to less than potential values and the crop evapotranspiration begins to decrease in proportion to the amount of water remaining in the root zone (Allen et. al., 1998).

17.3.5 Crop Production Functions under Deficit Irrigation

The mathematical relationships between crop yield and depth of irrigation water, evapotranspiration, or transpiration with or without consideration of the time of water deficit during crop growth period are defined as crop production functions (Barrett and Skogerboe, 1980). Crop production functions are useful in evaluating alternative irrigation strategies. The crop production functions are commonly used in irrigation-optimization models. Some well-known crop production functions are enlisted in Table 17.1 (Doorenbos and Kassam, 1979; Rao et al., 1988). Multiplicative form of crop production functions (different crop growth stages) are widely used for deficit irrigation modeling.

17.4 IRRIGATION SCHEDULING WITH DEFICIT IRRIGATION

Irrigation scheduling means when and how much water to be applied to the field at each irrigation. If the water supply is adequate, irrigation can be applied when the readily available moisture is depleted, that is, when plant begins to have an adverse effect on crop growth. The factors affect the irrigation scheduling when water is

TABLE 17.1
Crop Production Functions

Source	Crop Water Production Function	Notations	Remarks
De Wit (1958)	$$y = m\left(\frac{T}{E_0}\right)$$	y = total dry matter maas per unit area; E_0 = pan evaporation for the crop period; T = total transpiration for the crop period $$m = \frac{E_0}{WR}$$ WR = water requirement	Estimate yield and transpiration under field condition
Jensen (1968)	$$\frac{y_a}{y_m} = \prod_{i=1}^{n}\left(\frac{ET_a}{ET_m}\right)_i^{\lambda_i}$$	Y_a = actual yield (kg/ ha); y_m = maximum yield when water is no limit; n = number if growth stages; λ_i = sensitivity index of crops to water stress for i^{th} growth stage	Model accuracy depends on accuracy of the sensitivity of indices λ_t
Hiler and Clark (1971)	$$\frac{y_a}{y_m} = \sum_{i-1}^{n}\left(1 - \frac{ET_a}{ET_m}\right)\left(\frac{y_m - y_i}{y_m}\right)$$	Y_i = crop yield at i th growth stage.	Relative ET deficit
Stewart (1972)	$$\frac{y_a}{y_m} = \sum_{i-1}^{n}K_y\left[\frac{(ET_m - ET_a)_i}{ET_m}\right]$$	K_y = yield response factor	Used additive effect and relative ET deficit
Minhas et al. (1974)	$$\frac{y_a}{y_m} = \prod_{i=1}^{n}\left[1 - \left(1 - \frac{ET_a}{ET_m}\right)_i^2\right]^{\lambda_i}$$	λ_i = water deficit of yield for i th crop stage.	Relative ET deficit
Doorenbos and Kassam (1979)	$$\left(1 - \frac{y_a}{y_m}\right) = k_{yi}\left(1 - \frac{ET_a}{ET_m}\right)_i$$	Y_a = yield response factor for ith crop stage	Valid for most crop for water deficit in the range $$\left(1 - \frac{ET_a}{ET_m}\right) \leq 0.5$$
Rao et al. (1988)	$$\left(1 - \frac{y_a}{y_m}\right) = \sum_{i=1}^{n}K_{yi}\left(1 - \frac{ET_{ai}}{ET_{mi}}\right)$$	–	Additive approach
Rao et al. (1988)	$$\left(1 - \frac{y_a}{y_m}\right) = 1 - \prod_{i=1}^{gs}\left[1 - k_{yi}\left(1 - \frac{ET_{ai}}{ET_{mi}}\right)\right]$$	gs = crop growth stages	Multiplicative approach

plenty are agro-climatic parameters such as temperature, wind velocity, radiation, crop type, and so on. However, under deficit irrigation, plants are forced to certain levels of water stress during either a particular crop growth stage or throughout the whole growth season (crop period), without significant reduction in yields. Crop response to water deficit at different crop growth stages is not uniform. There are a number of critical periods when water stress on crop affect significantly. Water applied per irrigation and soil moisture contents before and after irrigation are monitored throughout the seasons, while crop is harvested at the end of season. Average daily crop water use (crop consumptive use) is estimated from the soil moisture content using the soil moisture depletion method, while daily reference evapotranspiration (ET_0) is computed from weather data using the FAO Penman–Montieth method. Crop coefficient values (K_c) are computed as the ratio of crop water use to ET_0. The water stress coefficients (K_s) are computed by relating crop coefficient of the fully irrigated treatments to the deficit irrigated treatments. The yield response factor (K_y) is obtained by relating relative yield reduction to relative crop water use deficits.

In deficit irrigation, plants are forced to certain levels of water stress during either a particular crop growth stage or throughout the whole growth season (crop period), without significant reduction in yields. In general, crop has been suffered mild water stress during the vegetative period and showed higher tolerance of water deficit (Thomas et al., 1976). Grimes and Dickens (1977) reported that either early or late irrigations reduced the cotton yield. Similar work on sugar beet (Okman, 1973 and Winter, 1980), sunflower (Jana et al., 1982; Rawson and Turner, 1983; Karaata, 1991), wheat (Day and Intalap, 1970), potato (Minhas and Bansal, 1991) and on many other crops has shown the possibility of achieving optimum crop yields under deficit irrigation practices by allowing a certain level of yield loss from a given crop with higher returns gained from the diversion of water for irrigation of other crops (Krida, 2000). The crop growth stages that were most sensitive to water stress were from flowering to boll formation in cotton and grain development stage in wheat (Jalota et al., 2006). Deficit irrigation is increased the WUE of a crop by reducing irrigations that have mild impact on yield of the crop. The resulting yield reduction may be small as compared the benefits gained through diverting the saved water to irrigate other crops for which water would normally be insufficient under traditional irrigation practices (Krida, 2000). A deficit irrigation program is a part of IWM and very useful for arid and semi-arid regions. It is necessary to know crop yield responses to water stress, either during defined growth stages or throughout the whole season (Kirda and Kanber, 1999). High-yielding varieties are more sensitive to water stress than low-yielding varieties (Doorenbos and Kassam, 1979). The sensitive growth stages of wheat to water stress are from stem elongation to booting, followed by anthesis and grain-filling. Crop yield linearly increased with increase in evapotranspiration. Quadratic crop production functions with the total applied water are developed and used to estimate the levels of irrigation water for maximizing yield, net profit, and levels to which the crops could be under irrigated. The time of irrigation was also suggested on the basis of crop sensitivity index to water stress taking rainfall probability and available soil water into account. It is necessary to consider the water retention capacity curve of the soil for deficit irrigation planning. In sandy soils, plants may undergo water stress quickly under deficit irrigation, whereas plants in deep soils of fine texture may have sufficient time to adjust to low soil water matric pressure and may remain unaffected by low soil water content (Krida, 2000). Therefore, deficit irrigation is more feasible in finely textured soils and precipitation and deficit irrigation regime and their interactions on crop water productivity.

17.5 OPTIMAL SCHEDULING UNDER DEFICIT IRRIGATION

Field studies are necessary for correct application of deficit irrigation for a particular crop in a particular region. In addition to optimization simulation models, simulating the soil water balance and related crop growth (crop water productivity modeling) can be a valuable decision support tool. By conjunctively simulating, the effects of different influencing factors (climate, soil, management, and crop characteristics) on crop production, models allow to better understand the mechanism behind improved WUE to schedule the necessary irrigation applications during the drought sensitive crop growth stages, considering the possible variability in climate, to test deficit irrigation strategies of specific crops in new regions and to investigate the effects of future climate scenarios or scenarios of altered management practices on crop production. It is a theoretical procedure to determine the savings in water and the economic benefit derived from deficit irrigation, using a water stress sensitivity index for winter wheat (Triticum aestivum L.) and spring barley (Hordeum vulgare L.) in a rainfed area and some water reduction is possible. The maximum allowable water reduction for spring barley is higher than that for winter wheat. Water reductions of 7% and 26% were allowed for winter wheat and spring barley, respectively, at a benefit to cost ratio of 1.5. This corresponded to an 8% and 35% increase in cultivated area, respectively. Haouaria and Azaiez (2001) proposed a mathematical programming model for optimal cropping patterns under water deficits in dry regions. Both annual and seasonal crops were examined in the same study. The model identified the optimal operating policy for each grower in the region having a given stock of irrigation water. Then, it allocated the water efficiently among growers and the model determined the global optimal cropping plan of the entire region. There are various optimization models are developed and a few models are described here as follows:

Gorantiwar and Smout (2003) proposed a three-stage simulation-optimization model for allocating water from a reservoir optimally based on a deficit irrigation approach. The allocation model was applied for a single crop (wheat) in an irrigation scheme in India and results with a deficit irrigation approach were compared with full irrigation (irrigation to fill the root zone to field capacity) and with the existing farmers' practices. It was found that deficit irrigation practicing enables the irrigated area and increased total crop production by 30%–45% and 20%–40%, respectively, over the farmers' practices and by 50% and 45%, respectively, over the full irrigation.

Babel et al. (2004) developed an integrated water allocation model (IWAM) for optimal allocation of limited water from a storage reservoir to different user sectors, considering socio-economic, environmental, and technical aspects. IWAM comprised three modules—a reservoir operation module, an economic analysis module, and a water allocation module. Weighting technique was used to convert the multi-objective decision-making problem into a single-objective function. The single-objective functions were optimized using linear programming (LP). The model applicability was demonstrated for various cases with a hypothetical example.

Ghahraman and Sepaskhah (2004) formulated a non-linear programming (NLP) optimization model with an integrated soil water balance. The proposed model was run for the Ardak area, Iran under single cropping cultivation (corn) as well as a multiple cropping pattern (wheat, barley, corn, and sugar beet). The water balance equation was manipulated with net applied irrigation water to overcome the difficulty encountered with incorrect deep percolation. The outputs of the model, under the imposed seasonal irrigation water shortages, were compared with the results obtained from a simple NLP model. The differences between these two models (simple and integrated) became more significant as irrigation water shortage increased.

Gorantiwar and Smout (2005) have developed a resource allocation model, area and water allocation model, with the concept of deficit irrigation through a variable depth irrigation approach. This model was applied to a medium irrigation scheme in India as a case study, to obtain the land and water allocation plans. These optimal allocation plans were compared to those obtained by using the model with the existing approach (full irrigation with a fixed irrigation interval of 21 days in Rabi and 14 days in the summer season). The allocation plans were obtained taking into account the different parameters that were included in the model, such as crops and cropping pattern, soils, irrigation interval, initial reservoir storage volumes, efficiencies, and the outlet and canal capacities. The total net benefits were compared for the two cases of fixed cropping distribution and free cropping distribution and a sensitivity analysis was conducted on other parameters. The total net benefits obtained with the variable depth irrigation approach introduced in the model were found to be 22% higher than those obtained with the existing approach.

Smout and Gorantiwar (2005) formulated the area and water allocation model (AWAM), which incorporates deficit irrigation for optimizing the use of water for irrigation. This model was developed for surface irrigation schemes in semiarid regions under rotational water supply. It allocated the land area and water optimally to the different crops grown in different types of soils up to the tertiary level or allocation unit. The model had four phases. In the first phase, all the possible irrigation strategies were generated for each crop-soil-region combination. The second phase prepared the irrigation program for each strategy, taking into account the response of the crop to the water deficit. The third phase selected optimal and efficient irrigation programs. In the fourth phase of the model, irrigation programs were modified by incorporating the conveyance and the distribution efficiencies.

Prasad et al. (2006) have developed optimal irrigation planning strategies for the Nagarjuna Sagar Right Canal command in the semiarid region of South India. Optimal cropping pattern and irrigation water allocations were made with full and deficit irrigation strategies for various levels of probability of exceedance of the expected annual water available. The results found that the optimization approach can significantly improve the annual net benefit with a deficit irrigation strategy under water scarcity.

Raes et al. (2006) proposed a soil water balance model BUDGET. The model determined root zone water on a daily basis by keeping track of incoming and outgoing water fluxes at its boundary and the yield decline was estimated with the K_y approach. To account for the effect of water stresses in the various growth stages, the multiplicative, seasonal, and minimal approach was integrated in the model. The model was applied for two crops (winter wheat and maize) under various levels of water stress in two different environments and the results were compared with observed yields. Simulated crop yields agreed well with observed yields for both locations using the multiplicative approach. The correlation value (R^2) between observed and simulated yields ranged from 0.87 to 0.94 with very high modeling efficiencies. The root mean square error values were relatively small and ranged between 7% and 9%. The minimal and seasonal approaches performed significantly less accurate in both of the study areas.

Smout et al. (2006) developed a modified area and water allocation model which uses simulation–optimization technique for optimum allocation of land and water resources to different crops cultivated in different allocation units of the irrigation scheme. It included both productivity and equity in the process of developing the allocation plans for optimum productivity or maximum equity. The modified model was applied on Nazare medium irrigation scheme in India. The results indicated that the two performance objectives productivity and equity conflict with each other and in this case, equitable water distribution may be preferred over free water distribution at the cost of a small loss in productivity.

Azaiez (2008) developed an integrated dynamic programming and LP model to solve for optimal land exploitation for a given crop. The model was particularly applicable for regions suffering from irrigation water scarcity, such as Saudi Arabia. The model applied deficit irrigation to increase the irrigated area at the expense of reducing the crop yield per unit area. The dynamic program guarantees that deficit irrigation was considered only when it is economically efficient. Moreover, it provided the best irrigation level for each growth stage of the crop, accounting for the varying impact of water stress overtime. The LP provided the best tradeoff between expanding the irrigated area and decreasing water share per hectare.

Raul et al. (2012) have considered a Hirakud canal command of Orissa in eastern India which has under severe threat of waterlogging in the monsoon season and acute shortage of irrigation water in the non monsoon season. An irrigation scheduling model (ISM) and a linear-programming optimization model (LPM) under hydrologic uncertainty were developed with the available land and water resources of the canal command area. The ISM was used to predict actual crop yield under different irrigation strategies, namely, full and deficit depths of irrigation. The crop yield obtained by the ISM under different irrigation management strategies was used in the LPM to optimize the land and water resources of the canal command at different probability of exceedances of net irrigation requirement and canal-water availability. The net annual return was found to decrease with the increase in the level of deficit with maximum return under full irrigation strategy. It was found that the uncertainty factor does not show any visible effect on the cropping pattern. From the sensitivity point of view, cropping area should be given emphasis, followed by the market price and cost of cultivation of different crops during the course of further study.

17.6 CONCLUSION

The adoption of deficit irrigation requires knowledge on the response of the different crops to water stress applied at various growth stages. The empirically derived crop yield response factors for a crop at a particular stage are estimated by considering the deficit irrigation at a particular stage while keeping the crop fully irrigated at other stages (Doorenbos and Kassam, 1979). But when deficit irrigation is applied throughout the growing period, then it would be difficult to determine the values of crop yield response factors at the different growth stages. These values of the crop yield response factors of various growth stages are used in the optimization models, considering either additive or multiplicative approach, to find out the overall response of the yield of the crops under deficit irrigation at the various stages. The crop yield response factors can also be determined at seasonal level by keeping the crops at deficit irrigation under whole growing period and finding out the relationship between relative yield reduction and seasonal relative evapotranspiration reduction (Doorenbos and Kassam, 1979). However, one finds significant mismatching between the values of yield reduction using the seasonal crop yield response factor and the yield reduction obtained by using the stage wise crop yield response factors under deficit irrigation. Overall, the yield reduction determined using seasonal crop yield response factor can be considered better under deficit irrigation as it uses the combined effect of the reductions in the yield at various stages due to water deficit at various stages. Optimization models used an economic criterion to find out the optimal cropping pattern to maximize the benefits for the optimal allocation of land and water resources of a command. Irrigation scheduling models (ISMs) are mostly not integrated into the optimization models under deficit irrigation. The actual yield is obtained by the ISM under fixed deficit levels and is then used in the optimization models to find out the optimal cropping pattern and resource allocation. Therefore, these optimization models may be optimized for a particular deficit level but may not give an overall optimal solution under deficit irrigation. The present study also aims to develop an optimization model to maximize the net economic returns by taking the deficit levels also as a variable in the model. The irrigation scheduling is integrated into the model and optimal deficit levels are worked out for different crops to give the optimal cropping pattern and optimal allocation of the water resources to maximize the benefits.

REFERENCES

Ali, A., 1995. Integrated multilevel irrigation management model for lower indus basin. Dissertation, Indian Institute of Technology Delhi, New Delhi, India.

Ali, M.A., Hoque, M.R., Hassan, A.A. and Khair, A., 2007. Effects of deficit irrigation on yield, water productivity, and economic return of wheat. *Agricultural Water Management*, 92: 151–161.

Ali, M.H. and Talukder, M.S.U., 2008. Increasing water productivity in crop production-a synthesis. *Agricultural Water Management*, 95: 1201–1213.

Allen, R.G., Pereira, L.S., Raes, D. and Smith, A., 1998. Crop evapotranspiration - guidelines for computing crop water requirements. Irrigation and drainage paper 56, Food and Agricultural Organization of the United Nations, Rome.

Azaiez, M.N., 2008. Modeling optimal allocation of deficit irrigation: application to crop production in Saudi Arabia. *Journal Mathematics Model Algorithm*, 7: 277–289.

Babel, M.S., Gupta, A.D. and Nayak, D. K., 2004. A model for optimal allocation of water to competing demands. *Water Resources Management*, 19: 693–712.

Barrett, J.W.H. and Skogerboe, V., 1980. Crop production function and the allocation and use of irrigation water. *Agricultural Water Management*, 3: 53.

Blank, H., 1975. Optimal irrigation decisions with limited water. Dissertation, Colorado State University, Fort Collins.

Cakir, R. and Ebi, U.C., 2010. Yield, water use and yield response factor of flue-cured Tobacco under different levels of water supply at various Growth stages. *Irrigation and Drainage*, 59: 453–464.

Calvache, M., Reichardt, K., Bacchp, O.O.S. and Neto, D.D., 1997. Deficit irrigation at different growth stages of the common bean (phaseolus vulgaris l., cv. Imbabello). *Science Agriculture, Piracicaba*, 54: 1–16.

Cresswell, H.P. and Paydar, Z., 1996. Water retention in Australian soils. I. Description and prediction using parametric functions. *Soil Research*, 34(2): 195–212.

Dastane, N.G., 1974. *A Practical Manual for Water Use Research*, Navbharat Publications, Puna, Maharashtra.

Day, A.D. and Intalap, S., 1970. Some effects of soil moisture stress on the growth of wheat. *Agronomy Journal*, 62: 27–29.

De Wit, C.T., 1958. Transpiration and crop yields, Versl. landbouk. Onderz. No. 64. Wageningen, the Netherlands.

Department of Economic and Social Affairs Population Division, 2017. World population prospects the 2017 revision. Retrieved from https://esa.un.org.

Dhawan, V., 2017. Water and agriculture in India. Published by Federal Ministry of Food and Agricultural. Retrieved from https://www.oav.de.

Doorenbos, J. and Kassam, A.H., 1979. Yield response to water. FAO Irrigation andDrainage Paper No. 33. Rome, Italy: 1–40.

English, M.J., 1990. Deficit irrigation. I. Analytical framework. *Journal of Irrigation and Drainage Engineering*, 116: 399–412.

English, M.J., Musick, J.T. and Murty, V.V., 1990. Deficit irrigation. In: G.J. Hoffman, T.A. Towell and K.H. Solomon, eds. *Management of Farm Irrigation Systems*, ASAE, St. Joseph, Michigan.

English, M. and Raja, S.N., 1996. Review perspectives on deficit irrigation. *Agricultural Water Management*, 32: 1–14.

Garg, N.K. and Ali, A., 1998. Two-level optimization model for lower Indus basin. *Agricultural Water Management*, 36: 1–21.

Garg, N.K. and Dadhich, S.M., 2014. Integrated non-linear model for optimal cropping pattern and irrigation scheduling under deficit irrigation. *Agricultural Water Management*, 140: 1–13.

Ghahraman, B. and Sepaskhah, A.R., 2004. Linear and non-linear optimization models for allocation of a limited water supply. *Irrigation and Drainage*, 53: 39–54.

Gorantiwar, S.D. and Smout, I.K., 2003. Allocation of scarce water resources using deficit irrigation in rotational systems. *Journal of Irrigation and Drainage Engineering- ASCE*, 129(3): 155–163.

Gorantiwar, S.D. and Smout, I.K., 2005. Multilevel approach for optimizing land and water resources and irrigation deliveries for tertiary units in large irrigation schemes.II: Application. *Journal of Irrigation and Drainage Engineering ASCE*, 131: 264–272.

Grimes, D.W. and Dickens, W.L., 1977. Cotton response to irrigation. California Agriculture 31, No. 5. Berkeley, California, United States of America, University of California.

Gupta, R., Tripathi, S.K. and Kumar, R., 2001. Modelling of sugarcane yield with soil and weather parameters in India. *Indian Journal of Sugarcane Technology*, 16(1): 20–26.

Hamdy, A., Ragab, R. and Scarascia-Mugnozza, E., 2003. Coping with water scarcity: water saving and increasing water productivity. *Irrigation and Drainage*, 52(1): 3–20.

Haouaria, M. and Azaiez, M.N., 2001. Theory and methodology optimal cropping patterns under water deficits. *European Journal of Operational Research*, 130: 133–146.

Hiler, E.A. and Clark, R.N., 1971. Stress day index to characterize effects of water stress on crop yields. *Transactions of Hydrology* (210-VI-NEH).

Jalota, S.K., Sood, A., Chahal, G.B.S., and Choudhury, B.U., 2006. Crop water productivity of cotton–wheat system as influenced by deficit irrigation, soil texture and precipitation. *Agricultural Water Management*, 84: 137–146.

Jensen, M.E., 1968. Water consumption by agricultural plants. Chapter 1. In: T.T. Kozlowski, ed. *Water Deficits and Plant Growth* Vol. II (pp. 1–22), Academic Press, New York.

Jana, P.K., Misra, B. and Kar, P.K. 1982. Effects of irrigation at different physiological stages of growth on yield attributes, yield, consumptive use, and water use efficiency of sunflower. *Indian Agriculturist*, 26: 39–42.

Karaata, H., 1991. Kirklareli kosullarinda aycicegi bitkisinin su-üretim fonksiyonlari. Köy Hizmetleri Arastirma Enst. Kirklareli, Turkey (Turkish), Report No 24 (PhD Thesis).

Kirda, C. and Kanber, R., 1999. Water, no longer a plentiful resource, should be used sparingly in irrigated agriculture. *Developments in Plant and Soil Sciences*, 84: 1–20.

Krida, C., 2000. Deficit irrigation scheduling based on plant growth stages showing water stress tolerance. Deficit irrigation practices. *FAO Water Reports*, 22: 3–10.

Minhas, J.S. and Bansal, K.C., 1991. Tuber yield in relation to water stress at stages of growth in potato (Solanum tuberosum L.). *Journal of the Indian Potato Association*, 18: 1–8.

Minhas, B.S., Parikh, K.S. and Srinivasan, T.N., 1974. Towards the structure of a production function for wheat yields with dated inputs of irrigation water. *Water Resource Research, David*, 10(3): 383–393.

Okman, C., 1973. Ankara Sartlarinda Seker Pancarinin su Istihlakinin Tayini Üzerinde bir Arastirma (Ph.D. Thesis). Ankara, Ankara University Publications.

Oweis, T., 1997. *Supplemental Irrigation: A Highly Efficient Water-Use Practice*, ICARDA, Aleppo, Syria: 16.

Paul, S., Panda, S.N. and Kumar, D.N., 2000. Optimal irrigation allocation: a multilevel approach. *Journal of Irrigation and Drainage Engineering, ASCE*, 126(3): 149–156.

Perry, C.J., 1999. The IIMI paradigm: Definitions and implications. *Agricultural Water Management*, 40(1): 45–50.

Prasad, A.S., Umamahesh, N.V. and Viswanath, G.K., 2006. Optimal irrigation planning under water scarcity. *Journal of Irrigation and Drainage Engineering*, 132(3): 228–237.

Raes, D., Geerts, S., Kipkorir, E., Wellens, J. and Sahli, A., 2006. Simulation of yield decline as a result of water stress with a robust soil water balance model. *Agricultural Water Management*, 81: 335–357.

Rao, N.H., Sarma, P.B.S. and Chander, S., 1988. A simple dated water-production function for use in irrigated agriculture. *Agricultural Water Management*, 13: 25–32.

Raul, S.K., Panda, S.N. and Inamdar, P.M., 2012. Sectoral conjunctive use planning for optimal cropping under hydrological uncertainty. *Journal of Irrigation and Drainage Engineering- ASCE*, 138: 145–155.

Rawson, H.M. and Turner, N.C., 1983. Irrigation timing and relationship between leaf area and yield in sunflower. *Irrigation Science*, 4: 167–175.

Smith, L.E.D., 2004. Assessment of the contribution of irrigation to poverty reduction and sustainable livelihoods. *Water Resources Development*, 20(2): 243–257.

Smout, I.K. and Gorantiwar, S.D., 2005. Multilevel approach for optimizing land and water resources and irrigation deliveries for tertiary units in large irrigation schemes. I: method. *Journal of Irrigation and Drainage Engineering*, 131: 254–263.

Smout, I., Gorantiwar, S. and Vairavamoorthy, K., 2006. Performance-based optimization of land and water resources within irrigation schemes. II: application. *Journal of Irrigation and DrainageEngineering*, 132(4): 341–348.

Stewart. J.I., 1972. Prediction of water production functions and associated irrigation programs to minimise crop yield and profit losses due to limited water (PhD Thesis). University of California-Davis, Univ microfilms 73–16, 1934.

Thomas, J.C., Brown, K.W. and Jordan, J.R., 1976. Stomatal response to leaf water potential as affected by preconditioning water stress in the field. *Agronomy Journal*, 68: 706–708. under water deficits. European Journal of Operational Research, 130, pp. 133–146.

Wang, E., Cresswell, H., Bryan, B., Glover, M. and King, D., 2009. Modelling farming systems performance at catchment and regional scales to support natural resource management. *NJAS-Wageningen Journal of Life Sciences*, 57: 101–108.

Wichelns, D., 2004. The policy relevance of virtual water can be enhanced by considering comparative advantages. *Agricultural Water Management*, 66(1): 49–63.

Winter, S.R., 1980. Suitability of sugar beets for limited irrigation in semi-arid climate. *Agronomy Journal*, 72: 649–653.

Zhang, H. and Oweis, T., 1999. Water–yield relations and optimal irrigation scheduling of wheat in the Mediterranean region. *Agricultural Water Management*, 38: 195–211.

Zhang, Y., Kendy, E., Qiang, Y., Changming, L., Yanjun, S., and Hongyong, S., 2004. Effect of soil water deficit on evapotranspiration, crop yield, and water use efficiency in North China Plain. *Agricultural Water Management*, 64: 107–122.

18 Water Resources Scenario of Indian Himalayan Region

Mir Bintul Huda and Rohitashw Kumar
National Institute of Technology

Nasir Ahmad Rather
Baba Ghulam Shah Badshah University

CONTENTS

18.1 INTRODUCTION

In the present times, changing climate is a burning issue. To deal with the various impacts of the changing climate, it needs to be defined first. Climate change is defined as a long-term change in the statistical distribution of weather patterns over periods of time that range from decades to millions of years. It may be a change in the average weather conditions or a change in the distribution of weather events with respect to an average, for example, greater or fewer extreme weather events. Climate change may be limited to a specific region or may occur across the earth. Climate Change is often synonymously used with global warming.

The rapid industrialization has led to increased levels of carbon dioxide, other greenhouse gases, and harmful components such as CFCs in the atmosphere. The increase in the levels of these compounds has led to an increase in the temperature of the earth; these have been associated with the change in number of components of hydrological cycle and systems such as increasing atmospheric water vapor content, increasing evaporation, changing precipitation patterns' intensity and extremes; reduced snow cover; and changes in soil moisture and runoff. The frequency of heavy precipitation events has increased over most areas. There have been significant decreases in water storage in mountain glaciers and northern hemisphere snow cover. Shifts in the amplitude and timing of runoff in glaciers and snowmelt fed rivers and ice-related phenomena in the rivers and lakes have been observed (Mir et al. 2015). Throughout the 21st century, the global climate has increased, which has projected the melting of glaciers and ice caps due to dominance of summer melting over winter precipitation increase. Simulations studies reported volume loss of 60% of the glaciers by 2,050. This, in turn, will reduce the water availability during warm and dry periods in regions supplied by meltwater from major mountain ranges. The IPCC Report (2007) has indicated the land classified as dry to have doubled since 1970. These alarming trends have been deduced on higher significance scales and are expected to accelerate in the coming years.

The scenario of climate change may not be clearly visible in an urban setup or the city or a country level, but the impacts can be assessed in the complex and fragile eco-systems of the world, which are sensitive to the changes in the natural balance of the hydro-meteorological trends. One such sensitive region is the Himalayas, also known as the Roof of the World. The imbalance or alteration will lead to a major disaster for the whole continent itself. Even the minor changes in the climate are magnified at substantial rates in such zones.

The Himalayas are the water reserves of Asia; it comprises of huge number of glacial reserves which are source for the world's biggest river basins, such as Indus, Ganga, Yangtze, and Yellow rivers. The population of most of South-Asia directly or indirectly depends on these river basins and the fragile balance of the region. The changing climate has an adverse impact on these water reserves, which will have deteriorating effects on the whole population of the continent, especially in terms of sustainability, food security, water crisis, livelihood, ecosystem imbalance, etc. The effects are serious in terms of their magnitude as well as their spatial and temporal extents. With time, these effects will be amplified and lead to irreversible scenarios.

The global warming effects in terms of increased snow melting in the Antarctic and Arctic region is largely reported by numerous studies, but the melting of the Himalayan glaciers does not receive its due concern. Beyond the Polar Regions, Himalayas forms the largest glacier concentration, about 9% of the Himalayas is covered under glaciers while 30%–40% is covered under snow. Hence, Himalayan glaciers are termed as the "Third Pole" and are the sources of the main rivers of Asia. But this third pole is under threat; the changing climate has led to less snow and less river discharges, shorter snowfall periods, lower levels of glacier consolidation, and weakening of glacier reserves increasing its susceptibility to melting in summers. The frequency of snowfall has decreased, and in most cause occurrence of rain is found more prominent in the higher reaches, where snowfall used to take place. The increased stream flows in the summer is a cause of the increase in the flood events in the plains of this region. In the desert regions of Ladakh, incidences of heavy rainfalls or cloudbursts have increased; these are the major cause of flash floods in the region causing huge devastation. In the shorter span, the visible effect of climate change in terms of water resources is the increase in the rainfall, leading to higher stream discharges and high flooding, but over the course of time it will be replaced by dryer spells, leading to barren lands and a widespread drought scenario in this region (Gautam et al., 2013). This will, in turn, cause a water crisis in the whole continent.

18.2 PHYSIOGRAPHIC FEATURES OF THE HIMALAYAS

India is bound by the Himalayas in the North and North East for a length of about 2,400 km. The mountains of Himalayas form the northern region of our country and are the highest in the world; Figure 18.1 gives the extent of the Himalayas in India. In Kashmir, the width of these mountains is about 400 km and they extend

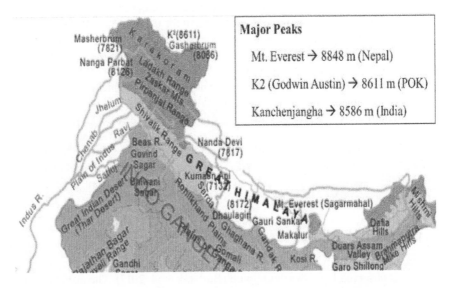

FIGURE 18.1   Indian Himalayas and its extend. (*Source*: World Book, 2018)

up to Arunachal Pradesh, where their width is about 160 km. Indian Himalayas are home to the world's highest mountain peaks, deep valleys, glacier reserves, and world-famous lakes and rivers. The Himalayan Mountains are broadly classified as, Trans Himalaya, Greater Himalayas, Middle Himalayas, Outer Himalayas, and the eastern hills. Zanskar Range, Godwin Austin, Ladakh, Kailash, and Karakoram Ranges form part of Trans-Himalayas; the major part of Trans-Himalayas lies in the Tibet. Kanchenjunga and Mount Everest lie in the Greater Himalayas, which is also known as Inner Himalayas; its average height is about 6,000 m and is always covered by snow and known as Himadri. Middle Himalayas have average height of about 3,000–4,500 m; Pir Panjal, Dhaula Dhar, and Mahabharat are a part of this range; the world-famous Kashmir Valley lies in this zone. Outer Himalayas are also known as Shivaliks; they have an average height of about 600–1,200 m; Dehradun and Patlidun are the prominent valleys that lie in this mountain range. Eastern hills run through thenorth-eastern India bound by the Brahmaputra in the east; Mishmi Hills, Naga Hills, Manipuri hills, and Mizo hills form a part of these.

The Himalayas form a natural border of India with China, Afghanistan, and Pakistan. They form a barrier for the westerlies (cold northern winds) that arise from the Mediterranean region and prevent them from penetrating into the central India. Also, it weakens the effect of the South-West monsoons in the northern India, especially the state of Jammu and Kashmir.

Himalayas are the source of many rivers which are mostly of glacial origin. They form the world's greatest fertile basins such as Indus and Ganga basins. These have huge potential of hydro-electric power generation and aquatic flora and fauna. The region as a whole is a fragile zone with wide ecological diversity.

18.3   CLIMATE

The climate of the Indian subcontinent is hugely influenced by the Himalayas. Himalayas act as a barrier to the cold air masses of the Central Asia. It plays a crucial range in the pattern of precipitation and monsoon. A wide variation in climate is found over the Himalayan region of India, which depends upon the topography and the location of a region. The higher altitudes (greater than 4,800 m) have very cold climate with temperatures below freezing point, which results in a permanent snow-covered zone; higher regions of Middle Himalayas have average summer temperature in the range of 15°C–18°C with winter temperature below 0°C, the climate in the higher elevations is cold alpine with cool summers and harsh winters; valleys of the Middle Himalayas have temperate climate with average summer temperature of about 25°C and cool winters; southern hills of the Himalayas have subtropical climate with average summer temperature of about 30°C and average winter temperature of about 18°C.

18.4   PRECIPITATION

The precipitation increases from the west to east on the Himalayan arc; a similar increase is observed from North to South (Burbank et al. 2012). The average annual rainfall on the southern slopes varies from 1,530 mm at Shimla, Himachal Pradesh,

TABLE 18.1

Precipitation in the Indian States of Himalayan Region

Parameter	States with Both MPR and NMPR						States with Only MPR	Parts of Eastern Himlayas Other Than AP
	Jammu & Kashmir		Himachal Pradesh		Arunachal Pradesh		Uttarakhand	
	Western		Western		Eastern			
Zone	MPR	NMPR	MPR	NMPR	MPR	NMPR	Western	
Annual Precipitation (mm)	1,413.2 ± 135.8	831.5 ± 99.2	1,476.2 ± 118.0	1,057.7 ± 120.0	2,937.5 ± 249.2	2,883.5 ± 330.2	1,558.03 ± 103.8	1,940.7 ± 113.3– 2,616.3 ± 71.4

Source: Singh et al. 2017.

and Mussoorie, Uttarakhand, in the western Himalayas to 3,050 mm at Darjiling, West Bengal, and eastern Himalayas. North of the Great Himalayas, at places such as Skardu, Gilgit, and Leh in the Kashmir portion of the Indus valley, only 75–150 mm of precipitation occurs. In relation to the monsoons, precipitation in the Indian Himalayas can be classified as Monsoonal Precipitation Regime (MPR) and Non-monsoonal Precipitation Regime (NMPR) (Singh et al. 2017). MPR has been defined as regime with >60% annual precipitation during the monsoons, whereas in regimes with <50% annual precipitation in monsoons is termed as NMPR. NPR precipitation is substantially influenced by monsoons, whereas NMPR is greatly influenced by western disturbances originating in the region of Mediterranean. The precipitation in the Indian Himalayas is given in Table 18.1.

18.5   RIVERS

In India, Himalayas is the origin of several important rivers. These are Indus, Jhelum, Chenab, Ravi, Beas, Sutlej, Ganga, Yamuna, Brahmaputra, and Spiti. These rivers form the lifeline of the country and have been substantial in the development of the fragile ecosystem of this zone. In India, rivers have been the lifelines of growth and culture. India is drained by 12 major river systems with a number of smaller rivers and streams. Major river systems in the north are the perennial Himalayan Rivers—Ganga, Yamuna, Indus, and Brahmaputra. The south has the non-perennial but rain fed Krishna, Godavari, and Cauvery while central India has the Narmada, Mahanadi, and Tapti.

Half of the country is drained by the Ganges-Brahmaputra and the Indus systems; they carry more than 40% of the utilizable surface water from the Himalayan watershed to the ocean. Flow in India's rivers is strongly influenced by monsoon resulting in an annual peak in most rivers. The northern rivers with sources in the Himalayas see an additional peak during the spring snowmelt. Because of this, water levels increase and flooding is a common phenomenon that also leads to yearly calamity in states like Bihar and Assam. During the dry season, the flow diminishes in most large rivers and even disappears entirely in smaller tributaries and streams. Due to low rains and dry rivers, drought is another common calamity across vast areas. Some parts of India suffer from flood and some parts from drought.

18.6 HIMALAYAN GLACIERS

The Himalayas are also known as the "Third Pole" because of the largest glacier reserves other than the Antarctica and Arctic. The world has approximately 1,98,000 glaciers, out of which 9,000 glaciers are in India in the Himalayan Region in the states of Sikkim, Jammu and Kashmir, Himachal Pradesh, Uttarakhand, and Arunachal Pradesh. The most important Himalayan Glacier is the Siachen (J&K), which is the largest glacier outside the Polar Region; it stretches to a length of about 72 km. It is located near India and Tibet border on the north-facing slopes of Karakoram Range. It feeds the Mutzgah and Shaksgam rivers. Other important glaciers in Jammu and Kashmir are Baltoro, Biafo, Nubra, Hispur, Nun-Kun, Machoi, and Rimo. Baltoro is second largest glacier in the Himalayan Region and feeds River Shigar, which is an important tributary of Indus River; Biafo is the source of the tributaries of Shigar River; Nubra is the source of Nubra River which joins Shyok River, a tributary of Indus. Besides Jammu and Kashmir, some important glaciers are in Uttarakhand (Gabgotri, Bandarpunch, Dokriani, and Doonagiri) Himachal Pradesh (Bara Shigri, Chhota Shigri), and Sikkim (Zemu, Rathong, Lonak).

18.7 GLOBAL WARMING

Studies show an increasing trend in the temperature data of the Himalayas. The rates of increase vary for different regions and periods of study depending upon the region, season, and the anthropogenic activities of the region. In the last decades, Himalayan and Tibetan Plateau has showed signs of warming at a rate higher than the previous century (Brohan et al. 2006; Diodato et al. 2011). An increase of 0.5°C in the annual average maximum temperature over 1971–2005 was stated. Dash et al. (2007) reported a rise of 0.9°C from 1901–2003 in the western Himalayas in the Indian region. Dimri and Dash (2011) reported an increase of 1.1°C–2.5°C over the western Himalayas in India for 1975–2006, that is, the western Indian Himalayas saw a 0.9°C rise over 102 years (1901–2003). Bhutiyani et al. (2007) found 1.6°C warming (0.16°C/decade) in the last century in this region. Singh et al. (2008) observed increasing trends in maximum temperatures and a seasonal average of daily maximum temperature for all seasons except monsoon over the lower Indus basin in the northwest Indian Himalaya. Fowler and Archer (2005) report increasing trends in winter temperature during 1961–2000 in the upper Indus basin (Pakistan) with warming rates of 0.07–0.51°C/decade in annual mean temperature and 0.1–0.55°C/decade in maximum temperatures. Khattak et al. (2011) reported increasing winter maximum temperature in the upper Indus basin. Jhajharia and Singh (2011) reported a 0.2–0.8°C/decade increase in average temperatures, 0.1–0.9°C/decade increase in maximum temperatures and a rise of 0.1–0.6°C/decade in minimum temperatures over eastern Indian Himalaya.

18.8 CHALLENGES OF WATER RESOURCES MANAGEMENT

The changing climate in the Himalayan zone over the period of time has led to changes in the critical balance of this ecologically diverse zone. Some of the major challenges of the Water Resources management due to global warming or climate change are discussed in the present section.

18.8.1 CLIMATIC EVENTS OF EXTREME NATURE

A study of the western Himalayas (Dimri and Dash 2011) reported an increased number of warm days and a decreased number of cold days during 1975–2006. The frequency of warmer years was observed to have increased from 1995 onwards in the Brahmaputra River Basin (Immerzeel 2008). Caesar et al. (2011) used daily data to compare trends in climatic extremes in Indo-Pacific regions including the Nepalese and Bhutanese Himalayas. The study revealed that the Nepalese and Bhutanese Himalayas show high rates of increase in maximum and minimum temperatures. These trends imply that temperature of the hottest day is increasing at a very high rate over the decades.

In addition to the increasing temperatures, extremes in the precipitation pattern and frequency were also observed. Sen Roy and Balling (2004) conducted a study on the north-western Indian Himalayas for 1910–2000 using daily data and reported increase in frequency of extreme precipitation events. The results were supported by Sen Roy (2009) using hourly data for 1980–2002. It was reported the northwestern Himalaya and northern parts of the Indo-Gangetic basin in the Himalayan foothills show increasing trends in precipitation extremes over all seasons. In Tibetan Plateau, You et al. (2008) reported rising trends in the southern and northern regions and decreasing trends in the central region for most extreme precipitation indices. Bookhagen (2010) reported that the Himalayas have almost twice as many extreme events as the Ganges Plain or the Tibetan Plateau, regardless of rainfall amount.

18.8.2 RETREATING GLACIERS

Owing to the threat from global warming, there have been many kinds of research on the impact of climate change on the glaciers. Most of these studies have reported retreating trends in the Himalayan glaciers except for the Karokoram glaciers, which differ due to the different orographic conditions, an all-year accumulation regime, concentration and role of avalanche, and ablation buffering due to thick debris cover (Hewitt 2011). One of the most important studies on glaciers was conducted by Pandey et al. (2011). 26 glaciers of the western Indian Himalayas were analyzed using satellite imageries over 1975–2007. The study reported fluctuating retreat rates with maximum retreat occurring during the 1989/1992–2001 period. Bajracharya and Mool (2010) studied several glaciers in Dudh Koshi basin in Nepal and reported their retreat in 1976–2007. Gangotri glacier was observed for its temporal variability and it was reported that although in earlier decades retreat was observed but no retreat was observed during 2006–2010 (Kargel et al. 2011). In 2011, Scherler et al. analyzed 286 mountain glaciers from the Hindu Kush, Karokaram, western Indian Himalaya,

Tibetan Plateau, West Kunlun Shan, and southern central Himalaya using satellite images from 2000 to 2008. It was found that 58% of these glaciers in the Karakoram region were either stable or advancing at slower rates; more than 65% of glaciers in the monsoon-influenced regions were retreating with several heavily debris-covered glaciers with low slope at the terminus being stable. Spatially, they found a higher concentration of retreating glaciers (79%) in the western Indian Himalaya and in the northern central Himalaya and West Kunlun Shan (86%) where debris-free glaciers were dominant. In comparison, they found 65% and 73% of sampled glaciers retreating at relatively slower rates, respectively, in Nepal and Bhutan Himalayas and Hindu Kush where debris-cover was common. A high rate of glacier retreat was widely observed in the Tibetan Plateau, which has the largest concentration of glaciers in China. Yao et al. (2007) analyzed 612 glaciers in the Tibetan Plateau and found large number of glaciers was retreating at rates increasing from 90% for 1980–1990 to 95% for 1990–2005. The overall area under glaciers in Nepal from 1962–2005 has reduced considerably (Bajracharya et al. 2011); overall shrinkage as well as fragmentation of glaciers was reported. Kulkarni et al. (2011) studied glacier shrinkage of 1,868 glaciers in 11 basins in the Indian Himalaya, where an overall reduction in glacier area of 16% from 1962–2002, ranging from 2.7–20% among different basins was reported.

18.8.3 IMPACT ON GLACIER DYNAMICS

Many scientists have agreed that unlike thickness of the glacier, flow rate and glacial length have longer response times to the changing climate (Kargel et al. 2011). The response times of the glaciers range from a couple of decades to about a century and these may be greater for large glacial reserves such as in the Himalayas (Armstrong 2010; Thompson et al. 2011). The data available for the glacial retreat is not sufficient enough to study and deduce the changes happening on the level of decades (Fujita and Nuimura 2011). There are various factors that influence the glacier retreat; these include precipitation amount, type of precipitation, albedo, temperature, debris cover, slope, aspect, elevation, etc. (Koul and Ganjoo 2010; Venkatesh et al. 2011; Scherler et al. 2011). Debris cover plays an important role in retarding the rate of retreat (Scherler et al. 2011; Bolch et al. 2011).

Fujita and Nuimura (2011) reported a comparative analysis of three small, relatively debris-free glaciers in Nepal, varying between 0.5 and 0.8 m water equivalent per annum; considerable thinning of the two glaciers in the humid climate in recent decades was observed. The higher mass-loss rates in the humid region were attributed to the glaciers' lower altitude, which makes them more sensitive to small changes in temperature and surface albedo.

Bolch et al. (2011) studied the glaciers in the Everest region with the help of high-resolution satellite data, and the results showed that these glaciers have been losing mass since 1970. Ten glaciers in the region lost mass at an average rate of -0.32 ± 0.08 m during 1970–2007. The greatest mass loss was observed at Imja Shar. In most glaciers, mass loss was observed at mid-ablation zones with negligible loss at glacier termini. Similarly, Kehrwald et al. (2008) concluded no net mass accumulation since 1950 in the Naimonanyi Glacier, located near northwest corner of Nepal in the Chinese Himalaya.

Most glaciers in the Himachal region show a considerable loss in their masses despite their debris cover (Berthier et al. 2007). An overall annual mass loss of −7 to −0.85 m was observed during 1999–2004. Negative mass balance was also observed in Himalayas in Uttarakhand. In contrast, some Karakorum glaciers have not shown a change in mass (Matsuo and Heki 2010), and some show positive mass balance (Hewitt 2005). Matsuo and Heki (2010) found that average ice loss from Asia's high mountain region during 2003–2009 has been twice as fast as the average loss rate over the previous four decades.

Ramanathan et al. (2007) attributed the warming trends in Asia to the black carbon. Kaspari et al. (2011) show that black carbon concentrations have increased approximately threefold from 1975–2000 relative to 1860–1975 in the high elevation regions of the Himalaya. While atmospheric black carbon causes glacial melting through warming related to light absorption, black carbon deposited on snow and ice accelerates melt through reduced surface albedo. Xu et al. (2009) suggest that black carbon deposited on Chinese glaciers was an important factor contributing to observed rapid glacial retreat. Lau et al. (2010) used numerical experiments to deduce that dust and black carbon was responsible for heating of troposphere which could lead to widespread enhanced land-atmosphere warming and accelerated snow-melt in the Himalayas. Yasunari et al. (2010) estimated black carbon concentration on the snow surface and performed numerical experiments suggesting accelerated glacier mass loss on Yala Glacier, Nepal and its related impact on water availability.

Glacier retreat does not occur due to yearly variations but occurs over decades and is an irreversible response to global warming. A large percentage of glaciers have been retreating with varying retreat rates, especially in the upper Indus basin. The studies available indicate that the reduction in glacier mass, shrinkage in its extent, increasing glacial lakes and the increasing temperatures over the Himalayas is a direct result of the changing climate.

18.8.4 EXPANSION OF GLACIAL LAKES

The melting and retreating glaciers due to the increasing temperatures give rise to the moraine-dammed glacial lakes (Evans and Clague 1994). The area under these glacial lakes has been increasing over the past decades (Bajracharya and Mool 2010). The number of glacier lakes is more in Eastern Himalayas than the western parts. In 2008, 50 of these glacial lakes on the Bhutan-China border were studied by Komori (2008) and deduced that 14 of these have been growing in the period 1960–2001. These growth rates were higher toward the southern side. It was found that from 1960–2000 (Bajracharya and Mool 2010), 245 small lakes disappeared from Nepal's Dudh Koshi Basin, while 24 new lakes were formed. Also, 11 supraglacial lakes had converted into moraine-dammed lakes and another 34 glacial lakes grown in size. A growth rate of 25–45 ha/year for glacial lakes in Nepal and Bhutan were suggested by Gardelle et al. (2010) during 1990–2009; whereas a relatively stable average growth rate of 4 ha/year in the western Indian Himalayas was reported. It was also observed that the growth rates of these lakes were higher in the Everest region during 2000–2009 than 1990–2000 periods, contrasting with Bhutan and Western Nepal Himalayas. Kulkarni et al. (2011) reported about five-fold increase in Lonak Lake from 23 to 110 ha between 1976 and 2007, in the eastern Himalayas. In China, Yong et al. (2010) found 64.7% increase in glacial lakes for 1976–2006, with a higher rate of change during the 1976–1988 period over the recent 1988–2006 period. The greatest risks imposed by these increasing glacial lakes is the Glacial Lake Outburst Floods (GLOF); these may incur risks that depend on the local topography, geological, and glaciological processes, but climate changes provide a favorable environment for their expansion (Reynolds and Taylor 2004; Watanabe et al. 2009).

18.8.5 Impact on Stream Flows

The variations in the streamflow have bigger implications on the water availability as well as the disasters such as floods and dam breaks. Climate change can alter the trend of streamflow. It can be influenced additionally by land-use changes as well. To analyze the streamflows, it is necessary to collect data of precipitation, temperature, and other climatic factors. A significant increase in the high magnitude floods in the rivers of Indian Himalayas was reported by Bhutiyani et al. (2007) in the last three decades. An increasing trend in the winter and spring stream flows in the upper Indus basin was observed (Khattak et al. 2011); these were attributed to increased temperatures and accelerated snowmelt. A higher percentage of observed upward trends in pre-monsoon and winter average flow was observed given the potential of snowmelt contribution in low flow periods, but no trends are observed in the post-monsoon season (Gautam and Acharya 2012).

Yao et al. (2007) noted an increase of 5.5% in river runoff attributable to glacial melting in the Tibetan Plateau and a higher increase (13%) in the surrounding Tarim River Basin. Zhang et al. (2011) attributed the increasing trends in flows of Niyang Basin, China to accelerated glacier melting. Lin et al. (2008) studied the streamflow trends in the Lhasa River during 1970s and 1980s; an increasing trend was observed in the last 20 years that was an effect of the increasing precipitation in summer and increasing temperature in winter.

It can be rightly concluded that the increasing trends in the streamflows in the Himalayan Basin can be attributed to the increase in the contribution from glacier melting and snowmelt as a response of increasing temperatures.

18.8.6 Natural Disasters

Among the natural disasters, floods account for 35% for 1975–2005 in South Asia (Shrestha 2008). Floods, landslides, and droughts are the most common disasters found in the Himalayan region. Owing to the vast river basins and their source being snowmelt and glaciers, Himalayas are more prone to flooding, and the magnitude and frequency of these floods are vastly influenced by the changing climates as already discussed. In Himalayas, probable types of floods are riverine floods, flash floods, glacial lake outburst floods (GLOF), and breached landslide-dam floods. The flash floods are common in the foothills, mountain borderlands, and steep coastal catchments.

The most prominent effect of global warming in the Himalayas is the glacial retreat and formation of moraine-dammed glacier lakes; these lakes are the most hazardous type as moraines can burst for several reasons such as sudden increased water volume; surge waves generated by glacier calving, snow, ice/rock avalanches;

earthquake; piping; and dam overtopping (Yamada and Sharma 1993; Kattelmann 2003). These events are common in the summer monsoon when temperature is high and inputs to the glacier lake can have multiple sources.

In 1981, the Zhang Zhangbo GLOF in China destroyed a large section of the China-Nepal road, a power station, and a bridge, with losses totaling more than USD 3 million (Bajracharya et al. 2007). The Dig Tso outburst in 1985 destroyed several infrastructures, land, shops, and the nearly completed Namche Hydropower plant worth USD 3 million (Vuichard and Zimmermann 1987). In 1994, the Luggye Tso GLOF in Bhutan caused loss of property (Richardson and Reynolds 2000) and more than 20 lives (Bajracharya et al. 2007). A huge landslide in 2000, resulting from snow and ice damming the Yigong River caused 30 deaths and more than USD 22 million of property damage in Arunachal Pradesh (ICIMOD 2010).

ICIMOD (2011) identified potentially dangerous glacial lakes in Nepal and these were categorized as low, medium, and high-risk lakes. Watanabe et al. (2009) studied the evolution of Imja Tsho from 1956–2007 and gave the details on the Imja GLOF risk. Risk assessment studies were carried out in the Bhutanese Himalayas (Ageta et al. 2000; Fujita et al. 2008; Komori 2008). Xu and Feng (1994) investigated middle section of the Chinese Himalaya and reported 139 moraine-dammed glacier lakes, of which 34 were identified as dangerous. They identified ice avalanches from advanced glacier tongues and ablation of dead ice beneath moraine ridges as potential GLOF triggers.

18.9 WATER HARVESTING

Harvesting is the technique of collection of water in periods of its abundance and then utilizing it for a varied number of purposes, especially irrigation uses. Water harvesting techniques may be classified as "short term and long term." The most common traditional techniques used in hilly regions for water harvesting are small ponds, dammed ponds, cemented ponds, etc. These traditional techniques have been replaced by modern techniques such as runoff harvesting tanks, semi-circular hoops, trapezoidal bunds, rock catchment, check dams, excavated tanks, etc.

The choice of the water harvesting structure depends upon the source of water, catchment area, storage required, usage requirements, etc. Some of the common water harvesting techniques that are suitable in the Indian Himalayan Region is as follows:

18.9.1 Hill Spring Outflow Harvesting

The zone consists of a number of natural springs, especially in the state of Jammu and Kashmir in India. The water from these springs can be tapped in tanks lined by polythene which can then be utilized for irrigation purposes (Figure 18.2).

18.9.2 Inter-Terrace Runoff Harvesting

Small dug out and lined tanks for intraterrace water harvesting can be constructed on the basis of the requirement of the farmers. Water can be easily stored in these

FIGURE 18.2   LDPE farm pond for spring water harvesting in J&K, India.

poly-lined tanks and be used for bringing additional land under cultivation, especially in the hilly regions of the Indian Himalayas.

18.9.3   Sub-Surface Water Stream Harvesting

Sub-surface water streams are very common in the watersheds of the Indian Himalayas. The water flows in the upper portion of the watersheds round the year and then gradually reduces from September to June. Water flows down on the beds of gullies or nallahs and its discharge reduces due to seepage in the porous beds as it moves down. It appears after few 100 m travel in the bed of Nala. The average flow rate varies between 0.01 and 0.1 m³/s. The water from these sources can be harvested in constructed tanks of ponds for irrigation during stress periods.

18.9.4   Contour Bunds

The Indian Himalayas are very suitable for the construction of contour bunds. In the contour bunds, bunds are constructed on the contours at a spacing of 5–10 m and provided with furrow upslope and cross ties. These structures are efficient in water storage and also easy in construction.

18.9.5   Artificial Glaciers

A new novel technology of construction of artificial glacier has been developed by Sonam Wangchuk known as Ice Stupas (Figure 18.3). The ice stupas are constructed through the gravity flow of water from a stream and utilization during water scarcity and shortage. At each dip/slope in the terrain, retaining walls (something like a mini

FIGURE 18.3   Ice stupas at Leh, India.

dam) is built which further slows down the water and facilitates the freezing of water in the form of steps, all along the slope into an "artificial glacier." All efforts are made to tap every drop of water; even the ones flowing below the frozen ice which would add to the surface runoff that they are harvesting. This artificial glacier then melts in April and supplies water to the fields of the few villages just in time when the barley needs its first water (locally known as Thachus).

18.9.6   FARM POND RENOVATION

The renovation of existing ponds in this region can be done at a number of sites. These ponds can be utilized for bringing the additional area under cultivation The renovation can be done mostly with locally available materials like boulders and also with sand and cement mixture for lining of the ponds. The use of poly lining can also be utilized for reducing seepage in these structures. The proper lining material of these ponds is very important for overall success of these structures.

18.9.7   ZING HARVESTING STRUCTURES

Water harvesting in the cold desert region of the Himalayas, especially Ladakh, can be done by construction of Zing (small pond or reservoir) diversion channel and artificial glacier. But, these ancient structures are in an extremely poor condition as the leakages are common. However, in artificial glacier the water from streams is diverted toward a shady area through a channel at the base of mountains. Dry stone masonry bunds are constructed in series, and water then distributed through a number of outlets and stored as a frozen form during the winter. It is formed between November to February and the stored frozen water is available in the month of spring

FIGURE 18.4   Zing water harvesting structure in Leh, India.

when there is an acute shortage of water. A number of sites have been identified where these structures can be constructed for reducing shortage of irrigation water (Figure 18.4).

The present scenario of changing climate poses a huge threat to the limited water reserves of the Indian Himalayan Region. This makes it compulsory to extensively adapt the water harvesting measures in the region. The harvesting of water serves two major purposes; it reduces the stress on the resources as well as fulfills the various uses, especially irrigation in the region.

18.10   CONCLUSION

Water Resources is an essential natural resource present on the earth, which makes it suitable for the different types of species to live. It is in a critical balance with the entire eco-system. The world's biggest water reserves originate in the Himalayas in the forms of huge glaciers apart from the Antarctic and Arctic. Also, large river basins such as Ganges, Indus, and Yellow river originate from the Himalayas and are a means of sustenance for the whole of Asia in its different impacts in the human lives. But, this balance has been hugely disturbed by the changes in the climate as a result of various interferences in the hydrological cycle and its components. The changes in the climate commonly known as global warming accelerates the melting of snow and glaciers in the "Third Pole," which increases the stream flows in the rivers originating in this region. But this increase in the flows is not a favorable change as scientists are of the opinion that it will be followed by drought in the coming times. Presently, a huge increase in the extreme climatic and hydrological events has been observed, which leads to a huge loss of lives and property. One of the prominent scenarios observed in the Himalayan region is of the effect of climate change on its glaciers in terms of the glacier dynamics as well as the glacier retreats. This increases the extent of the glacial lakes that have been associated with the GLOFs and are hazardous. The review of the Himalayan Region in terms of its water resources potentially paves the way for the proper and extensive planning, conservation, and management of such priceless reserves that mankind has been bestowed with. Also, the review thrusts upon the adoption of suitable water harvesting measures in the region.

REFERENCES

Ageta, Y., S. Iwata, H. Yabuki, N. Naito, A. Sakai, C. Narama, and C. Karma, 2000. Expansion of Glacier Lakes in Recent Decades in the Bhutan Himalayas. IAHS Publication No. 264. pp. 165–176.

Armstrong, R. L., 2010. *The Glaciers of the Hindu Kush-Himalayan Region: A Summary of the Science Regarding Glacier/Melt Retreat in the Himalayan, Hindu Kush, Karakoram, Pamir and Tien Shan Mountain Ranges*. Kathmandu, Nepal, ICIMOD. ISBN: 978-92-9115-176-9.

Bajracharya, S. R., S. B. Maharjan, and F. Shrestha, 2011. Glaciers Shrinking in Nepal Himalaya. In J. Blanco and H. Kheradmand. (Eds.) *Climate Change - Geophysical Foundations and Ecological Effects*, InTech, pp. 445–457. ISBN: 978-953-307-419-1. Available: www.intechopen.com/books/climate-change-geophysical-foundationsand-ecological-effects/glaciers-shrinking-in-nepal-himalaya

Bajracharya, S. R. and P. Mool, 2010. Glaciers, Glacial Lakes and Glacial Lake Outburst Floods in the Mount Everest Region, Nepal. *Annals of Glaciology* 50: 81–86.

Bajracharya, S. R., P. K. Mool, and B. R. Shrestha, 2007. *Impact of Climate Change on Himalayan Glaciers and Glacial Lakes*. Kathmandu, Nepal, ICIMOD, p. 119. ISBN: 978-929-115-032-8.

Berthier, E., Y. Arnaud, R. Kumar, S. Ahmad, P. Wagnon, and P. Chevallier, 2007. Remote Sensing Estimates of Glacier Mass Balances in the Himachal Pradesh (Western Himalaya, India). *Remote Sensing of Environment* 108: 327–338.

Bhutiyani, M. R., V. S. Kale, and N. J. Pawar, 2007. Long-Term Trends in Maximum, Minimum and Mean Annual Air Temperatures across the Northwestern Himalaya during the Twentieth Century. *Climatic Change* 85: 159–177.

Bolch, T., T. Pieczonka, and D. I. Benn, 2011. Multi-Decadal Mass Loss of Glaciers in the Everest Area (Nepal Himalaya) Derived from Stereo Imagery. *The Cryosphere* 5: 349–358.

Bookhagen, B., 2010. Appearance of Extreme Monsoonal Rainfall Events and Their Impact on Erosion in the Himalaya. *Geomatics, Natural Hazards and Risk* 1(1): 37–50.

Brohan, P., J. J. Kennedy, I. Harris, S. F. B. Tett, and P. D. Jones, 2006. Uncertainty Estimates in Regional and Global Observed Temperature Changes: A New Dataset from 1850. *Journal of Geophysical Research* 111: D12106. doi: 10.1029/2005JD006548.

Burbank, D. W., B. Bookhagen, E. J. Gabet, and J. Putkonen, 2012. Modern Climate and Erosion in the Himalaya. *Comptes Rendus Geoscience* 344: 610–626.

Caesar, J., L. V. Alexander, B. Trewin, K. Tse ring, L. Sorany, V. Vuniyayawa, N. Keosavang, A. Shimana, M. M. Htay, and J. Karmacharya, 2011. Changes in Temperature and Precipitation Extremes over the Indo Pacific Region from 1971 to 2005. *International Journal of Climatology* 31: 791–801.

Dash, S. K., R. K. Jenamani, S. R. Kalsi, and S. K. Panda, 2007. Some Evidence of Climate Change in Twentieth-Century India. *Climatic Change* 85: 299–321.

Dimri, A. P. and S. K. Dash, 2011. Wintertime Climatic Trends in the Western Himalayas. *Climatic Change* 111(3–4): 775–800. doi: 10.1007/s10584-011-0201-y.

Diodato, N., G. Bellocchi, and G. Tartari, 2011. How do Himalayan Areas Respond to Global Warming. *International Journal of Climatology* 32(7): 975–982. doi: 10.1002/joc.2340.

World Book. 2018. World Book Encyclopedia 2018, 22 Volumes. By Paul Kobasa (Editor).

Evans, S. G. and J. J. Clague, 1994. Recent Climatic Change and Catastrophic Geomorphic Processes in Mountain Environments. *Geomorphology* 10: 107–128.

Fowler, H. J. and D. R. Archer, 2005. Hydro-Climatological Variability in the Upper Indus Basin and Implications for Water Resources. *Regional Hydrological Impacts of Climatic Change—Impact Assessment and Decision Making* 295: 131–138.

Fujita, K. and T. Nuimura, 2011. Spatially Heterogeneous Wastage of Himalayan Glaciers. *Proceedings of the National Academy of Sciences* 108: 14011–14014.

Fujita, K., R. Suzuki, T. Nuimura, and A. Sakai, 2008. Performance of ASTER and SRTM DEMs, and Their Potential for Assessing Glacial Lakes in the Lunana Region, Bhutan Himalaya. *Journal of Glaciology* 54: 220–228.

Gardelle, J., Y. Arnaud, and E. Berthier, 2010. Contrasted Evolution of Glacial Lakes Along the Hindu Kush Himalaya Mountain Range Between 1990 and 2009. *Global and Planetary Change* 75: 47–55.

Gautam, M. R. and K. Acharya, 2012. Streamflow Trends in Nepal. *Hydrological Sciences Journal* 57: 344–357.

Gautam, M. R., G. R. Timilsina, and K. Acharya, 2013. Climate Change in the Himalayas: Current State of Knowledge. Policy Research Working Paper, 6516, The World Bank Development Research Group, Environment and Energy Team, June 2013.

Hewitt, K., 2005. The Karakoram Anomaly. Glacier Expansion and the 'Elevation Effect', Karakoram Himalaya. *Mountain Research and Development* 25: 332–340.

Hewitt, K., 2011. Glacier Change, Concentration, and Elevation Effects in the Karakoram Himalaya, Upper Indus Basin. *Mountain Research and Development* 31: 188–200.

International Center for Integrated Mountain Development (ICIMOD), 2010. Managing Flash Flood Risk in the Himalayas. Information sheet # 1/10. Available:www.preventionweb. net/files/13252_icimodmanagingflshflo odriskinthehim.pdf.

ICIMOD, 2011. *Glacial Lakes and Glacial Lake Outburst Floods in Nepal.* Kathmandu, Nepal, International Center for Integrated Mountain Development. ISBN: 978-92-9115-194-3.

Immerzeel, W., 2008. Historical Trends and Future Predictions of Climate Variability in the Brahmaputra Basin. *International Journal of Climatology* 28: 243–254.

Inter Governmental Panel on Climate Change (IPCC), 2007. Contribution of Working Group II to the Fourth Assessment Report of the Intergovernmental Panel on Climate Change, 2007. In *Climate Change 2007: Impacts, Adaptation and Vulnerability*, M. L. Parry, O. F. Canziani, J. P. Palutikof, P. J. van der Linden and C. E. Hanson (Eds). Cambridge, United Kingdom and New York, NY, Cambridge University Press.

Jhajharia, D. and V. P. Singh, 2011. Trends in Temperature, Diurnal Temperature Range and Sunshine Duration in Northeast India. *International Journal of Climatology* 31: 1353–1367.

Kargel, J. S., J. G. Cogley, G. J. Leonard, U. Haritashya, and A. Byers, 2011. Himalayan Glaciers: The Big Picture is a Montage. *Proceedings of the National Academy of Sciences* 108: 14709–14710.

Kaspari, S. D., M. Schwikowski, M. Gysel, M. G. Flanner, S. Kang, S. Hou, and P. A. Mayewski, 2011. Recent Increase in Black Carbon Concentrations from a Mt. Everest Ice Core Spanning 1860–2000 AD. *Geophysical Research Letters* 38: L04703. doi: 10.1029/2010GL046096.

Kattelmann, R., 2003. Glacial Lake Outburst Floods in the Nepal Himalaya: A Manageable Hazard. *Natural Hazards* 28: 145–154.

Kehrwald, N. M., L. Thompson, Y. Tandong, E. Mosley-Thompson, U. Schotterer, V. Alfimov, J. Beer, J. Eikenberg, and M. Davis, 2008. Mass Loss on Himalayan Glacier Endangers Water Resources. *Geophysical Research Letters* 35: L22503. doi: 10.1029/2008GL035556.

Khattak, M. S., M. S. Babel, and M. Sharif, 2011. Hydro-Meteorological Trends in the Upper Indus River Basin in Pakistan. *Climate Research* 46: 103–119.

Komori, J., 2008. Recent Expansions of Glacial Lakes in the Bhutan Himalayas. *Quaternary International* 184: 177–186.

Koul, M. N. and R. K. Ganjoo, 2010. Impact of Inter-And Intra-Annual Variation in Weather Parameters on Mass Balance and Equilibrium Line Altitude of Naradu Glacier (Himachal Pradesh), NW Himalaya, India. *Climatic Change* 99: 119–139.

Kulkarni, A. V., B. P. Rathore, S. K. Singh, and I. M. Bahuguna, 2011. Understanding Changes in the Himalayan Cryosphere Using Remote Sensing Techniques. *International Journal of Remote Sensing* 32: 601–615.

Lau, W. K., M. K. Kim, K. M. Kim, and W. S. Lee, 2010. Enhanced Surface Warming and Accelerated Snow Melt in the Himalayas and Tibetan Plateau Induced by Absorbing Aerosols. *Environmental Research Letters* 5: 025204. doi: 10.1088/1748-9326/5/2/025204.

Lin, X., Y. Zhang, Z. Yao, T. Gong, H. Wang, D. Chu, L. Liu, and F. Zhang, 2008. The Trend on Runoff Variations in the Lhasa River Basin. *Journal of Geographical Sciences* 18: 95–106.

Matsuo, K. and K. Heki, 2010. Time-Variable Ice Loss in Asian High Mountains from Satellite Gravimetry. *Earth and Planetary Science Letters* 290: 30–36.

Mir, B. H., A. Akhoon, and N. A. Rather, 2015. Assessing Impact of Climate Change on Surface Water Resources in Dal Catchment. *Civil Engineering* 83: 33247–33254.

Pandey, A. C., S. Ghosh, and M. S. Nathawat, 2011. Evaluating Patterns of Temporal Glacier Changes in Greater Himalayan Range, Jammu & Kashmir, India. *Geocarto International* 26: 321–338.

Ramanathan, V., M. V. Ramana, G. Roberts, D. Kim, C. Corrigan, C. Chung, and D. Winker, 2007. Warming Trends in Asia Amplified by Brown Cloud Solar Absorption. *Nature* 448: 575–578.

Richardson, S. D. and J. M. Reynolds, 2000. An Overview of Glacial Hazards in the Himalayas. *Quaternary International* 65: 31–47.

Reynolds, J. M. and P. J. Taylor, 2004. Review on: Inventory of Glaciers, Glacial Lakes and Glacial Lake Outburst Floods, Monitoring and Early Warning Systems in the Hindu Kush Himalaya Region: Bhutan. *Mountain Research and Development* 24(3): 272–274.

Scherler, D., B. Bookhagen, and M. R. Strecker, 2011. Spatially Variable Response of Himalayan Glaciers to Climate Change Affected by Debris Cover. *Nature Geoscience* 4: 156–159.

Sen Roy, S., 2009. A Spatial Analysis of Extreme Hourly Precipitation Patterns in India. *International Journal of Climatology* 29: 345–355.

Sen Roy, S. and R. C. Balling Jr, 2004. Trends in Extreme Daily Precipitation Indices in India. *International Journal of Climatology* 24: 457–466.

Singh, P., K. H. Umesh, and N. Kumar, 2008. Modelling and Estimation of Different Components of Streamflow for Gangotri Glacier Basin, Himalayas/Modélisation Et Estimation Des Différentes Composantes De L'écoulement Fluviatile Du Bassin Du Glacier Gangotri, Himalaya. *Hydrological Sciences Journal* 53: 309–322.

Singh, S. P., R. D. Singh, S. Gumber, and S. Bhatt, 2017. Two Principal Precipitation Regimes in Himalayas and Their Influence on Tree Distribution. *Tropical Ecology*, 54(4): 679–691.

Shrestha, M. S., 2008. Impacts of Floods in South Asia. *Journal of South Asia Disaster Study* 1(1): 85–106.

Thompson, L. G., E. Mosley-Thompson, M. E. Davis, and H. H. Brecher, 2011. Tropical Glaciers, Recorders and Indicators of Climate Change, are Disappearing Globally. *Annals of Glaciology* 52: 23–34.

Venkatesh, T. N., A. V. Kulkarni, and J. Srinivasan, 2011. Relative Effect of Slope and Equilibrium Line Altitude on the Retreat of Himalayan Glaciers. *The Cryosphere Discuss* 5: 2571–2604.

Vuichard, D. and M. Zimmermann, 1987. The 1985 Catastrophic Drainage of a Moraine-dammed Lake, Khumbu Himal, Nepal: Cause and Consequences. *Mountain Research and Development* 7(2): 91–110.

Watanabe, T., D. Lamsal, and J. D. Ives, 2009. Evaluating the Growth Characteristics of a Glacial Lake and its Degree of Danger of Outburst Flooding: Imja Glacier, Khumbu Himal, Nepal. *Norsk Geografisk Tidsskrift-Norwegian Journal of Geography* 63: 255–267.

Xu, D. and Q. Feng, 1994. Dangerous Glacier Lakes and Their Outburst Features in the Tibetan Himalayas. *Bulletin of Glacier Research* 12: 1–8.

Xu, J., R. E. Grumbine, A. Shrestha, M. Eriksson, X. Yang, Y. Wang, and A. Wilkes, 2009. The Melting Himalayas: Cascading Effects of Climate Change on Water, Biodiversity, and Livelihoods. *Conservation Biology* 23: 520–530.

Yamada, T. and C. K. Sharma, 1993. Glacier Lakes and Outburst Floods in the Nepal Himalaya. *IAHS Publications-Publications of the International Association of Hydrological Sciences* 218: 319–330.

Yao, T., J. Pu, A. Lu, Y. Wang, and W. Yu, 2007. Recent Glacial Retreat and its Impact on Hydrological Processes on the Tibetan Plateau, China, and Surrounding Regions. *Arctic, Antarctic, and Alpine Research* 39: 642–650.

Yasunari, T. J., P. Bonasoni, P. Laj, K. Fujita, E. Vuillermoz, A. Marinoni, P. Cristofanelli, R. Duchi, G. Tartari, and K. M. Lau, 2010. Estimated Impact of Black Carbon Deposition during Pre-Monsoon Season from Nepal Climate Observatory–Pyramid Data and Snow Albedo Changes Over Himalayan Glaciers. *Atmospheric Chemistry and Physics* 10: 6603–6615.

You, Q., S. Kang, E. Aguilar, and Y. Yan, 2008. Changes in Daily Climate Extremes in the Eastern and Central Tibetan Plateau during 1961–2005. *Journal of Geophysical Research* 113: D07101, doi: 10.1029/2007JD009389.

Zhang, M., Q. Ren, X. Wei, J. Wang, X. Yang, and Z. Jiang, 2011.Climate Change, Glacier Melting and Streamflow in the Niyang River Basin, Southeast Tibet, China. *Ecohydrology* 4(2): 288–298.

Index

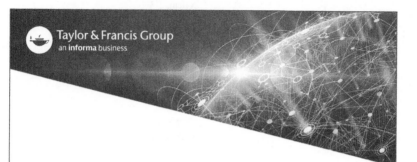